1986

BIOLOGY OF GALL INSECTS

BIOLOGY OF GALL INSECTS

Editor

T. N. ANANTHAKRISHNAN

EDWARD ARNOLD

© T. N. Ananthakrishnan 1984

First published in India by Oxford & IBH Publishing Co.
66 Janpath, New Delhi 110001

First published in Great Britain 1984 by
Edward Arnold (Publishers) Ltd., 41 Bedford Square, London WC1B 3DQ

Edward Arnold (Australia) Pty Ltd., 80 Waverley Road, Caulfield East, Victoria 3145,
    Australia

Edward Arnold, 300 North Charles Street, Baltimore, Maryland 21201, U.S.A.

ISBN 0 7131 2906 9

Printed in India by Printsman Press, Faridabad.

Introduction

Galls are adaptations of plants to attack by gall-insects, endowing them with more favourable conditions of micro-climate, providing an optimal environment for rapid reproduction and abundant food for the duration of the season. Many gall insects show a high degree of specificity with respect to the site of gall formation—leaf, flower, stem or root. Closely related species of the same genus also tend to produce galls quite different in appearance. Considerable information is available on gall morphology projecting the diverse patterns of galling by different groups of insects, as also on the anatomy and developmental morphology. However comparatively little information exists on the bioecological aspects of gall insects in terms of their polymorphism, voltinism, fecundity, life cycles, population dynamics, predator/parasite-gall insect interactions as well as dispersal trends, to cite some of the more important parameters. This multi-author volume aims at projecting a vivid picture of several of the above-mentioned aspects of the biology of gall insects.

The opening Chapter on the adaptational strategies of cecidogenous insects emphasises the role of hyperplasy, hypertrophy in the establishment of gall form, transformation of differentiated tissues into meristematic tissues, the organizational complexities of the nutritive zone, resource partitioning, synchronization of the time of active growth with the appearance of the gall insect as well as implications of polymorphism.

The second Chapter on 'Gall-forming Aphids' presents a vivid picture highlighting such aspects as the taxonomic position of gall-forming aphids, geographic distribution, gall morphology and induction, evolution of life cycles, host-parasite relationships, population ecology of aphids and parasite/predator-aphid interactions, adaptive significance of galls, behavioural patterns and long distance dispersal. Brief mention has also been made on dimorphism in galls due to time of gall initiation, with biological differences seeming to result from nutritional conditions within the gall, as well as on the study of clonal reproduction in the gall, enabling detection of isozyme patterns leading to genetic differentiation within and among populations.

Chapter III on the 'Biology and ecology of gall-forming Psylloidea' discusses the complexities in the organization of different types of psyllid galls, followed by their morphological and physiological adaptations. The adaptive significance of psyllid galls in terms of the advantages of gall formation—nutritional as well as protection from climatic extremes and natural enemies. Besides the role of the gall as a means of dispersal, the economic significance of gall-forming psylloids is also discussed.

Chapter IV deals with 'Gall-forming coccids' with emphasis on the taxonomy, distribution and host plants of the gall formers. In this connection a tabular statement of all the gall-forming coccids with their hosts and distribution has been provided and the Chapter also seeks to discuss in detail the host-plant relationships of gallicolous coccids and morphological specialization associated with them. The role of the newly settled first instar 'crawler stage', or the dispersal stage of the life cycle, which initiates gall formation has also been discussed.

Chapter V on the 'Biology of gall thrips' seeks to highlight the distributional patterns, as well as trends in gall induction involving preference, susceptibility and sequence of gall formation. Biological studies in particular, pertain to aspects of intraspecific diversity, oviposition, fecundity and life cycle patterns. Emphasis has also been laid on such ecological parameters as interactions with biotic and abiotic factors, and successional patterns in thrips gall systems.

Chapter VI on 'Gall Tephritidae' highlights the relationships between the cecidogenous tephritids and their host plants, the natural history of gall tephritids and gall structure and formation. In this attempt emphasis has been laid on the host plants, location and structure of tephritid galls, biogeography, bioecology involving life histories, adaptive specialization and phenology. The two principal avenues of evolution of gall tephritids— through taxonomy and phylogeny, and through the types of association between gall tephritids and the host plants—have also been discussed.

Chapter VII on the 'Biology of gall midges' attempts to discuss at length the morphology and variability of diagnostic characters of cecidomyiids supported by illustrations, followed by aspects of life cycles involving two to several generations per year depending upon the species. Three major biological groups are distinguished on the basis larval feeding habits— mycophagous, phytophagous and zoophagous and the frequency of these gall midges of the Palaearctic region indicated. Interesting information is provided on the inquiline gall midges, since 90% of the inquilines live in gall cecidomyiids. An incisive discussion is on the modes of gall formation in cecidomyiids and patterns of population changes, economically important gall midges as well as the number of relationships of common gall midge species among the biogeographical regions of the world, with the evolutionary trends in gall midges is presented on the basis of several maps.

Chapter VIII on the 'Biology of gall-wasps' attempts to unravel the intricate web of mutual interrelationships between gall-wasps and their host plants, projecting the gall-wasp as a focal organism of the gall community, involving the inquilines and parasites since they are dependent upon the galls produced. Starting with a relevant account of the classification of gall-wasps, discussion is made of the adult morphology followed by gall structure and gall formation. The host plants of gall-wasps have been discussed along with the distribution of galls on the host plants providing an indepth picture. In discussing life cycles and reproduction, unisexual, bisexual and heterogamous species have been involved, with several diagrammatic representations of the life cycles. Inquiline gall-wasps and gall-wasp parasites are also discussed, besides the trophic interrelationships of genera of cynipid oak gall inhabitants.

Chapter IX on the 'Biosystematics of chalcids and sawflies' associated with plant galls attempts to discuss the interrelationships between cecidicoles and cecidozoa which tend to vary in different galls. In view of the fact that an inquiline species in one gall system may be a gall-former in another plant species, the need for an incisive study of the biosystematics of these major groups has been emphasised. Chalcids and sawflies also having indirect associations with plant galls are also discussed.

Chapter X on the 'Geography of gall insects' including a survey of the major patterns of geographic distribution of gall forms, appear to be a fitting contribution to this book on the biology of gall insects. Besides discussing current ideas on zoogeography, this Chapter also attempts to interpret problems due to recent immigration and parallel evolution besides indicating that the diverse geological changes are sufficiently reflected in the present distribution of plants and insects and are major factors in our understanding of the evolution of gall insects.

The last Chapter on the 'Biology of gall mites' is devoted to cecidogenous acarines mostly belonging to the Eriophyidae and Tenuipalpidae and provides a brief account of the diagnosis and bioecology of gall mites. Emphasis is made on the types of life cycles involving protogyne and deutogyne types. Host relationships relating to the diversity of galls induced on diverse hosts are also discussed.

I wish to acknowledge the assistance rendered by Dr. A. Raman in rearranging some of the Chapters. I am also thankful to Mr. K. Raman for helping me in preparing the index.

T.N. ANANTHAKRISHNAN

Contents

Contents

The Contributors

1. Ananthakrishnan, T.N. Entomology Research Institute, Loyola College, Madras 600034, India
2. Askew, R.R. Department of Zoology, University of Manchester, Williamson Building, Manchester M139 PL
3. Beardsley, J.W. Department of Entomology, University of Hawaii at Manoa, Honolulu, Hawaii
4. Brewer, J.W. Colorado State University, Fort Collins, Colorado, U.S.A.
5. Channabasavanna, G.P. Department of Entomology, University of Agricultural Sciences, Hebbal, Bangalore 560024, India
6. Freidberg, Amnon, Department of Zoology, The George S. Wise Faculty of Life Sciences, Tel Aviv University, Tel Aviv 69978, Israel
7. Gagné, Raymond J. Systematic Entomology Laboratory, C/o U.S. National Museum NHB 168, Washington D.C. 20560, U.S.A.
8. Hodkinson, I.D. Department of Biology, Liverpool Polytechnic Byrom Street, Liverpool L3 3AF England
9. Nangia, Neelu, Department of Entomology, University of Agricultural Sciences, Hebbal, Bangalore 560024, India
10. Narendran, T.C. Department of Zoology, University of Calicut, Kerala 673635, India
11. Raman, A. Entomology Research Institute, Loyola College, Madras 600034, India
12. Skuhravá, M. Institute of Encyclopedy, Czechoslovak Academy of Sciences, Prague, Czechoslovakia
13. Skuhravý, V. Institute of Entomology, Czechoslovak Academy of Sciences, Prague, Czechoslovakia
14. Wool, David, Department of Zoology, George S. Wise Faculty of Life Sciences, Tel Aviv University, Tel Aviv 69978, Israel

1. Adaptive Strategies in Cecidogenous Insects

T.N. Ananthakrishnan

Many plants respond to the stimulus mostly induced by the feeding of phytophagous insects by producing abnormal growths often resulting in the pathological disturbance of tissues, culminating in gall production. Though the tendency to produce disorganised growth and malformation in plants by insects is common, true galls are always positive, directional responses resulting in disharmonic growth effects, with the polarity related to the insect rather than the rest of the plants (Ananthakrishnan, 1978). Gall formation or cecidogenesis is a complex phenomenon involving the recanalisation and reorientation of plant development and such growth activities result in the insects becoming partially or completely enclosed, so that the gall insects grow, mature and reproduce within the galls. An interaction between the offensive stimuli involving growth substances released by insects and the defensive response by plants (Rosenthal and Janzen, 1978) appears to be the hall mark of gall production. The precise mechanisms involved however, are not as yet clearly understood, though diverse views are currently available. Several groups of insects are recognised gall makers, and among the more dominant groups are the Hymenoptera including the gall wasps (chalcids and cynipids) and saw-flies (tenthredinids). Diptera including the cecidomyiids or gall-midges and the occasional gall-formers such as the agromyizids, trypetids and anthomyiids, Homoptera including the aphids, aleurodids, coccids and psyllids, and the Thysanoptera or thrips, besides the acarines or the eriophyid mites. Gall-forming Coleoptera (curculionids and chrysomelids) and Lepidoptera are also known.

Initiation and exploitation of plant tissues leading to the formation of plant galls by insects is considered as a highly developed form of phytophagy. One of the earliest groups of insects to develop this trait as early as the late Cretaceous are the saw-flies, while all the other gall insect groups

1

acquired this trait only during the Tertiary (Zwolfer, 1978), the presumed period when the stabilisation and diversification of angiosperms occurred. What is of particular interest is their ability to redirect the growth processes of the host plant, developed independently in many taxa, enabling them to obtain proper nourishment and shelter. The principle of host dominance applies well to gall insects in view of the dominance of certain plant taxa getting adapted to feeding by specific insect species, so well demonstrated by *Quercus*, *Salix*, and *Betula*, becoming the mainstay of many gall midges, (Zwolfer, 1978) because of their dominance in recent geological epochs. Many more examples are available from the different groups of gall-forming insects indicating the association of particular insect generic complexes with specific plant generic complexes, paving the way for a remarkably high degree of host-specificity. This is invariably coupled with the reciprocal action of the plant, forming different types of gall involving various morphogenetic patterns (Ananthakrishnan, 1984). The specific response of plants is the result of an altered metabolism due to the action of an alien chemical stimulus together with the wounding effects as a result of feeding or oviposition, inducing alterations in the cellular and metabolic environment initially around the feeding area. Different kinds of tissue reorganisation including the major cecidogenetic phenomena like hypertrophy and hyperplasia ensue leading to gall formation (Ananthakrishnan, 1982).

The evolution of the gall-forming capacity also depends on the habits and behaviour of the cecidozoans and this is very obvious among chalcids, gall midges, some gall wasps and tephritids living in higher plants, where not only sufficient nutritive substances are available, but also undifferentiated meristematic tissues, ideal for the development of a gall (Zwolfer, 1978). For instance, the gall wasp larvae have a remarkable capacity to regulate the growth of the host plants in order to produce a gall. The plant parts attacked very often being in a suitable physiological status, also facilitates oviposition, enabling the formation of a gall under the influence of the developing larvae. Heterogony or alternation of generations is very characteristic of some gall insects like aphids and cynipids involving galling of different organs on the same plant. A typical example is the gall of *Biorrhiza pallida* Oliv. on *Quercus*, where the bisexual generation in the bud galls alternate with the unisexual generation in the root galls. In the cynipid *Diplolepis Quercus-folii* (Linn), the bisexual generation represented by the bud gall alternates with the unisexual generation of the leaf galls, so that in each generation, the galls of the uni- and bisexual generations appear morphologically different.

While it is true that plant galls represent a unique and complex phenomenon involving the mutual adaptation of the gall maker and the plant, the fact that gall formation is a specific, highly organised reaction of the plant which tends to isolate the gall former in space and time (Mani, 1964) cannot be overlooked. It is generally believed that it is the gall maker that derives all the benefits in the form of nutrition and shelter for reproduction, but the plant tends to suffer through loss of essential substances as well as changes in the direction of growth. However consideration of the process of gall formation as an adaptation to nullify the pathogenic effects of gall insects needs also to be recognised; and the host plant through gall formation has possibly compelled the gall maker to become an extremely specialised feeder (Mani, 1964).

Gall-forming insects have developed several strategies for successful survival and both hyperplasia and hypertrophy are necessarily adaptive phenomena, the cellular realignments caused by hypertrophy in the early stages of feeding often resulting in the establishment of gall-form. Transformation of differentiated tissues into meristematic tissues is another adaptive trend, there often being a precise regulatory mechanism behind the shape and form of every insect-induced gall. Besides the genetic expression evident in the galling patterns of insects like thrips, a convergence of external form and divergence of internal structure also occur, so that galls of highly host specific species reveal a strikingly identical pattern of galling, mostly as rolls, their subtle variations being revealed only by their structure. The very large number of such galls produced by species of *Liothrips* is a typical instance. The opposite condition is equally true wherein different species belonging to one genus show striking complexities in galling patterns. A typical example is afforded by the different species of *Crotonothrips* forming diverse types of galls on different species of *Memecylon* (Melastomaceae) (Fig. 1). Instances are also known wherein different thrips genera or generic complexes are associated with plant generic complexes, tending to lend support to the phenomenon of coevolution (Ananthakrishnan, 1983).

It is not the intention here to examine the modalities of gall formation by different groups of gall insects, but only to emphasise the adaptational strategies involved. While gall insects provide sufficient evidence for age and geographical history of phytophagy, they also form excellent material for the study of insect migration and dispersal and attempts have been made in this volume by Gagnè to explain the continental connections and discontinuities of gall formers based on close proximity and recent connections between continents. The influence of geographic and climatic processes such as the break-up of continents, drifting of land masses, advance and retreat of glaciers have been major factors in the evolution of gall insects.

Gall formation is often considered as an adaptation for exploitation of nutritionally inferior plant tissues. The organisation of a nutritive zone in galls is an adaptation *par excellence* and this specialised zone may be in the form of isolated patches of cells as in many thrips-induced galls (Raman and Ananthakrishnan, 1983) or a highly organised one surrounding the gall insect as in many cecidomyiid and cynipid galls (Meyer, 1969). In any case, under the influence of the cecidozoan, the course of morphogenetic events is altered so that a new physiological environment around the gall insect(s)

Figure 1. Patterns of variability among the leaf galls on species of *Memecylon* induced by different species of *Crotonothrips*

is constituted by specialised cells providing nutrition (Fig. 2). The nutritive cells display very characteristic morphological and cytochemical patterns (Bronner, 1978) reflecting the modes of feeding of the gall maker, the sites of incidence, the pattern of organisation and the subcellular morphology appearing significant to each type of gall insect. Of particular interest is the capability of the minute mites to attack individual cells of the host plant, and cause pronounced hyperplasy giving rise to abnormal outgrowth of erinial hairs. Highly evolved cecidia like the galls of cecidomyiids and cynipids show a radial vascular irrigation pattern centred around the larva and diverging from the nutritive zone to the outer parenchyma, where interestingly enough, the vascular differentiation develops before the differentiation of sclereid tissues. This radial irrigation develops from a pre-existing meristematic 'cambium' tissue and becomes arranged in a centripetal fashion (Meyer, 1969).

The intense protein synthesis in the actively growing plant tissues appears to constitute an important factor in cecidogenesis, since the time

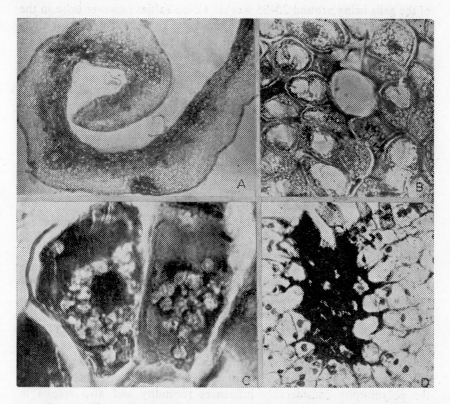

Figure 2. The profile of the nutritive zone in thrips induced gall (*Loranthus elasticus—Phorinothrips loranthi*)

A—Gall (t.s.). Nutritive zone extending along the upper mesophyll.
B & C—Cells of the nutritive zone—magnified.
D—Necrosis along the feeding spot.

of active growth of the plant parts relates to the time of emergence of adult gall insects, oviposition or hatching of young larvae, so that the time of appearance of most gall insects synchronizes with the time of active growth. (Mani, 1964) indicates the existence of an early phase, trophic phase and a post-trophic phase. While in some gall insects the life history terminates with the trophic phase which marks the maturation of the gall and a planned escape of the cecidozoan conditioned further by the change in the climatic conditions, as in *Oligotrophus-Tilia* system (Rohfritsch, 1964). In most others termination of the trophic phase does not mark the end of the life history, but is followed by a series of post-trophic changes. The trophic phase may be very short or may extend to two to three years. Very rarely, gall insects with a short annual life cycle occur in herbaceous plants. In spite of the short life cycle patterns of around 20–25 days, thrips induce galls generally in spring or summer, perishing the same year, the life-span of the galls being around 20–25 weeks. Complexities however arise in the case of the life cycles of heterogonic cecidozoa by the occurrence of alternation of generations. The oak gall cynipids for instance are the most complex involving bisexual univoltine and bivoltine species, unisexual univoltine and bivoltine species and heterogonic species (Askew, this volume). The selective process leading to the development of the various life cycles are of significance, Thelytoky and bivoltinism increasing the reproductive potential of a species.

Resource partitioning is often evident in some galls, the concerned gall insect-like aphids tending to keep the habitat quality constant, so as to maintain their reproductive fitness. Witham (1980) showed that with one leaf/gall, as leaf size nearly doubled, the average number of aphids per gall more than tripled. Galling even in various positions on a leaf are known to produce significant differences in reproductive success. The presence of a second gall on a leaf blade has been reported to negatively affect the fitness of the basal leaf gall. In the aphid gall of *Pempherulus betulae* there is only one gall per leaf so that gall position has a significant effect on fitness. Galls positioned distally on the leaf blade contained fewer progeny which developed to maturity at a slower rate.

Another aspect of reproductive fitness relates to the occurrence of polymorphism, a feature very common in gall thrips, where one finds major and minor females, oedymerous and gynaecoid males, besides normal individuals. Such polymorphism introduces intra- and inter-specific competition in population regulation within galls. The reproductive behaviour of the polymorphs considerably influences fecundity and also results in patterns of mating which lead to recognisable variation in fecundity and subsequent production of morphs. The impact of polymorphism in the *Mimusops* gall thrips, *Arrhenothrips ramakrishnae* Hood on fecundity as well as population regulation within galls is indicated in figure 3.

A further aspect of adaptive strategy relates to the ability of some gall thrips to switch over to the formation of leaf galls and inflorescence gall on the same plant with accompanying changes in the duration of their life cycles. This is exemplified by the rosette gall thrips of *Thilakothrips babuli* Ramakrishna on *Acacia leucophloea* Willd. Alary polymorphism also occurs and with the drying of these galls some of the apterous adults undergo diapause and with *Acacia* subsequently sprouting flower buds, they leave the diapausing site, migrate to flowers to form the floret galls within which both the macropterous and apterous populations develop. What is of significance from the view point of the strategy adopted by the thrips species, is a shorter life cycle in the inflorescence galls (12–14 days) as compared to that of the rosette galls (20–28 days), in view of the comparative short life of the inflorescence (Varadarasan and Ananthakrishnan, 1982).

Regarding the question of resource utilisation in galls by members of the gall communities, two aspects appear to be involved—feeding by the immature and adults of the gall-inducing species as well as by inquiline species which often live in close association with gall inducers, being generally incapable of gall induction. However the nature of the inquiline association is so diverse that it is not possible to arrive at a satisfactory definition, though Askew (1971) suggests that inquilinism is a sort of commensalism. These inquilines, mostly species belonging to the same group as the gall formers, may feed on the gall tissues utilising the provisioned food or actively predate on the eggs, immatures and adults of the gall species or modify the galls made by the primary gall inducer or in rare

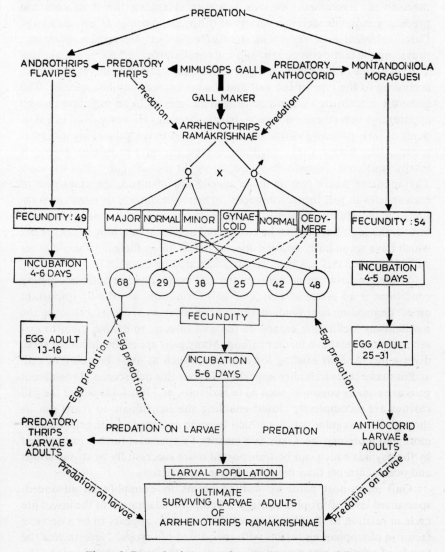

Figure 3. Reproductive strategies of *Mimusops* gall thrips.

cases play a dual role as an inquiline in one gall and as a gall inducer in another. There can be no better examples than the rose galls made by the cynipid *Diplolepis polita* (Ashmead) wherein the cynipid inquiline species *Periclistus pirata* (Osten Sacken) causes additional cell proliferation thereby becoming enclosed so that the galls become structurally modified (Shorthouse, 1973). Oak gall inquilines are also known to form irregular chambers inside the thick walls of their host, but not interfering with gall formation. It is in cynipid galls that the most astonishing number and variety of inquilines are found, sometimes as many as 25 species, mostly cynipids themselves. *Androthrips flavipes* Schmutz is among the most common predatory inquiline seen in a variety of thrips galls such as *Mimusops elengi, Casearia tomentosa, Ficus retusa, Acacia leucophloea, Memecylon edule,* etc., providing the biotic component and the regulation of gall thrips population. The functions of inquilines in a gall as indicated earlier is therefore varied, according to the type of the gall and the concerned inquiline species. The question of transition from a cecidogenetic species to an inquiline though apparent, is nevertheless not very sharply defined. However from the view point of their inducing structural modifications in the galls of higher insect groups, they play a vital role, though their capacity to initiate galling is totally restricted. Sometimes as occurring in the company galls of some Thysanoptera where two or more species are involved, the exact role of each species in gall formation needs further elucidation, through observations on the ability of each to induce gall initiation. As Mani (1964) observed, the inquilines are to be considered as primary cecidogenous species which have secondarily become specialized to life in the gall of some other gall species and this has been amply illustrated by the galls of cynipids.

While synchronization between the gall maker and the optimum galling conditions is an essential aspect of gall formation, an equally important aspect demanding coordination and synchronisation of events, relates to the mechanisms facilitating escape of the gall insects, to enable them to disseminate and establish further galling. Many galls are of the dehiscent type, diverse mechanisms existing for dehiscence such as loose breaking of gall, and in many cases humidity appears to favour this dehiscence. Indehiscent galls are equally common, such as in leaf rolls, and leaf folds where the gall cavities are incompletely closed enabling the individuals to crawl out or through their regular ostioles which become enlarged as the galls become more and more mature. Galls as a means of dispersal is further exemplified by the psyllids which can be transported more successfully by strong winds and ocean currents than their free-living counterparts.

Gall insect-host plant association seems to exemplify an advanced/ specialised level of 'trophic strategy'. Temporal adaptation of the insect life cycle in relation to the phenology of the host plant appears to be a decisive factor in phytophagous insects with specialised life cycles. Nevertheless the trends of relationships among gall insects on the one hand and among the gall-susceptible hosts on the other indicative of 'coevolution', require further elucidation. The recognition of 'open gall systems' as induced by thrips and mites where the maturation and development of the galls are largely dependent on the build-up of populations, and closed 'gall systems' where cecidogenesis is controlled by the action of a single gall maker as in cecidomyiids and cynipid galls is undoubtedly a further step in the understanding of galling systems in general. Curiously enough, the Homoptera-induced galls suggest an intermediate system wherein some of the gall type are induced by populations and others by single organism. The gamut of adaptive exploitation of gall insects in terms of the chemistry of the galls is an aspect deserving consideration to be able to present a more comprehensive picture of the dynamism of the gall insects.

REFERENCES

1. Ananthakrishnan, T.N. 1978. Thrips galls and gall thrips. *Zool. Surv. Ind. Tech. Monogr.* 1, 69 pp.
2. Ananthakrishnan, T.N. 1981. Thrips-plant gall association with special reference to patterns of gall diversity in relation to varying thrips population. *Proc. Ind. Natl. Sci. Acad.* B47 1: 41–46.
3. Ananthakrishnan, T.N. 1983. Adaptive strategies in cecidogenous Thysanoptera. *Curr. Sci.* 52: 77–79.
4. Askew, R.R. 1971. *Parasitic Insects.* London: Heinemann Educational Books Ltd. 316 pp.
5. Bronner, R. 1978. Contribution a l'etude histochimique des tissus nourriciers de zoocecidies. *Marcellia* 40: 1–134.
6. Mani, M.S. 1964. *Ecology of Plant Galls.* The Hague, W. Junk Publ. 434 pp.
7. Meyer, J. 1969. Problems actuels de la Cecidologie. *Bull. Soc. bot. Fr.* 116: 445–481.
8. Raman, A., Ananthakrishnan, T.N. 1983. Morphological studies on some thrips (Thysanoptera : Insecta) induced galls. 2. Fine structure *Proc. Ind. nat. Sci. Acad.* (In Press).
9. Rohfritsch, O. 1967. Ejection de la coque cecidienne d'*Oligotrophus reamurianus* de la feuille de tilleul. *Marcellia* 34: 136–171.
10. Rosenthal, G.A., Janzen, D.H. 1979. *Herbivores: Their interaction with secondary plant metabalites.* New York: Academic Press.
11. Shorthouse, J.D. 1973. The insect community associated with rose galls of *Diplolepis polita* (Cynipidae, Hymenoptera). *Quest. Ent.* 9: 55–98.
12. Whitham, T.J. 1980. The theory of habitat selection: examined and extended using *Pemphigus* aphids. *Amer. Naturalisi* 115: 449–466.
13. Zwölfer, H. 1978. Mechanismen und Ergebnisse der Co-evolution von phytophagen und entomophagen Insekten und den höheren Pflanzen. *Veröffentl. naturw. Verein Hamburg, Sonderbd.* 2: 7–50.

2. Gall-forming Aphids

David Wool

Introduction

Among gall-forming insects, the aphids are unique in many ways. They have evolved unusually complex life cycles (a character they share with free-living aphids). They often are impressively host-specific. And, unlike many other gall-forming insects, they not only use the gall as a nutritional and protective environment for their immatures, but also as a reproductive incubator in which individual genotypes are multiplied parthenogenetically hundreds, or even thousands of times before dispersal. These features, together with the complex host-insect interactions involved in producing the gall, make the gall-forming aphids a fascinating research material for entomologists, ecologists, and evolutionary biologists.

Information on gall-forming aphids has been collected since Linnè, and reviewed early in the 20th century, in two different ways. Some entomologists provided extensive reviews of gall-forming organisms, of which only a small minority are aphids (Buhr, 1964, Felt, 1940; Goot, 1915, 1917; Houard, 1913, 1922; Mani, 1964, 1973; Mimeur, 1949, Ross, 1911). Others produced extensive taxonomic reviews of aphids, among which only a small minority produce galls (Bodenheimer and Swirski, 1957; Börner and Heintze, 1957; Cottier, 1953; Eastop and Hille Ris Lambers, 1976, Palmer, 1952; Theobald, 1926).

The purpose in the present review is to bring together the available information. The information up to the 1950's was taken mainly from older reviews. In addition, the volumes of Biological Abstracts for the years 1950–1969 were searched, and a computerized search of literature was ordered in the Biosis data bank for 1969/1982. The literature reviewed includes most of the known facts. One important topic which is not included in this review is the histological structure of the galls. [This subject has

11

been omitted because a chapter on gall structure is included in this volume].

Taxonomy

Gall-forming aphids occur in three families of the superfamily Aphidoidea (Table 1), although not all species, genera or even tribes within those families are gall-formes. Of the three families, the *Adelgidae* and *Phylloxeridae* are taxonomically rather poorly known (they are only briefly mentioned in the most recent text on aphid taxonomy by Eastop and Hille Ris Lambers, 1976). This fact is necessarily reflected in the present review.

Aphid taxonomy is very difficult due, among other reasons, to their complex life history. Different morphological forms of the same species (in particular the gall-inhabitants and the free-living root forms of the insects), described over the years on different hosts and in different parts of the world, were often given species' and even generic status. Some of the better known species have many invalid synonyms in the literature. For example, the genus *Smynthurodes* had four invalid synomyms, and within the genus, *S. betae* alone has 11 invalid synonyms. Within the genus *Forda* (three invalid synonyms), *F. marginata* has 16 invalid synonyms and *F. formicaria* has nine (Eastop and Hille Ris Lambers, 1976). This situation is probably, true for the great majority of the species, which are insufficiently known. It is possible therefore, that the figure given by Mani (Mani, 1964) of over 700 species of gall-forming aphids, may be grossly inflated.

Another complication is the definition of a gall. Many aphid species cause leaf or stem deformations which in the older literature are referred to as galls, even when all the aphids are not enclosed within the plant structure. The term 'pseudogalls' is used in such cases by some authors. In the Appendix, confirmed species of gall-forming aphids have been listed, with references to descriptions and illustrations of the galls and/or the aphids. The list includes only species for which the primary host is known. To decide on the correct species name among all synonyms, the recent text by Eastop and Hille Ris Lambers has been relied upon (Eastop and Hille Ris Lambers, 1976). The list is restricted to those species which form 'true' galls, although the term is not always clearly defined. In cases of doubt (such as in the *Phylloxeridae* and some *Prociphilini* in the *Pemphigidae*) some examples have been included in the list.

Not included in the list are the reports of Moristu (Moristu, 1947, 1956) on four species of *Aphididae* forming galls on cherry trees (one species *Tuberocephalus momonis* (Matsumura), is reported as forming a true gall). This is the only report of a gall aphid in this family (one of the largest in the Aphidoidea), and no further supporting evidence is known.

The list should be regarded as conservative, since many more species taxonomically belong to the listed genera (Eastop and Hille Ris Lambers, 1976) but are not mentioned in the literature on galls.

Table 1. Taxonomic position of gall-forming aphids
(generally after Bodenheimer and Swirski, 1957).

Superfamily Aphidoidea

Family	*Subfamily*	*Tribe*	*Some genera*
Eriosomatidae (=Pemphigidae)	Eriosomatinae	Byrsocryptini	*Eriosoma, Tetraneura*
	Pemphiginae	Pemphigini	*Pemphigus*
		Pachypappini	*Pachypappa, Mordwilkoja*
		Prociphilini	*Thecabius*
	Melaphinae		*Melaphis, Schlechtendalia*
	Hormaphidinae		*Hormaphis, Hamamelistes, Astergopteryx*
	Fordinae	Fordini	*Forda, Paracletus, Smynthurodes*
		Baizongiini	*Baizongia, Geoica, Slavum, Rectinasus, Aploneura*
Adelgidae (=Chermesidae)	Adelginae		*Adelges, Sacchiphantes, Dreyfusia*
Phylloxeridae	Phylloxerinae		*Phylloxera, Viteus*

Geographic distribution

Species of gall-forming aphids are known from all continents except the Antarctica; their geographic distributions, however, present considerable problems, some of them having important evolutionary implications.

The complete life cycle (holocycle) of these aphids generally involves two hosts: The galls occur only on the primary host, and the distribution of the gall forms is necessarily limited by the distribution of the host trees. The forms living on the secondary hosts, however, often have a much wider distribution (Hille Ris Lambers, 1957) and where the primary host does not exist, may live without ever forming galls (anholocycle).

The problems of describing the geographical distribution of these aphids are complicated by the voluntary or involuntary intervention of man. Transfer of ornamental and other trees between countries and continents resulted in transfer of the aphids to areas in which they have never existed naturally. For example, the genus *Tetraneura* is now known from all continents, but this distribution is almost certainly not natural (Hille Ris Lambers, 1970). *T. ulmi* and other forms are almost certainly introduced in North America. Three species are known in Australia (on secondary hosts) but they probably originated in Eastern Asia (Japan) where they produce galls (Hille

Ris Lambers, 1970). The primary hosts of the *Fordinae* do not grow naturally in the American continent, and in Europe they are limited to the south, but several aphid species live anholocyclically on roots of grasses in America and in Central and Northern Europe.

Primary hosts of gall-forming aphids are listed in Table 2. Each family, or subfamily, of aphids is usually associated with a single plant family or even genus. This is true for aphids in general (Eastop, 1973). Species of *Picea, Quercus, Ulmus, Rhus* and *Populus* typically grow in the northern and central parts of the Old and the New World. *Carya* is limited to the southern and central United States. *Pistacia* grows in the Mediterranean region, east to Iran and Iraq, and reaches as far south and west as North Africa and the Canary Islands. No primary hosts (and therefore no aphid galls) occur in the tropics of Africa, Asia and America, although anholocyclic forms were collected in East Africa (Eastop, 1958) and in Rhodesia (Mordvilko, 1934).

More details on the geographic distribution of gall-forming species are given in the Appendix.

Table 2. Primary hosts of gall-forming aphids

Family (or Subfamily)	Primary host
Adelgidae	Picea (Pinaceae)
Phylloxeridae	Quercus (Fagaceae)
	Carya (Juglandaceae)
	Vitis (Vitaceae)
Eriosomatidae (=Pemphigidae):	
Sub fam. Eriosomatinae:	Ulmus (Ulmaceae)
„ Pemphiginae:	Populus (Salicaceae)
„ Hormaphidinae:	Styrax (Styracaceae)
	Hamamelis (Hamamelidaceae)
	Distylium (Hamamelidaceae)
Sub fam. Melaphinae:	Rhus (Anacardiaceae)
„ Fordinae:	Pistacia (Anacardiaceae)

Gall morphology and induction

Aphid galls vary greatly in size and shape, (see references to illustrations in the Appendix) as well as in aphid clone size. *Phylloxeridae* galls look like small, round pockets on leaf blades, often only a few millimetres in diameter. Galls produced by fundatrices of *Fordini* on *Pistacia atlantica*, and by species of *Thecabius* on *Populus* spp., are also pocket-like (in the former only a single aphid is present in the gall). *Adelgidae* form cone-like galls (sometimes referred to as 'Pineapple galls') made of many individual chambers. *Fordini* galls are flat, bag-like, sometimes looking like leaf-folds (which they are not!). *Baizongiini* galls are elongate, globular or branched, structures, often completely sealed with no opening slit (for example:

Figure 1. A branch of *Pistacia palaestina* heavily infested with
galls of *Forda formicaria*.

Figure 2. Two galls of *Geoica utricularia* on one leaf of *Pistacia palaestina*.
The leaflets carrying the galls are dead, but the rest of the leaf is unaffected.

Geoica, Baizongia). The latter may be 10–20 cm long. Most *Pemphigus* species make globular or otherwise rounded galls on petioles of *Populus* spp. Of outstanding shapes are the galls of *Mordwilkoja vagabunda* (on *populus*) and *Slavum wertheimae* ('Cauliflower galls' on *Pistacia*), looking like corals (Grigarick and Lange, 1968; Hille Ris Lambers, 1957; Wertheim and Linder, 1961). The numbers of individuals per mature gall vary from one to thousands. The ball-like gall of *Astegopteryx styracicola*, 7–8 cm in diameter, is estimated to house more than 200,000 aphids (Aoki, Yamane and Kiuchi, 1977).

In almost all cases, the aphid galls are formed on growing twigs or on young, unfolding leaves. Some authors suggest that the complexity of possible structures to be formed is reduced with increasing age of the tissue (Mani, 1964).

There is no doubt that the stimulus inducing gall formation in the host is contributed by the aphids. It appears to be species-specific, because different aphid species produce distinctly different galls on the same host plant (For example: Koach and Wool, 1977). The exact agent inducing the reaction in the plant remains, so far, unknown.

In the aphid, *Hormaphis hamamelidis*, investigators (Lewis and Walton, 1958) claimed to have identified a 'cecidogenic substance' transmitted by the aphids into the host plant. The substance was detected in aphid myce-tomes and salivary glands as well as in plant tissues penetrated by the aphid mouth parts. But there is no proof so far that this material is in fact cecidogenic nor is there any clue to its chemical nature. The saliva of Homoptera contains a large number of compounds, at least some of which could potentially induce galls (Miles, 1972).

Laboratory experiments using the mouthparts of a Lygaeid bug as a model for aphid-feeding apparatus, suggested that the two most likely can-didates for cecidogenesis are the plant growth-hormone IAA (either transmitted with the saliva, or formed in the plant by oxidative transami-nation of Tryptophan by salivary enzymes), and free amino acids in the saliva, as suggested by earlier workers (Miles, 1968 and 1972). The former mechanism seems more likely (Miles, 1968). The cecidogenic agent may be the same in all Hemiptera, and the specificity of gall forms may depend on the type and stage of plant tissue attacked, and/or feeding behaviour of the insect (loc. cit. and see Dunn, 1960). In at least one case, the affected tissue is located further away apically from the point of an aphid attack (Akimoto, 1981).

In the presence of the insect, the condition of the plant tissue at the time of insect attack is important for gall formation. In the *Pemphigidae*, often only primordial leaves in the bursting bud respond to the aphid stimulus, and timing must be very precise for successful gall formation. In the Adelgid *Adelges abietis* (and other species), the aphid nymph attacks the terminal bud of the host plant (*Picea* sp.) in autumn and overwinters there with its mouth parts in the tissue: only those surviving aphids which have reached the correct part of the bud tissue early enough succeed in producing a gall in the following spring (Rohfritsch, 1981).

The importance of the physiological and developmental state of the leaf for gall formation is demonstrated by recent work on *Pemphigus*. Two forms of galls of *Pemphigus populitransversus* on *Populus deltoides*, which were suspected of belonging to two subspecies (Bird, Faith, Rhomberg, Riska and Sokal, 1979), were shown to be formed by the same species but at different times: elongate galls were formed early in the season, when the leaf petioles were just beginning to unfold. Globular galls were formed later in the season. The difference affected not only the gall shape but also biological parameters of the aphid clone inside (Faith, 1979; Rhomberg, 1980).

In some *Pemphigus species*, normal gall development ceases when the fundatrix dies or is parasitised, and often the gall disappears and only a scar remains (Harper, 1959b). *Baizongia pistaciae* galls continue to grow for several weeks before they attain final size. The feeding of increasing numbers of aphids seems to be necessary for continued gall development. On the other hand, many *Fordinae* and *Pemphiginae* galls attain the final size and form very quickly, when only the fundatrix, and perhaps one to two offspring, are present. Clone size then continues to increase without obvious changes in gall dimensions.

Life cycles

Many of the detailed descriptions of gall-aphid life cycles are very old. Some relatively recent publications describe the life cycles of a few aphids in great detail, in an attempt to fill the gaps in the understanding of their biology: *Pemphigus betae* doane (Grigarick and Lange, 1962, 1968; Harper, 1963) *Mordwilkoja vagabunda* Walsh (Ignoffo and Granovsky, 1961a and 1961b); *Pemphigus borealis* Tullgr. (Dolgova, 1970). These studies are interesting because an entire life cycle was produced under controlled conditions and experiments on host preferences could be carried out. A contribution to the little-known biology of the *Phylloxeridae* is a recent paper on *Phylloxera caryaecaulis* (Caldwell and Schuder, 1979). The biology of the Adelgid aphids *Gilleteela cooleyi* (under a synonym and erroneously assigned to the *Phylloxeridae*) (Cumming, 1959) and *Adelges abietis* (Plumb, 1953) were also described in great detail.

A typical, complete life cycle (holocycle) of gall-forming *Eriosomatidae* (*Pemphigidae*) involves migration between a primary host, on which the gall is formed, and a secondary host on which the aphids do not form a gall. Sexual reproduction occurs only on the primary host.

The cycle begins when a nymph, emerging from an overwintering fertilised egg on the primary host, finds its way to a newly unfolding leaf in early spring and begins to feed (Dunn, 1960). In the resulting gall the *fundatrix* nymph, after several molts, matures into an apterous adult, which reproduced parthenogenetically. In most *Pemphigus* species, the offspring *fundatrigeniae* (all female) mature into alate forms (*alienicolae*) which disperse from the galls in midsummer and larviposit near the secondary hosts.

The new generation nymphs (*radicicolae*) penetrate the soil and feed freely on grass roots. Toward the end of the summer, a new alate form is produced, the *sexuparae* (fall migrants) which fly back to the primary host and give birth to male and female *sexuales*, tiny organisms which do not feed. These produce, after mating and fertilization, the overwintering eggs to complete the cycle. Variations of this general cycle occur in several groups.

In many Fordinae, two or more generations of apterous fundatrigeniae are produced in the gall during the summer before the alate stage. The final generation of alienicolae then appears in the fall, and the radicicolae overwinter on the grass roots. The sexuparae are formed the following spring, and the entire cycle lasts two years rather than one (Bodenheimer and Swirski, 1957; Wertheim, 1950).

Populations of aphids may survive and reproduce continuously (parthenogenetically) as radicicolae on the secondary hosts without producing sexuparae (anholocycly). This phenomenon is known in many species of *Eriosomatinae, Pemphiginae,* and *Fordinae* in areas where their primary hosts do not exist, and may occur elsewhere in parallel with the normal holocycle. A. Mordvilko based his ideas about life cycle evolution in aphids on the presence of anholocycly (see section on Life Cycle Evolution) (Mordvilko, 1924, 1929 and 1934).

Some gall-forming aphids do not have a secondary host (monoecious species). The alate fundatrigeniae in such species are the sexuparae and give rise to the sexuales on the primary (only) host. Among these species are *Slavum wertheimae* (Fordinae) (Hille Ris Lambers, 1957), *Pemphigus monophagus, P. spirothecae* and *Thecabius populimonilis* (Pemphiginae).

In some species, two types of galls are produced on the primary host. In some Fordini on *Pistacia atlantica*, the fundatrix produces a small primary gall along the midrib of the leaflet, but her apterous offspring migrate from the gall to the leaflet margins where they produce the final, typical galls of the species (Bodenheimer and Swirski, 1957; Koach and Wool, 1977; Wertheim, 1950). The fundatrix gall was erroneously recognized as belonging to a different species or genus by some of the early cecidologists (Houard, 1922; Mimeur, 1949). A similar behaviour is known in *Thecabius populimonilis* (Maxson and Knowlton, 1929) and *Th. affinis* (Palmer, 1952) of the Prociphilini.

In most Pemphigidae, there is no doubt that the gall is produced by a single fundatrix. There are two species in which several nymphs attack the growing bud simultaneously. In *Slavum wertheimae*, many aphids may settle on one bud or even one leaf primordium in the early stages of gall formation, but eventually only one remains and the others perish (Wertheim and Linder, 1961). Similarly, in *Mordwilkoja vagabunda*, in the early stages of feeding, the young fundatrices are gregarious and 10 to 15 may feed on the apical region of the same bud (Ignoffo and Granovsky, 1961a and 1961b). Feeding by at least two nymphs seems to be required to arrest terminal growth and initiate gall formation (Ignoffo and Granovsky, 1961b). After the buds reaches a 'rose' stage, however, they move from the apical meristem to the leaf stipules and each nymph locates itself individually on a stipule. The affected stipule encloses the aphid and develops into a gall (Ignoffo and Granovsky, 1961b). All the fundatrigeniae are, then, offspring of a single aphid.

Life cycles of *Adelgidae* and *Phylloxeridae* differ from the patterns already described, in some important details. First and foremost, they do not larviposit their parthenogenetic offspring as do the *Eriosomatidae*, but lay parthenogenetic eggs. Second, *Adelgidae* overwinter as nymphs (and not as eggs) on the primary host. A general description of their life cycle is given in (Carter, 1971).

The life cycle of *Gilletteela* (=Adelges) *cooleyi* may be typical (Cumming, 1959). The fertilized eggs, deposited by the sexuales on *Picea* in the fall, hatch three to four weeks later. The nymphs settle on the stem below the dormant buds and overwinter there. Survivors begin feeding in early spring, become adult, and lay parthenogenetic eggs. The emerging nymphs (gallicolae) produce the galls in the summer, but the early feeding of the fundatrix seems to be required to initiate gall formation (Cumming, 1959; Rohfritsch, 1981). Galls are formed by swelling of the stem and needle stalkets, forming separate chambers at the base of needles, enclosing the aphids. The entire gall is pineapple-shaped, the numbers of chambers (and aphids within chambers) are variable.

The adults of this summer generation are alates, and in late summer they migrate to their secondary host (*Pseudotsuga*), where they oviposit parthenogenetically. The emerging fall generation ('sistens') live and overwinter on the needles of the secondary host. When they mature, next summer, they produce two kinds of parthenogenetic offspring: alate sexuparae, which fly to *Picea* and produce sexuales to complete the cycle, and apterous oviparae which remain on the secondary host. Some 'sistens' may be produced anholocyclically on *Pseudotsuga* (Cumming, 1959).

One variation of this pattern is reported in *Pineus floccus* in Virginia, U.S.A. (Walton, 1980), where all generations on both hosts reproduce parthenogenetically and no sexual reproduction exists. There is one generation a year on the primary host (*picea rubens*) forming a gall, and three or more generations on the secondary host (*Pinus strobus*). The latter produce two kinds of offspring-alates, which fly to *Picea* and oviposit to produce the gallicolae, and aptera remaining on the pine.

Life cycles of *Phylloxeridae* seem to be similar to the *Adelgidae* but apparently, in most cases, with no alteration of hosts. The recent description of *Phylloxera russellae* may be used as an example (Stoetzel, 1981). Overwintering eggs on the host (*Carya illinoinensis*) hatch in the spring. The nymphs settle on unfolding leaves and initiate gall formation. Within a few weeks they mature and begin ovipositing in the galls. The fundatrigeniae develop into *aperous* sexuparae, which lay two kinds of eggs inside the galls: large eggs produce females, small eggs give males. These sexuales emerge from the galls, mate and each female lays one overwintering egg, to complete the cycle on the same host. Another species on the same host, *Ph. notabilis*, has two or more summer generations per year on the leaves, and produces, in addition to the apterous sexuales, alate oviparae as well which may spread to other trees (Cumming, 1959; Whitehead and Eastop, 1937). In the notorious Phylloxerid, *Viteus vitifolii*, the aphids alternate between the leaves (the gall stage) and the roots (radicicolae) of the same host plant, the grape vine (*Vitis*). In Europe and the Middle East, the radicicolae occur anholocyclically (Bodenheimer and Swirski, 1957).

The primary hosts of most species of gall-forming aphids are deciduous trees (excluding the *Adelgidae*, infesting evergreen Pinaceae). The galls are therefore shed with the leaves in the fall. However, some exceptions are interesting to note. Among the Pachypappini, *Morwilkoja vagabunda* galls on *Populus* xylify and remain attached to the branches when the leaves are shed. They seem to attract the migrating sexuparae, which aggregate in them and produce the sexuales within this shelter (Ignoffo and Granovsky, 1961a and 1961b). Similarly the old galls of two species of Fordinae (*Baizongia pistaciae* on *Pistacia palaestina* and *Slavum wertheimae* on *P. atlantica*) remain attached to the branches (Koach and Wool, 1977; Wool, 1977). One species of *Pistacia* in the Mediterranean region, *P. lentiscus*, is exceptional in being evergreen. The empty galls of the single gall-producing aphid on this host, *Aploneura lentisci*, are not shed in the end of the season and support a variety of inquilines, including another, free-living aphid (*Aphis* sp.).

A number of questions about the life cycles of gall aphids remain unanswered so far, in particular the environmental triggering of the formation of alates (alienicolae and sexuparae), and the induction of the sexual cycle. In free-living aphids, the stimulus for the appearance of winged forms and sexuales seems to be a combination of low temperatures and short daylength (Lees, 1966; Blackman, 1975). These direct stimuli can hardly reach the aphids inside the completely sealed galls of *Baizongia*, *Geoica* or *Slavum* (Fordinae). It seems that the induction may be indirect, through the changes in the condition of the host tree towards leaf fall.

Life cycle evolution

Aphids are a very old group of insects, known already from the Permian (Mordvilko, 1934). The fossil forms resemble some extant species of *Phylloxeridae* and *Pemphigidae* in wing venation and other characters. The extant species may not be directly derived from the extinct forms but probably had some common ancestors (Mordvilko, 1934).

In a series of papers, dating from 1924 to 1934, A.K. Mordvilko suggested a detailed hypothesis to relate known facts about the distribution of aphids and their hosts to the evolution of their life cycles. Although his hypothesis may not be the only explanation for the available data, it certainly deserves attention.

According to Mordvilko, the primitive aphids had a sexual cycle, perhaps several sexual generations per year, on a single host. Then, alate female individuals of the summer generations changed from sexual to parthenogenetic, *oviparous* reproduction (this is still the case in the summer generation of the *Adelgidae*), but still on the same host. Evolution proceeded by the reduction and disappearance of wings in the parthenogenetic females, but retention of wings in the males.

Alternation of hosts (*heteroecy*) could evolve, according to Mordvilko, only in moderate climates. The secondary hosts, mostly herbaceous plants, became available much later in the evolutionary history of the aphids. Heteroecy must have evolved after the life cycle reached the stage of alternating sexual and parthenogenetic generations on the same host (otherwise the entire cycle, including the sexual part, would have to be found on the alternate host as well).

The first step in the evolution of heteroecy was probably the facultative establishment of some of the summer, parthenogenetic generations on alternative hosts as they became available. Only gradually it developed into the present obligate heteroecy.

The reason why the sexual cycle does not occur on the secondary hosts seems to be different adaptations of the fundatrix from the fundatrigeniae or radicicolae. In gall aphids, the emerging fundatrix larva resulting from a sexual cycle on a secondary host would be unable to induce a gall and will not survive.

Finally, anholocycly evolved following some drastic environmental change (for example warm to cold, and humid to dry climate) which caused extinction of the primary hosts, but left the parthenogenetic forms on the secondary hosts unharmed.

This hypothesis provides an explanation for some puzzling observations. Some anholocyclic forms of *Eriosomatidae*, for example, are found in areas where no primary hosts exist today. Anholocyclic forms of *Fordinae* are known from Central and Northern Europe and North America, where *Pistacia* trees do not grow. Mordvilko (1924 and 1929) deduces from such patterns of distribution that *Ulmus* trees had existed during the glacial periods in places such as the Philippines, Java and South Africa, and that *Pistacia* trees grew in Europe during some preglacial period.

Mordvilko's hypothesis, stimulating as it is, is not universally accepted. The present distribution of anholocyclic forms may not reflect the past distribution of primary hosts. Mordvilko may have been unaware of the wide distances over which alate fundatrigeniae (alienicolae) can be transported by wind even in one 'leg', let alone the distance that could be travelled during centuries (Hille Ris Lambers, 1970). An alate fundatrigenia from a *Pistacia* gall, presumably from the Mediterranean region, was trapped in the Netherlands at the time of their dispersal from the galls on their primary host (Hille Ris Lambers, 1957).

Despite this objection, which might very well be justified in the majority of cases, Mordvilko's ideas are still a stimulating contribution to the study of the evolution of aphids and aphid-host plant relationships.

Host-parasite relationships

HOST SPECIFICITY

Examination of known primary hosts for many gall-forming aphids indicates that each family or even subfamily of aphids is restricted to a single family or genus of host trees (see Table 2 and Appendix). This is true for many free-living aphids as well (Eastop, 1973). Within each plant genus (for example: *Populus*, *Pistacia*, *Ulmus*), when considered on a world scale, it appears that most gall-forming species have several primary hosts (see Appendix). However, descriptions of both aphids and plants suffer from problems of misclassification, which increase the range of hosts reported for host specific species (Eastop, 1973). My own observations of the Fordinae in the limited geographical area of Israel, and some indirect reports in the literature, suggest that upon detailed investigation these aphids will be found to be strictly host specific within limited geographical areas.

This seems to be the case for many herbivorous insects (Fox and Morrow, 1981). There may be a variety of reasons for a polyphagous species (on a world scale) to be strictly monophagous locally, ranging from the availability of only a small subset of potential hosts (perhaps only one) in any one locality, to genetic variation in the plant or the herbivore or both (Fox and Morrow, 1981). What appears to be polyphagy may, in many cases, turn out to be the result of morphological similarity of genetically different strains or races of aphids, adapted to morphologically similar but genetically distinct races of host plants.

The indirect evidence pointing in this direction is that, for example, *Populus* species are known to hybridise often in nature, but not all hybrids are equally susceptible to aphid attack (Grigarick and Lange, 1962). Similarly, although all spruces are susceptible to attack by *Adelgidae*, some plants never develop galls. One of eight trees observed for over 20 years has never developed galls, although it was 'covered with pseudofundatrices' of *Adelges abietis* every fall. The same tree was a good host for another Adelgid species, *A. laricis* (= *Chermes strobilobius*) (Rohfritsch, 1981).

Differences in susceptibility to aphid attack were reported among varieties of pecan (*Carya illinoinensis*) against species of *Phylloxera*. First-instar nymphs hatching on such varieties are unable to induce galls and die (Stoetzel and Tedders, 1981). In Israel, *Pistacia* trees were often found to be heavily infested with galls of *Forda formicaria*, *Slavum wertheimae*, *Geoica utricularia* and other species of *Fordinae*—side by side with trees on which not a single gall could be found. Whether a plant will or will not be successfully colonized by aphids may depend on physiological, nutritional, ecological, or genetical factors (on the part of both the plant and the aphid).

In the *Picea excelsa* tree that did not develop galls, stinging by the aphids did cause cellular response, but the attacked cells died after a short period of activation and a gall was not formed. This may be an inherited defence mechanism of some plant genotypes (Rohfritsch, 1981).

In Israel all but one species are very strictly host specific, and the one exception—*Geoica utricularia*—when investigated in detail, seems to provide strong support for this generalization (Koach and Wool, 1977). Considering only the common species (so that the chance occurrence of a single, rare gall on a particular host will not introduce bias), it is found that *Aploneura* lentisci occurs only on *Pistacia lentiscus*; *Baizongia pistaciae*, *Forda formicaria*, and *F. marginata*, occur strictly on *P. palaestina*; *Slavum wertheimae*, *Forda riccobonii*, and *Smynthurodes betae* only occur on *P. atlantica*. This is true even when the different hosts grow side by side. On a single tree considered by botanists to be a hybrid (*palaestina* × *atlantica*) *Smynthurodes betae* and *Forda formicaria* were found—the only case of galls typical for the two hosts occurring together. The most interesting case was the exceptional *Geoica utricularia* (unresolved species complex according to Eastop and Hille Ris Lambers (Eastop and Hille Ris Lambers, 1976). Galls identified as *G. utricularia* were collected on three hosts: *P. palaestina*, *P. atlantica*, and *P. khinjouk* (in the Sinai). Examination of 19 morphological characters did not show any difference between alates from the three hosts. However, electrophoretic examination of esterase isozymes in aptera from 700 galls on *P. palaestina* and 400 on *P. atlantica* showed that they are clearly distinct (Koach and Wool, 1977).

DAMAGE TO THE PRIMARY AND SECONDARY HOSTS

The infestation rates of some gall-forming species on particular trees may be extremely high. It is reasonable to suppose that the host must divert nutrients for gall information and insect growth and reproduction, which could otherwise be used for growth and reproduction of the plant itself (Ignoffo and Granovsky, 1961b). Galls are sometimes described as 'nutritional sinks', attracting nutrients from surrounding tissues (Antipov, 1980; Rilling and Steffan, 1978). In cases where the gall is formed on the apical bud (such as *Baizongia pistaciae* or *Mordwilkoja vagabunda*), the branch will not grow (Ignoffo and Granovsky, 1961b). *Pemphigus populivenae* galls on *Populus fremontii* cause chlorosis of the leaf area distal to the gall, and damaged leaves usually fall off the tree one to two months earlier than unparasitised leaves (Whitham, 1978). *Phylloxera* species may cause severe damage to cultivated pecan varieties, and heavy infestation may result in defoliations and loss of the nut yield (Stoetzel and Tedders, 1981) or even death of branches (Caldwell and Schuder, 1979). There are scattered reports of the use of insecticides against gall aphids, indicating that they may cause damage of commercial dimensions to ornamental and other tree species (For example: Seutin and Baurant, 1968). But, it most known cases, *the commercial damage—sometimes severe—is caused by the free-living forms on the secondary host*, not the gall-form on the primary host.

Pemphigus betae and *P. bursarius* (Primary host: *Populus*) sometimes cause severe damage to beets in Canada. The use of insecticides against *P. betae* early in the season increased beet yield, but destroyed an effective natural enemy (Harper, 1961). There were attempts to breed beet varieties resistant to the aphid (Harper, 1964). *P. bursarius* is an important pest of lettuce in England (cited in Harper, 1961). The root forms of *P. populitransversus* are known to cause damage to cabbages in Texas. In Yugoslavia, gall aphids of the *Eriosomatinae* (Primary host: *Ulmus*) cause damage to pear nurseries (their secondary hosts) by the buildup of very large colonies feeding on the roots (Zivanovic, 1978). The most notorious of all is the grape Phylloxera, *viteus vitifolii*. This aphid, of American origin, infested France, Germany and other European countries and caused the destruction of millions of acres of grape vines. The aphid makes galls on vine leaves, but it is the root form (radicicolae) which destroys the plant. The danger of future catastrophes still exists (Schaefer, 1976).

ADAPTIVE SIGNIFICANCE OF GALLS

These facts support the suggestion that the gall represents an adaptation of the plant to the attack by the aphid. The damage to the foliage of the primary host by the aphids, when confined to the gall, is presumably much smaller than the potential damage by the same number of free-living aphids. A good example of damage limited to a small portion of total leaf area is provided by *Geoica utricularia* galls on *Pistacia palaestina* (photograph in Wool, 1977). Toward the end of the season, the leaflets carrying galls die and dry, but the rest of the leaf remains intact and green. Untimely shedding of leaves carrying galls has not been noticed. The reaction of the plant to insect attack may therefore be advantageous for the host.

There is no doubt that gall formation is advantageous for the aphids. Their food supply is guaranteed for the duration of the season, while the gall-free leaf tissues quickly become hard and impenetrable by the aphid mouth parts. Humidity in the galls, particularly in the completely-sealed gall of the *Baizongiini*, must be close to air saturation—a very important advantage for small insects, for example in the hot and dry Mediterranean summer (gall-forming aphids are found on *Pistacia* even in the Negev and Sinai deserts). Temperatures inside the galls are milder than outside them. In short, the gall provides the aphids with what appears to be an optimal environment for rapid reproduction.

An interesting adaptation for life in galls appears to be the excretory system in gall aphids. Rather than excrete honeydew, as do many free-living aphids, the gall forms excrete white wax threads, which are found in the empty galls (together with the exuviae) when the alates had left. If droplets of liquid are excreted, they are covered by a thin wax layer, and accumulate in the galls like small glass beads. In one case, aphids reportedly actively remove these droplets through the gall opening (see the section on behaviour).

The specific reaction of the primary host to attack by each gall aphid, and the considerations listed above, suggest that the aphids and primary hosts have evolved a complex host-parasite system to their mutual advantage. No such relationships apparently exist with the secondary hosts, and damage to the plants may be heavy. (In monoecious species such as the Phylloxerids on pecan, it is possible that the damage is mainly due to the free-living generations, but this is not clearly stated in the references cited above.)

Population ecology, parasites and predators

The ecology of some free-living aphids, in particular those of agricultural importance, has been studied in great detail (for example: van Emden et al., 1969). The effects of daylength, temperature, humidity, host-plant condition, crowding, parasites and predators on some species is well known (review in Bodenheimer and Swirski, 1957), and see (Tomiuk and Wöhrmann, 1980b; Wool, van Emden and Bunting, 1978).

Very few gall-forming species were studied ecologically in any detail, apart from occasional attempts to collect basic ecological data related to the reproductive potential, seasonal abundance and alate migration (for example: Dolgova, 1970, Ignoffo and Granovsky, 1961a; Wertheim, 1950). Exceptions are recent publications on *Pemphigus betae* (Whitham, 1978, 1979 and 1980) which indicate that more interesting facts are there to be discovered by ecologists.

Pemphigus betae fundatrices actively compete for the 'best' location on the leaf for gall production (Whitham, 1979). Colonisation rate in early spring is very high—over 80 per cent of all fundatrices arrive on the leaves during a three-day period, just after bud burst. The fundatrices prefer the largest leaves, although it is unclear how they can judge which leaf is about to become the largest, while still in the bud. Perhaps leaves differ in quality or quantity of nutrients in the sap. It seems that aphids in the galls are resource-limited: the number of offspring per fundatrix, their mean developmental time, and the mean number of embryos per migratory alate fundatrigenia were all positively correlated with leaf size and show a downward trend with an increasing number of galls per leaf. The percentage of aborted galls declined with increasing leaf size (Whitham, 1978 and 1980).

Many fundatrices fail to colonise suitable leaves and die within a few days of arrival. Detailed observations revealed that late-arriving fundatrices are far more likely to fail in the colonisation attempt. An increasing density of fundatrices apparently forces them to utilise smaller, less desirable leaves.

An interesting observation supporting the suggestion of resource limitation on gall aphids is that applying foliar fertiliser to leaves of *Populus fremontii* infested with galls of *Pemphigus populivenae* significantly increased the numbers of aphids per gall and embryos per alate compared with untreated branches (but no such effect was found in *P. betae*) (Whitham, 1978).

The gall environment may protect the aphids from some potential predators. On the other hand, the concentration of aphid biomass in the gall may be attractive to predators. In my studies of *Fordinae* in the Negev desert, I noticed in the summer of 1973 that a large number of *Geoica utricularia* galls on *Pistacia atlantica* were punctured (almost half of the gall was missing) and empty. One suspected that birds were responsible but there was no evidence to support this suspicion. Recently sparrows (*Passer montanus* L.) in Japan were observed to be preying upon galls of two aphid species—a *Tetraneura* sp. on *Ulmus davidiana* and an introduced *Colopha moriokaensis* (Monzen) on *Zelkova serrata*—by tearing the top part of the gall and feeding on the aphids (Sunose, 1980). The birds do not feed on the gall tissue, and the author speculates that they first learned that food is available inside the closed gall by mistaking the gall for fruit. Occasional bird predation on *Pemphigus* galls is mentioned in Grigarick and Lange (1962).

A number of predators find their way into the galls—either through the opening, if there is one, or by boring into them—and feed on the aphids. The most common predator on *Fordinae* galls in Israel and Iran is a small pyralid moth, *Alophia* (perhaps *A. combustella*) (Bodenheimer and Swirski, 1957; Davatchi, 1958). This moth was found infesting galls of *Baizongia pistaciae*, *Geoica utricularia*, *Aploneura lentisci*, *Forda marginata* and *Smynthurodes betae*. While the two former galls are large and contain enough biomass to support even several moth larvae, the others are rather small and perhaps the caterpillar—always one per gall—moves from one gall to another before it completes it development and pupates in the gall.

A. combustella is listed as a common predator of *Fordinae* galls on *Pistacia vera* in Russia (Fet, 1979).

Another group of predators associated with gall forming aphids are Diptera, especially *Chamaemyidae* and *Syrphidae*. At least two Chamaemyid flies feed on, and pupate in, galls of *Fordinae* in Israel—*Leucopis steinbergi* Tanasitsjchunk and *L. palumbi* Rodani. They are common in galls on *Forda marginata* and *Aploneura lentisci* and may infest up to 80 per cent of the galls in late season (Wool, unpubl.). *Leucopis pemphigae* Mall. is a predator of *Pemphigus* aphids in Canada (Harper, 1963), and *L. Simplex* Loew predates on *Phylloxera vitifolii* aphids in the leaf galls (Stevenson, 1967).

Recently maggots of the Syrphid *Meliscaeva* (=*Syrphus*) *auricollis* Meigen were found in galls of *Aploneura lentisci*. This species feeds on free-living aphids and it is unknown whether its occurrence in the galls is accidental or not.

Another predator is an Anthocorid bug, quite common in galls of *Smynthurodes betae* and *Aploneura lentisci* in Israel (and see Devatchi, 1958). *Anthocoris antevolens* white preys on *Pemphigus* aphids in Canada (Harper, 1963). Another bug, *Ceratocaspus modestus* (Miridae) is a predator of the grape phylloxera, *Viteus vitifolii* (Wheeler and Henry, 1979).

Some Coccinellid beetles also prey on gall aphids—notably *Scymnus cervicalis* on grape *phylloxera* galls (Wheeler and Jubb, 1979), and *Scymnus* sp. on *Pemphigus* galls (Harper, 1963). All these predators, however common, seem to have little effect on aphid population size. Since reproduction within the gall is parthenogenetic, even few survivors of predation are sufficient to ensure the survival and propagation of the gall 'genotype' on the secondary hosts (It is the author's experience that in galls containing *Alophia* or *Leucopis* larvae, not all aphids are destroyed and some alates manage to complete their development and disperse). Exceptions are parasites of the gall-forming fundatrices. 'Mummified' fundatrices have been found, as a result of parasitisation by wasps, in galls on several occasions, but no wasps emerged (or the 'mummies' were empty when collected). Such galls are, of course, destroyed.

The gall 'community' also includes inquilines (such as spiders, mites, psyllids, pseudococcids and scale insects). In empty galls of *Aploneura lentisci* an *Aphis* sp. was found surviving and reproducing (Wool unpubl.). This is unique, and can happen only in *Aploneura*, since this species lives on the evergreen *Pistacia lentiscus* and the old galls are not shed with the leaves, but remain for months on the plant and are apparently succulent enough for the aphids to live on.

It should also be mentioned that a large Tachinid fly, *Fischeria bicolor*, is a parasite of *Alophia* larvae in the galls, and two Braconid wasps (one of them identified as *Bracon hebetor*) are perhaps also parasites of the same predator species.

The foregoing account of predators does not include spiders, ants, lizards and birds, which are known to consume the free-living forms—in particular, the migrating alates—of many aphids, gall-formers included.

Behaviour

The behaviour of gall-forming aphids is of course difficult to observe directly. Free-living forms, such as alate fundatrigeniae, of several species have been observed in nature as well as in the laboratory, for example, the attraction of Fordinae alates to light, the arrival and behaviour of sexuparae on the primary host (Wertheim, 1950). It is more difficult to observe behaviour patterns inside the galls, or that of the subterranean radicicolae, without disturbing normal activity. Careful work has uncovered some very complex patterns of behaviour in gall-forming aphids.

a) Apterous first-instar larvae of *Pemphigus dorocola* actively push honeydew globules (coated with wax) as well as exuviae, out through the gall opening (Aoki, 1980). This is the only report of active gall-cleaning behaviour, although cases of honeydew globules dropping out of galls were recorded earlier.

b) In the same species, first instar-larvae were observed to use their hind legs in aggressive behaviour against predatory intruders (Aoki, 1978). Similar behaviour was observed in other *Pemphigidae* which form pseudo-galls or are free-living. In *Colopha* and *Colophina* species, the aggressive larvae do not reproduce (Aoki, 1976).

c) In *Astegopteryx styracicola* (Hormaphidinae) in Japan, an apterous, presumably sterile, 'soldier' caste was observed (Aoki, 1977 and 1979). When the large, ball-shaped gall is disturbed, numerous apterous individuals fall out which may bite human skin and cause severe irritation. They have been shown to attack Syrphid maggots, but are assumed to have evolved as a defense mechanism against predatory mammals. They were shown to be specialised second instar aptera, morphologically different from the normal ones (Aoki, 1979a).

d) Territorial behaviour of fundatrices of *Pemphigus betae*, suggesting competition for the best position on the leaf, was observed (Whitham, 1979). The nymphs emerge from overwintering eggs and arrive at the bursting buds during a period of a few days in early spring. When two or more settle on the same leaf, position at the base of the leaf is actively defended by kicking and shoving with the hind legs. The largest nymph usually wins.

e) In *Epipemphigus niisimae* in Japan, first instar fundatrices intrude into galls made by other conspecific nymphs and fight for possession of the gall. Again the larger nymph usually wins and the other dies (Aoki and Makino, 1982). Nymphs of *Eriosoma yangi* parasitize and usurp galls of other *Eriosoma* species (Akimoto, 1981).

f) Long-distance dispersal in gall-forming aphids is generally limited to the winged forms (fundatrigeniae and sexuparae). Fundatrix larvae migrate from their hatching sites to the buds of the host, and in some species of *Fordinae*, as well as in *Thecabius*, apterous nymphs leave the fundatrix galls to produce the final ones on the leaf. These patterns may be considered short-range dispersal behaviour. Morphological differences were found between migratory and non-migratory first instar larvae of *Pachypappa marsupialis*, the former dispersing into other, empty or occupied conspecific galls (Aoki, 1979b). In contrast, dispersal of apterous *Viteus vitifolii* from the leaf galls seems to be passive and undirected (Stevenson, 1975).

Genetics

CYTOGENETICS

Aphids present the geneticist and developmental biologist with several interesting questions. The different morphological forms which a single clone (genotype) assumes during its life cycle must be controlled by environmental factors (Lees, 1966; Blackman, 1975; van Emden et al., 1969) which trigger the activation and inactivation of different genes at different times in ontogeny. There are indications that response to environmental stimuli varies genetically among clones (Blackman, 1971 and 1972). Differences in esterase activity among clones (Wool, Bunting and van Emden, 1978) are also genetically determined (Blackman, 1979). All these facts were collected in free-living aphids, but should be expected to occur in gall-formers as well.

Most aspects of aphid genetics are very difficult to study, since sexuales are formed under very specialised conditions in most species. The exception is the study of aphid chromosomes, which does not require controlled mating. Gall-aphid chromosome numbers are listed by Blackman (Blackman, 1980) as a part of a study of *Aphidoidea* chromosomes. Chromosome numbers are not considered of reliable taxonomic value, and it is impossible to suggest a 'primitive' number (Blackman, 1980). In Table 3 are listed chromosome numbers for some gall-forming aphids extracted from several sources including personal communication with R. Blackman. The Fordinae, in particular anholocyclic populations, show within-species variation in karyotypes: in the genus *Forda*, cells with different numbers of chromosomes were found in single embryos (Blackman, per. com.). The reasons for this are not clear.

POPULATION GENETICS AND MORPHOLOGICAL VARIATION

Gall-forming aphids are attractive for population genetical and geographic variation research because clonal reproduction within the gall permits partitioning of variation in measurable characters into genetical and environmental components. Galls are often quite conspicuous and easy to collect—an important attribute in large-scale studies.

Table 3. Chromosome numbers of gall-forming aphids.
(Eriosomatidae = Pemphigidae). Root forms are indicated, others were
collected on the primary host. B = Blackman. pers. comm.* = Variable
numbers of chromosomes, see text.

Subfam. *Pemphiginae*	2n	Collected in	Ref.
Epipemphigus imaicus	18	India	Khuda-Bukhsh, 1980
Pemphigus populitransversus	20	Canada	Harper & MacDonald, 1966
Mordwilkoja vagabunda	20	Canada	Harper & MacDonald, 1966
Thecabius affinis	38	Britain	Blackman, 1980
Thecabius populiconduplifolius	28	Canada	Harper & MacDonald, 1966
Pachypappa sacculi	10	Canada	Harper & MacDonald, 1966
Kaltenbachiella pallida	28	Britain	Blackman, 1980
Subfam. *Hormaphidinae*			
Hamamelistes spinosus	50	Canada	Blackman, 1980
Subfam. *Fordinae*			
Aploneura lentisci	16	Britain (roots)	Blackman, 1980
Baizongia pistaciae	24	Britain (roots)	Blackman, 1980
Forda formicaria	20 (20–23*)	Britain (roots)	B
Forda formicaria	20	Israel	B
Forda dactylidis	30	Iran	Blackman, 1980
Forda marginata	28 (17–32*)	Britain, North America (roots)	B
Forda marginata	20/26*	Israel	B
Forda marginata	18	California	B
Forda hirsuta	18	Iran	Blackman, 1980
Geoica eragrostidis	16–18*	Britain (roots)	Blackman, 1980
Geoica utricularia	18	Israel	Blackman, 1980
Geoica eragrostidis	18	Italy	Blackman, 1980
Geoica setulosa	20–31*	Britain (roots)	Blackman, 1980
Geoica setulosa	20	Iran	Blackman, 1980
Paracletus cimiciformis	16	Israel	Blackman, 1980
Rectinasus buxtoni	26	Iran	Blackman, 1980
Smynthurodes betae	8	Britain, Iran, Israel	Blackman, 1980 B

In the last 30 years, considerable efforts were invested in studies of morphological variation of gall-forming aphids. The most intensively-studied species is *Pemphigus populitransversus*; populations of this species were collected throughout the United States and Canada, in studies related to local as well as continent-wide morphological variation (Faith, 1979; Heryford and Sokal, 1971; Rinkel, 1965; Senner and Sokal, 1974; Sokal, 1952; 1962; and 1963; Sokal and Thomas, 1965; Sokal et al. 1971; Sokal and Riska, 1981). Another *Pemphigus* species, *P. populicaulis*, was investigated recently (Sokal et al., 1980). On a smaller scale, and with a different emphasis, three species of *Fordinae* were investigated in Israel (Wool and Koach, 1976; Wool, 1977).

The work of Sokal and his students on *Pemphigus* is perhaps the most intensive morphometric geographical variation study ever carried out on a single organism. Because difficult methodological problems were encountered, new quantitative approaches to the evaluation of and testing for geographical trends were developed (Gabriel and Sokal, 1969; Royaltey et al., 1975; Sokal and Oden, 1978a and 1978b) which contribute considerably to geographic variation research.

The basic analytical methods involve measuring large numbers of characters (15–35 in different studies) on some aphids from each gall—usually in *Pemphigus* the fundatrix and some alate fundatrigeniae (in *Fordinae* the fundatrix is dead and shrivelled at the time of alate emergence). Each character is then subjected to 'nested' analysis of variance (Sokal and Rohlf, 1969) to partition the variation into components: among aphids within galls (nongenetic, environmentally caused variation plus measurement error); among gall means on the same tree (presumably, mostly genetic variation among fundatrices); among tree means within localities, and among locality means (resulting from direct environmental effects, random historical events ('founder effect'), or selection for adaptation to local conditions). The magnitude of these components may vary geographically. Multivariate statistical analyses of covariance, principal components and factor analyses are used to discover patterns of covariation among characters, and these may again vary geographically. Multiple regression analyses may be used to relate morphometric to climatic measurements.

The magnitude of the within-gall variance component varied considerably among different characters in the studies mentioned above. The components of variation among galls within localities, and among localities, were large and significant in most characters both in *Pemphigus* and in the *Fordinae*. In many of the studies on *Pemphigus*, as well as in the *Fordinae*, a general 'size' factor explaining a large proportion of the variation emerged from the multivariate analysis of the characters. Characters associated with this factor apparently do not represent independent parts of the genome, but rather seem to be affected as a group by whatever controls aphid body size (genetic and/or environmental). In *Geoicautricularia* (*Fordinae*), this size factor appeared to be affected by locality temperature—aphid size was larger, the colder the mean locality temperatures (Wool and Koach, 1976; Wool, 1977). This pattern was independent of the geographical proximity of the localities.

Despite their genetic identity, there often was no correlation between stem mother (fundatrix) and alate characters within localities in *Pemphigus*. *Environmental* factors seem to have a different effect on the ontogeny of the two morphs (the fundatrices, of course, reach maturity weeks before their alate offspring, and locality (and gall) conditions may not remain the same).

Repeated sampling of Pemphigus from the same group of localities over a six-year span (Sokal et al., 1971) revealed differences in variation patterns.

These, and results of later studies, were interpreted to mean that the sexuparae, arriving at the primary host, come from different secondary hosts (on which different aphid genotypes may be selected—accounting for differences in within-locality variation), but generally from the same nearby geographical area, which maintains the stability of among-locality variation (Sokal et al., 1980).

A dimorphism in *Pemphigus populitransversus* gall morphology was detected in several localities—some galls were globular and others, elongate. This difference was first thought to represent two different aphid strains or even species, since biological differences were also associated with this dimorphism (Senner and Sokal, 1974). However, careful examination (Faith, 1979) indicated that the difference was due to the time of gall initiation: early galls become elongate, late galls are globular. The biological differences (mainly higher clone sizes in globular galls) seem to result from the nutritional conditions in the gall (Rhomberg, 1980). The early galls may be initiated by nymphs emerging from overwintering eggs on *Populus*, while the globular ones—are issued by individuals overwintering as radicicolae on the secondary hosts, arriving on *Populus* as sexuparae in the spring and producing sexuales whose eggs hatch shortly after fertilisation (Faith, 1979).

In later studies, the geographic variation patterns of aphids from elongate and globular galls were investigated separately (Sokal and Riska, 1981). Again no correlation of stem mother and alate characters were detected within localities, but locality *means* of the two life history stages were strongly correlated; geographical differences in climate may affect both stem mothers and alates in the same direction, masking the switching on and off of different groups of genes during ontogeny and the different directional response of the two morphs in the same gall to the environment (Sokal et al., 1980), while local differences may not be strong enough to produce a noticeable directional effect.

Environmental effects on the geographic patterns of morphological characters were detected in *Geoica utricularia* (*Fordinae*) as mentioned above. This species occurs in very different environments (from cool and humid, to hot and dry desert conditions) in the small geographical area of Israel (Koach and Wool, 1977; Wool and Koach, 1976; Wool, 1977). The best correlation of alate size-related characters with climate was with winter temperatures, at which time galls and alates do not exist, ruling out a direct environmental effect on aphid size and suggesting a possible selective effect operating at some other stage in the life cycle. The number of sensoria on the antennal segments behaved independently of size and seemed to be positively related to locality rainfall, although it is not clear in what way (Wool, 1980). Another species, *Baizongia pistaciae*, is limited only to the cool and humid part of the range (possibly due to the absence of its primary host in the desert) and no climatically-related patterns were detected in any character (Wool, 1977).

It was suggested that the shape of the distribution of canalised morphological characters may be used to estimate the niche breadth of the species (Fraser, 1977). In an attempt to apply this idea to gall-aphid characters, a number of practical difficulties were encountered (Wool, 1980).

ISOZYME VARIATION

The clonal reproduction in the gall permits the application of electrophoretic techniques (now commonly used in population genetical research) to study the genetic differentiation within and among populations, as well as the extent of somatic mutations and recombination and developmental effects on gene expression during ontogeny of a single gall 'organism'. This depends on the presence of suitable genetic markers.

Electrophoresis is now widely used in studies of free-living aphids and their parasites (recent examples: Baker, 1977; Bunting and van Emden, 1980; Tomiuk et al., 1979; Tomiuk and Wöhrmann, 1980a; 1980b, and 1981; Wool et al., 1978a and 1978b), but so far little work was done on gall-forming aphids. In an electrophoretic study of esterases in 400 galls of *Geoica utricularia* from *Pistacia atlantica* and 700 from *P. palaestina*, a clear distinction was revealed, possibly indicating that these forms belong to different species (Koach and Wool, 1977).

The role of the parthenogenetic reproduction in the galls has been the basis for some evolutionary genetic speculation (Wool, 1980), considering that it permits an enormous exact multiplication of single genotypes. The completion of the entire cycle involves two kinds of risks: selection for the best adapted genotypes, and random, non-selective mortality. Selection may operate at any stage in the life cycle, but it is probably minimized in the gall stage when parthenogenetic reproduction begins, and cases of extinction of entire broods are rare (beyond parasitisation of fundatrices by wasps). Random mortality effects must be very strong during the two migratory phases, in which the alates are predated upon by many organisms (for example: spiders, ants, reptiles and birds). Those which survive may not find a suitable host. Parthenogenetic reproduction in the gall permits multiplication of each genotype many times, thus considerably reducing the risk of its being lost due to chance mortality factors.

Concluding remarks

Reviewing the literature on gall-forming aphids convinces one that, after more than 200 years of observation and research, basic entomological information on the biology of the majority of the species is still scarce. Knowledge of the complete life cycle of many more species is required, in order to clarify some of the taxonomic confusion and validate, or refute, the evolutionary theories about the origin of their life cycles and host specificity. Laboratory techniques for rearing the aphids through the complete cycle should be most welcome, since following the aphids in nature is very difficult.

Despite these limitations, gall-forming aphids are excellent research objects in the fields of evolutionary biology and entomology, and many interesting problems are there to be solved.

Acknowledgements

Grateful thanks are due to colleagues, whose contributions to this review were most important: Prof. E. Swirski of the Volcani Institute, Bet Degan, Israel, who made his library available, and Dr. V. Eastop of the Department of Entomology, British Museum (Natural History), London, who advised on several taxonomic difficulties. Correspondence with Dr. R. Blackman (British Museum) about the chromosomes of gall forming aphids was very helpful. Mrs. O. Manheim's assistance with the preparation of the literature list and the Appendix is gratefully acknowledged.

REFERENCES

1. Akimoto, S. 1981. Gall formation by *Eriosoma* fundatrices and gall parasitism in *Eriosoma yangi* (Homoptera, Pemphigidae). *Kontyû* 49: 426–436.
2. Alleyne, E.H., Morrison, F.O. 1977. Some Canadian poplar aphid galls. *Can. Entomol.* 109: 321–328.
3. Antipov, N.I. 1980. Water relations in leaf galls of various plant species. *Fiziol. Rast* (Mosc) 27: 140–144 (in Russian, English summary).
4. Aoki, S. 1976. Occurrence of dimorphism in the first-instar larva of *Colophina clematis* (Homoptera, Aphidoidea). *Kontyû* 44: 130–137.
5. Aoki, S. 1977. *Colophina clematis* (Homoptera, Pemphigidae) an aphid species with 'Soldiers'. *Kontyû* 45: 276–282.
6. Aoki, S., Yamane, S., Kiuchi, M. 1977. On the biters of *Astegopteryx styracicola* (Homoptera, Aphidoidea). *Kontyû* 45 : 563–570.
7. Aoki, S. 1978. Two pemphigids with first instar larvae attacking predatory intruders (Homoptera, Aphidoidea). *New Entomol.* 27 : 7–12.
8. Aoki, S. 1979a. Further observations on *Astegopteryx styracicola* (Homoptera, Pemphigidae)., an aphid species with soldiers biting man. *Kontyû* 47 : 99–104.
9. Aoki, S. 1979b. Dimorphic 1st instar larvae produced by the fundatrix of *Pachypappa marsupialis* (Homoptera, Aphidoidea). *Kontyû* 47 : 390–398.
10. Aoki, S. 1980. Occurrence of simple labour in a gall aphid, *Pemphigus dorocola* (Homoptera, Pemphigidae). *Kontyû* 48 : 71–73.
11. Aoki, S., Makino, S. 1982. Gall usurpation and lethal fighting among fundatrices of the aphid *Epipemphigus niisimae*. (Homoptera, Pemphigidae). *Kontyû* 50: 365–376.
12. Baker, J.P. 1977. Assessment of the potential for and development of organophosphorous resistance in field populations of *Myzus persicae*. *Ann. Appl. Biol.* 86 : 1–9.
13. Bird, J., Faith, D.P., Rhomberg, L., Riska, B., Sokal, R.R. 1979. The morphs of *Pemphigus populi-transversus*: allocation methods, morphometrics, and distribution patterns. *Ann. Ent. Soc. Amer.* 72 : 767–774.

14. Blackman, R.L. 1971. Variation in the photoperiodic response within natural populations of *Myzys persicae* Sulz. *Bull. ent. Res.* 60 : 533–546.
15. Blackman, R.L. 1972. The inheritance of life-cycle differences in *Myzus persicae* (Sulz) (Hem., Aphididae). *Bull. ent. Res.* 62 : 281–294.
16. Blackman, R.L. 1975. Photoperiodic determination of the male and female sexual morphs of *Myzys persicae*. *J. Inseet Physiol.* 21 : 435–453.
17. Blackman, R.L. 1979. Stability and variation in aphid clonal lineages. *Biol. J. Linn. Soc.* 11 : 259–277.
18. Blackman, R.L. 1980. Chromosome numbers in the *Aphididae* and their taxonomic significance. *Syst. Entomol.* 5 : 7–25.
19. Bodenheimer, F.S., Swirski, E. 1957. The *Aphidoidea* of the Middle East. Israel, The Weizmann Science Press of Israel.
20. Börner, C., Heintze, K. 1957. *Aphidina-Aphidoidea. Handbuch der Pflanzenkrankheiten* Vol. 5. Berlin & Hamburg, P. Sorauer.
21. Buhr, H. 1964. Bestimmung Tabellen der Gallen (Zoo-und Phytocecidien) an Pflanzen Mittel und Nordeuropas. Vol. I & II. Jena, G. Fischer Verlag.
22. Bunting, S. van Emden, H.F. 1980. Rapid response to selection for increased esterase activity in small populations of an apomictic clone of *Myzus persicae*. *Nature* 285 : 502–503.
23. Caldwell, D.L., Schuder, D.L. 1979. The life history and description of *Phylloxera caryaecaulis* on shagbark hickory. *Ann. Ent. Soc. Amer.* 72 : 384–390.
24. Carter, C.I. 1971. Conifer woolly aphids (Adelgidae) in Britain. London, H.M. Stationery Office, Forestry Commission Bulletin 42.
25. Carter, C.I. 1975. A gall forming Adelgid, *Pineus similis*, new record to Britain. With a key to the Adelgid galls on Sitka spruce. *Entom. Mon. Mag.* 111 : 29–32.
26. Cottier, W. 1953. Aphids of New Zealand. *N.Z. Dept. of Scientific & Industrial Research Bull.* 106.
27. Cumming, M.E.P. 1959. The biology of *Adelges cooleyi* (Gill). (Homoptera, Phylloxeridae). *Can Entomol.* 41 : 601–617.
28. Davatchi, G.A. 1958. Étude Biologique de la Faune Entomologique des *Pistacia* sauvages et cultivès. *Rev. Path. Veg. Ent. Agr. Fr.* 37 : 3–166.
29. Dolgova, L.P. 1970. The heteroecious life cycle of *Pemphigus borealis* Tullg. (Homoptera, Aphidoidea) *Entomol. Rev.* 49 : 17–23.
30. Dunn, J.A. 1960. The formation of galls by some species of *Pemphigus* (Homoptera, Aphididae) *Marcellia* 30 (Suppl.) 155–167.
31. Eastop, V.F. 1958. A study of the *Aphididae* (Homoptera) of East Africa. London, Colonial Office.
32. Eastop, V.F. 1973. Deductions from the present day host plants of aphids and related insects. In *Insect-Plant relationships*, ed. H.F. van Emden, Oxford : Blackwell Publ. Co. pp. 157–178.
33. Eastop, V.F., Hille Ris Lambers, D. 1976. *Survey of the World's Aphids*. The Hague, W. Junk.
34. Emden, H.F. van, Eastop. V.F., Hughes, R.D., Way, M.J. 1969. The ecology of *Myzus persicae*. *Ann. Rev. Entomol.* 14 : 197–270.
35. Faith, D.P. 1979. Strategies of gall formation in *Pemphigus* aphids. *J.N.Y. Entomol. Soc.* 87 : 21–37.
36. Felt, E.P. 1940. *Plant galls and gall makers*. Comstock Publ. Co.
37. Fet, V. Ya. 1979. Ecology of gall forming aphids (Homoptera, Aphidoidea) and a complex of invertebrates linked to them on the Pistachio *Pistacia vera*. *Izv. Akad. Nauk. Turkm. SSR, SER BIOL.* (3) 67–70 (in Russian, English summary).
38. Fox, L.R., Morrow, P.A. 1981. Specialization : species property or local phenomenon? *Science* 211 : 887–893.

39. Fraser, A. 1977. The use of canalized characters to specify the limits of ecological niches. *Amer. Natur.* 111 : 196–198.

40. Gabriel, K.R., Sokal, R.R. 1969. A new statistical approach to geographic variation analysis. *Syst. Zool.* 18 : 259–278.

41. Goot, P. van der, 1915. *Beitrage zur kenntnis der Hollandischen Blattlause.* Haarlem, H.D. Tjeenk Willink & Zoon.

42. Goot, P. van der, 1917. *Zur Kenntnis der Blattlause Java's.* Contributions a'la Faune des Indes Neederlandaises. Vol. 1 (III) : 1–301.

43. Grigarick, A.A., Lange, W.H. 1962. Host relationships of the sugar-beet root aphid in California. *J. Econ. Entomol.* 55 : 760–764.

44. Grigarick, A.A., Lange, W.H. 1968. Seasonal development and emergence of two species of gall-forming aphids, *Pemphigus bursarius* and *P. nortoni*, associated with poplar trees in California. *Ann. Ent. Soc. Amer.* 61 : 509–514.

45. Harper, A.M. 1958. Notes on behavior of *Pemphigus betae* Doane infected with *Entomophthoraphidis* Hoffm. *Can. Entomol.* 90 : 439–440.

46. Harper, A.M. 1959a. Gall aphids on poplar in Alberta. I. descriptions of galls and distributions of aphids. *Can. Entomol.* 91 : 489–496.

47. Harper, A.M. 1959b. Gall aphids on poplar in Alberta. II. periods of emergence from galls, reproductive capacities, and predators of aphids in galls. *Can. Entomol.* 92 : 680–685.

48. Harper, A.M. 1961. Effect of insecticides on the sugar-beet root aphid, *Pemphigus betae. J. Econ. Ent.* 54 : 1151–1153.

49. Harper, A.M. 1963. Sugar-beet root aphid, *Pemphigus betae* Doane (Homoptera, Aphididae) in Southern Alberta. *Can. Entomol.* 95 : 863–873.

50. Harper, A.M. 1964. Varietal resistance of sugar beets to the sugar-beet root aphid, *Pemphigus betae* Doane (Homoptera, Aphididae). *Can. Ent.* 96 : 520–522.

51. Harper, A.M. 1966. Three additional gall aphids from Southern Alberta. *Can. Entomol.* 98 : 1212–1214.

52. Harper, A.M., MacDonald, M.D. 1966. Chromosomes of gall-aphids in the subfamily *Eriosomatinae* (Homoptera : Aphididae). *Can. J. Genet. Cytol.* 8 : 788–791.

53. Heryford, N.N., Sokal, R.R. 1971. Seasonal morphometric variation in *Pemphigus populi-transversus. J. Kansas Ent. Soc.* 44 : 384–390.

54. Hille Ris Lambers, D. 1957. On some Pistacia aphids (Homotp., Aphididae) from Israel. *Bull. Res. Counc. Israel* 6B : 170–175.

55. Hille Ris Lambers, D. 1970. A study of *Tetraneura* Hartig, 1841 (Homoptera, Aphididae) with descriptions of a new subgenus and new species. *Boll. Zool. Agr. Bachicolt* 2 : 21–101.

56 Hottes, F.C., Frison, T.H. 1931. The plant lice, or Aphiidae, of Illinois. Illinois Natural History Survey Bulletin 19 : 123–447.

57. Houard, C. 1908. Les Zoocécidies des Plantes d'Europe et al Bassin de la Mediterranée, Vol. I, II. Paris, A. Hermann Librairie Scientifique.

58. Houard, C. 1913. Les Zoocécidies des Plantes d'Europe et la Bassin de la Mediterrannée (Supplément, Vol. III). Paris: A. Hermann Librairie Scientifique.

59. Houard, C. 1922. Les Zoocécidies des Plantes d'Afrique, d'Asie et d'Oceanie. Vol. I, II. Paris: Libraire Scientifique Jules Hermann.

60. Ignoffo, C.M., Granovsky, A.A. 1961a. Life history and gall development of *Mordwilkoja vagabunda* (Homoptera, Aphidae) on *Populus deltoides. Ann. Ent. Soc. Amer.* 54 : 486–499.

61. Ignoffo, C.M., Granovsky, A.A. 1961b. Life history and gall development of *Mordwilkoja vagabunda* (Homoptera: Aphidae) on *Populus deltoides.* Part II: gall development. *Ann. Ent. Soc. Amer.* 54 : 635–641.

62. Khuda-Bukhsh, A.R. 1980. Karyotypic studies in 6 species of aphids (Homop-

tera, Aphididae) from the Garhwal Himalayas, India. *Entomon* 5 : 247–250.

63. Koach, J., Wool, D. 1977. Geographic distribution and host specificity of gall-forming aphids (Homoptera, Fordinae) on *Pistacia* trees in Israel. *Marcellia* 40 : 207–216.

64. Lees, A.D. 1966. The control of polymorphism in Aphids. *Adv. Insects Physiol.* 3 : 207–277.

65. Lewis, I.F., Walton, L. 1958. Gall formation on *Hamamelis virginiana* resulting from material injected by the aphid *Hormaphis hamamelidis*. *Trans. Amer. Microsc. Soc.* 77 : 147–200.

66. Mani, M.S. 1964. *The Ecology of Plant Galls*. The Hague: W. Junk.

67. Mani, M.S. 1973. *The Plant Galls of India*. MacMillan.

68. Maxson, A.S., Knowlton, G.F. 1929. The tribe *Pemphigini* (Aphididae) in Utah. *Ann. Ent. Soc. Amer.* 22 : 251–271.

69. Miles, P.W. 1968. Studies on the salivary physiology of plant bugs: experimental induction of galls. *J. Insect Physiol.* 14 : 97–106.

70. Miles, P.W. 1972. The saliva of Hemiptera. *Adv. Insect Physiol.* 9 : 183–255.

71. Mimeur, J.M. 1949. *Contribution a l'étude des Zoocécidies du Maroc*. Encyclopedie Entomologique, P. Lechavalier (ed.).

72. Mordvilko, A. 1924. On the theory of plant lice migrations. *USSR, C.R. Acad. Sci.* (1924) 141–144.

73. Mordvilko, A. 1928. On the evolution of cycles and the origin of Heteraecy (migrations) in plant lice. *Ann. Mag. Nat. Hist.* (10th series) 2 : 570–583.

74. Mordvilko, A. 1929. Anolocyclic elm aphids *Eriosomea* and the distribution of elms during the Tertiary and Glacial periods. URSS, C.R. Acad. Sci. (1929) 197–202.

75. Mordvilko, A. 1934. On the evolution of aphids. *Archiv. f. Naturgeschichte, N.F.,* 3 : 1–60.

76. Moristu, M. 1947. Four gall-forming aphids on cherry trees in Japan. *Mushi* 18 : 39–48.

77. Moristu, M. 1956. Notes on galls of cherry trees. *Bull. Fac. Agric.*, Yamaguti University 7 : 293–294.

78. Palmer, M.A. 1952. *Aphids of the rocky mountain region*. Thomas Say Foundation Vol. 5.

79. Plumb, G.H. 1953. The formation and development of the Norway spruce gall caused by *Adelges abieties*. *The Connect. Expl. Sta. Bull.* 566 : 1–77.

80. Rhomberg, L. 1980. Causes of life-history differences between the morphs of *Pemphigus populitransversus*. J.N.Y. *Entomol. Soc.* 88 : 106–112.

81. Rilling, G., Steffan, H. 1978. Experiments on the carbon dioxide fixation and the assimilate import by leaf galls of *Phylloxera* (Dactylosphaera) *vitifolii* on grapevine (*Vitis rupestris*) (in German, Engl. summ.) *Angew. Bot.* 52 : 343–354.

82. Rinkel, R.C. 1965. Microgeographic variation and covariation in *Pemphigus populi-transversus*. *Univ. Kansas. Sci. Bull.* 46 : 167–200.

83. Rohfritsch, O. 1981. A 'defense' mechanism of *Picea excelsa* L. against the gall former, *Chermes abietis* L. (Homoptera, Adelgidae). *Z. Angew. Ent.* 92 : 18–26.

84. Ross, H. 1911. *Die Pflanzengallen (Cecidien) mittel-und Nordeuropas*. Jena, G. Fischer Verlag.

85. Royaltey, H.H., Estrachan, E., Sokal, R.R. 1975. Tests for patterns in geographic variation. *Geographical Analysis* 7 : 369–395.

86. Schaefer, H. 1976. 1st die Reblaus tot? *Deutches Weinbau Jahrbuch* (1976) 69–74.

87. Senner, J.W., Sokal, R.R. 1974. Analysis of dimorphism in natural populations with consideration of methodological and epistemological problems. *Syst. Zool.* 23 : 363–386.

88. Seutin, E., Baurant, R. 1968. Essai de lutte pratique contre le *Chermes* de la

Picea. Parasitica. 23 : 49–55.

89. Sokal, R.R. 1952. Variation in a local population of Pemphigus. *Evolution* 6 : 296–315.

90. Sokal, R.R. 1962. Variation and covariation of characters of alate *Pemphigus populi-transversus* in eastern North America. *Evolution* 16 : 227–245.

91. Sokal, R.R., Rinkel, R.C. 1963. Geographic variation of alate *Pemphigus populi-transversus* in Eastern North America. *Univ. Kansas Sci. Bull.* 10 : 467–507.

92. Sokal, R.R., Thomas, P.A. 1965. Geographic variation of *Pemphigus Populi-transversus* in Eastern North America: stem mothers and new data on alates. *Univ. Kansas Sci. Bull.* 46 : 201–252.

93. Sokal, R.R , Rohlf, F.J. 1969. Biometry. Freeman & Co.

94. Sokal, R.R., Heryford, N.N., Kishpaugh, J.R.L. 1971. Changes in microgeographic variation patterns of *Pemphigus populitransversus* over a six-year span. *Evolution* 25 : 584–590.

95. Sokal, R.R., Oden, N.L. 1978a. Spatial autocorrelations in biology I. Methodology. *Biol. J. Linn. Soc.* 10 : 199–228.

96. Sokal, R.R., Oden, N.L. 1978b. Spatial autocorrelations in biology. 2. Some biological implications and four applications of evolutionary and ecological interest. *Biol. J. Linn. Soc.* 10 : 229–249.

97. Sokal, R.R., Bird, J., Riska, B. 1980. Geographic variation in *Pemphigus populicaulis* (Insecta, Aphididae) in Eastern North America. *Biol. J. Linn. Soc.* 14 : 163–200.

98. Sokal, R.R., Riska, B. 1981. Geographic variation in *Pemphigus populitransversus* (Insecta, Aphididae). *Biol. J. Linn. Soc.* 15 : 201–234.

99. Stevenson, A.B. 1967. *Leucopis simplex* (Diptera, Chamaemyidae) and other species occurring in galls of *Phylloxera vitifoliae* in Ontario. *Can. Entomol.* 99 : 815–820.

100. Stevenson, A.B. 1975. The grape phylloxera *Daktulosphaira vitifoliae* (Homoptera, Phylloxeridae) in Ontario, Canada: Dispersal behavior of 1st stage apterae from leaf galls. *Proc. Ent. Soc. Ontario* 106 : 24–28.

101. Stoetzel, M.B., Tedders, W.L. 1981. Investigations of two species of *phylloxera* on pecan in Georgia, *J. Ga. Ent. Soc.* 16 : 144–151.

102. Stoetzel, M.B. 1981. Two new species of *Phylloxera* (Phylloxeridae, Homoptera) on pecan. *J. Ga. Ent. Soc.* 16 : 127–144.

103. Sunose, T. 1980. Predation by tree sparrow (*Passer montanus* L.) on gall making aphids. *Kontyû* 48 : 362–369.

104. Swain, A.F. 1919. A synopsis of the *Aphididae* of California. Univ. California publications. *Entomology (Technical Bulletin)* 3 : 1–221.

105. Theobald, F.V. 1929. *The plant lice or Aphididae of Great Britain.* Vol. 3. London: Headley Bros.

106. Tomiuk, J., Wöhrmann, K., Eggers-Schumacher, H.A. 1979. Enzyme patterns as a characteristic for the identification of aphids. *Z. Ang. Ent.* 88 : 440–446.

107. Tomiuk, J., Wöhrmann, K. 1980a. Enzyme variability in populations of aphids. *Theor. Appl. Genet.* 57 : 125–127.

108. Tomiuk, J., Wöhrmann, K. 1980b. Population growth and population structure of natural populations of *Macrosiphum rosae* (L.) (Hemiptera, Aphididae). *Z. Ang. Ent.* 90 : 464–473.

109. Tomiuk, J., Wöhrmann, K. 1981. Changes of the genotype frequencies at the MDH locus in populations of *Macrosiphum rosae* (L.) (Hemiptera, Aphididae). *Biol. Zbl.* 100 : 631–640.

110. Walton, L. 1980. Gall formation and life history of *Pineus floccus* (Patch) (Homoptera, Adelgidae) in Virginia. *Virginia J. Sci.* 31 : 55–60.

111. Wertheim, G. 1950. Studies on the biology, ecology and taxonomy of the tribe *Fordini* in Israel. Jerusalem: Hebrew University (in Hebrew, English summ.) Ph.D. Thesis.

112. Wertheim, G. 1954. Studies of the biology and ecology of the gall producing aphids of the tribe *Fordini* (Homoptera: Aphidoidea) in Israel. *Trans. R. Ent. Soc. Lond.* 105 : 79–97.

113. Wertheim, G. 1955. New Observations on gall-producing aphids on *Pistacia atlantica* in Israel. *Bull. Res. Counc. Israel* 4 : 392–394.

114. Wertheim, G., Linder, J. 1961. The early development of the cauliflower gall. *Bull. Res. Counc. Israel* IOB : 133–136.

115. Wheeler, A.G. Jr., Jubb, G.L. Jr. 1979. *Scymnus cervicalis* Mulsant, a predator of grape phylloxera *Daktulosphaira vitifoliae* with notes on *S. brullei* Mulsant as a predator on woolly aphids on elm (Coleoptera, Coccinellidae). *Coleopt. Bull.* 33 : 199–204.

116. Wheeler, A.G. Jr., Henry, T.J. 1978. *Ceratocaspus modestus* (Hemiptera, Miridae) a predator of grape phylloxera: Seasonal history and description of fifth instar. *Melsheimer Ent. Ser.* 25 : 6–10.

117. Whitehead, F.E., Eastep, O. 1937. The seasonal cycle of *Phylloxera notabilis* Pergande (Phylloxeridae, Homoptera). *Ann. Ent. Soc. Amer.* 30 : 71–74.

118. Whitham, T.G. 1978. Habitat selection by *Pemphigus* aphids in response to resource limitations and competition. *Ecology* 59 : 1164–1176.

119. Whitham, T.G. 1979. Territorial behavior of *Pemphigus* gall aphids. *Nature* 279 : 324–325.

120. Whitham, T.G. 1980. The theory of habitat selection: Examined and extended using *Pemphigus* aphids. *Amer. Natur.* 115 : 449–466.

121. Wool, D., Koach, J. 1976. Morphological variation of the gall-forming aphid, *Geoica utricularia* (Homoptera) in relation to environmental variation. In *Population Genetics and Ecology*, ed. Karlin, S. & Nevo, E, Academic Press, PP. 239–272.

122. Wool, D. 1977. Genetic and environmental components of morphological variation in gall-forming aphids (Homoptera, Aphididae, Fordinae) in relation to climate. *J. Anim. Ecol.* 46 : 875–889.

123. Wool, D., Bunting, S., van Emden, H.F. 1978. Electrophoretic study of genetic variation in British *Myzus persicae* (Sulz) (Hemiptera, Aphididae). *Biovhem. Genet.* 16 : 987–1006.

124. Wool, D., van Emden, H.F., Bunting, S.W. 1978. Electrophoretic detection of the internal parasite, *Aphidius matricariae*, in *Myzus persicae*. *Ann. Appl. Biol.* 90 : 21–26.

125. Wool, D. 1980. On ecological inference from Kurtosis and skewness of morphological characters. *Res. Popul. Ecol.* 22 : 263–272.

126. Wool, D. 1980. Galls and Genetic variation in the *Fordinae* (Aphididae, Homoptera). *Bull. Soc. Bot. Fr.* 127 (Actual. Bot. 1) : 197–198.

127. Zivanovic, V. 1978. Gall aphids of the genus *Schizoneura* from elm as a pest in pear nurseries. *Zast Bilja* 29 : 257–264 (in Serbo-Croatian, Eng. summ.).

ADDENDUM

Notes added in print

Since this article went to press, there came to the author's attention a number of recent publications on gall-forming aphids which he thought should be mentioned in this review.

Population ecology

A series of four papers (Parry, 1978a, 1978b, 1980; Parry and Spiers, 1982) describe in detail the ecological factors affecting the population size of the Adelgid *Gilletteela cooleyi* on its secondary host in Scotland. The author points at winter mortality as the principal regulator of population numbers. Winter mortality is caused by dislodgement from the host, freezing, and perhaps changes in the host plant (Parry 1978a, 1980). Those aphids which survive the winter migrate, as sexuparae, back to the primary host, but some fail to migrate and die on the secondary host, perhaps due to changes of soluble nitrogen levels in the host (Parry, 1978b). Predators do not seem to play an important role at these stages in the life cycle.

Predators. The larvae of a gelechid moth, *Nola innocua* Butler, feeds on the galls of two aphid species in Japan (Ito and Hattori, 1982). The larvae feed on the gall tissue and are not aphid predators. The maggots of a syrphid, *Cnemodon* sp., and an Itonidid, *Aphidoletes abietis* Kieff., were found in Switzerland preying on the aphids inside tightly closed galls of three species of Adelgidae (Mitchell and Maksymov, 1977).

Host selection. One factor determining the suitability of *Populus angustifolia* trees, or leaves within trees, to gall formation by the fundatrices of *Pemphigus betae* may be the phenol content of the leaves. Phenol gradients exist within leaves. Trees with higher gall density per leaf, had lower phenol concentration (Zucker, 1982). It is possible that *Pemphigus* fundatrices are seeking feeding sites low in phenol concentration.

ADDITIONAL REFERENCES

128. Ito, Y., and Hattori, I. 1982. A kleptoparasitic moth, *Nola innocua*, attacking aphid galls. *Ecological Entomology* 7: 475–478.
129. Parry, W.H. 1978a. Studies on the factors affecting the population levels of *Adelges cooleyi* (Gillette) on Douglas fir. I. Sistens on mature needles. *Z. Ang. Ent.* 85: 365–378.
130. Parry, W.H. 1978b. Studies on the factors affecting the population levels of *Adelges cooleyi* (Gillette) on Douglas fir. II. Progredientes and sistens on current year needles. *Z. Ang. Ent.* 86: 8–18.
131. Parry, W.H. 1980. Studies on the factors affecting the population levels of *Adelges cooleyi* (Gillette) on Douglas fir. III. Low temperature mortality. *Z. Ang. Ent.* 90: 133–141.

132. Parry, W.H., and Spiers, S. 1982. Studies on the factors affecting the population levels of *Adelges cooleyi* (Gillette) on Douglas fir. IV. Polymorphism in progredientes. *Z. Ang. Ent.* 94: 253–263.

133. Mitchell, R.G., and Maksymov, J.K. 1977. Observations of predation on spruce gall aphids within the gall. *Entomophaga* 22: 179–186.

134. Zucker, W.V. 1982. How aphids choose leaves: the roles of phenolics in host selection by a galling aphid. *Ecology* 63: 972–981.

APPENDIX

The following list contains only species known to form galls (occasionally pseudogalls) on a primary host. Readers interested in a description or illustration of any species may consult the publication code (numbered in columns 3 and 4 of the table) in the reference list. The code gives the publication number, followed (in parentheses) by either page numbers or index numbers used by the author of that publication to refer to that species.

Hosts (P=primary, S=secondary)	Distribution	Ref: Description G=gall, A=aphid	Ref: Illustration G=gall, A=aphid
1	2	3	4
Family phylloxeridae			
Acanthochermes quercus Kollar	Central Europe	G: **21** (5515)	G:
P: Quercus (Q. petrae, Q. ruber)		A:	A:
S:			
Phylloxera caryae-avellana Riley	South & Central USA	G: **36** (p. 74)	G: **36** (14.4)
P: Carya		A:	A:
S:			
Phylloxera caryaefallax Riley	South & Central USA	G: **36** (p. 72)	G: **36** (14.5)
P: Carya		A:	A:
S:			
Phylloxera caryaeglobuli Walsh	South & Central USA	G: **36** (p. 72)	G: **36** (14.6,46)
P: Carya		A:	A:
S:			
Phylloxera caryaesemen Walsh	South & Central USA	G: **36** (p. 72)	G: **36** (14.8)
P: Carya		A:	A:
S:			
Phylloxera rimosalis Perg.	South & Central USA	G: **36** (p. 72)	G: **36** (14.7)
P: Carya		A:	A:
S:			

Phylloxera caryaecaulis (Fitch) P: Carya (C. ovata, C. laciniosa, C. glabra, C. illinoinensis) S: none (monoecious)	South & Central USA	G: **36** (p. 75), **23** A: **23**	G: **23** (1) A: **23** (3–10)
Phylloxera devastatrix Pergande P: Carya illinoinensis (Pecan) S: none (monoecious)	South & Central USA	G: **36** (p. 75) A:	G: A:
Phylloxera notabilis Pergande P: Carya illinoinensis (Pecan) S: none (monoecious)	South & Central USA	G: **36** (p. 74) **101** A:	G: **101** (1) A:
Phylloxera russellae Stoetzel P: Carya illinoinensis (Pecan) S: none (monoecious)	South & Central USA	G: **101** A: **102**	G: **101** (2, 3), **102** (9, 10) A: **102** (1–8)
Phylloxera texana Stoetzel P: Carya illinoinensis (Pecan) S: none (monoecious)	South & Central USA	G: **102** A: **102**	G: **102** (12, 13) A: **102** (11)
Phylloxera (20 more species) P: Carya	South & Central USA	G: **36** (p. 71–5)	
Viteus vitifolii Fitch = Daktylosphaera vitifolii (Fitch) = Pemphigus vitifoliae Fitch = Phylloxera vastatrix Planch. = Phylloxera vitifoliae Fitch P: Vitis vinifera (leaves) S: Vitis vinifera (roots)	North America, Europe, N. Africa, Middle East, Australia	G: **20** (p. 366–70) **21** (7625) **59** (1884–5) **71** (p. 63) A: **104** (180)	G: **20** (189), **36** (3.5), **71** (68), **116** (1) A: **20** (186–8)

Family Adelgidae (*Chermesidae*)

Adelges coccineus Cholodk. P: Picea excelsa S: Larix	Europe	G: **57** (96) A:	G: **57** (17–20) A:

1	2	3	4
Adelges laricis Vallot =Cnaphalodes strobilobius (Kltb.) =:Adelges strobilobius Kalt. P: Picea (P. excelsa, P. glaucus, P. alga, P. abies, P. orientalis) S: Larix L. decidua, L. europaea)	Europe, N. America	G: **21** (4791), **41** (p. 536–41), **57** (94), **84** (1162) A: **24**	G: **20** (170) **21** (163) **57** (16), **84** (69) **24** (2) A: **24** (4, 5, 6)
Adelges (7 more species) P: Picea sp.	N. America		G: **36**
Aphrastasia pectinatae (Cholodk.) P: Picea (P. alba, P. excelsa) S: Abies (A. alba, A. balsamea, A. nordmannianae)	N. & Central Europe	G: **20** (p. 341), **21** (4790) A:	G: **20** (169) A:
Dreyfusia nordmannianae (Eckst.) =Adelges normandiannae (Eckst.) P: Picea orientalis S: Abies (A. alba, A. nordmannianae)	Europe, N. America	G: **20** (p. 332), **21** (4788) A: **24**	G: **20** (159–161), **24** (13) A: **24** (20–23)
Dreyfusia prelli Grossmann P: Picea orientalis S: Abies nordmannianae	Switzerland, Caucasus	G: **21** (4789) A:	G: A:
Gilletteella cooleyi (Gill.) =Adelges cooleyi Gill. P: Picea (P. glauca, P. engelmanni) S: Pseudotsuga menziesii	N. Europe, N. America	G: **21** (4783), **36** (p. 45) A: **27, 36**	G: **27, 36** (14.1), **24** (7) A: **27, 24** (15–19)
Pineus cembrae (Cholodk.) =Adelges sibiricus Cholodk. =Chermes sibiricus Cholodk. P: Picea excelsa S: Pinus cembrae	N. Europe	G: **21** (4784), **57** (103) A:	G: **57** (23) A:

Pineus orientalis (Dreyfus) P: Picea orientalis S: Pinus montana, P. silvestris	Caucasus	G: **21** (4785) A: **24**	G: **24** (20) A: **24** (31, 32)
Sacchiphantes abietis (L.) =Chermes abietis Kalt. P: Picea (P. alba, P. glauca, P. pungens, P. rubra, P. excelsa, P. sitchensis) S: Larix	Central & N. Europe Morocco, Lebanon, Turkey	G: **19** (203), **21** (4787) **57** (101), **71** (p. 60), **84** (1161) A: **41** (p. 525–8), **24**	G: **21** (160–2), **71** (61–64), **24** (516) A: **20** (177), **24** (13, 14)
Sacchiphantes viridis (Rtzb). =Chermes geniculatus Rtzb. =Chermes laricis Htg. =Adelges viridis Ratz. P: Picea (P. alba, P. morinda, P. nigra, P. orientalis, P. sitchensis) S: Larix	N. Europe	G: **21** (4786), **57** (102) A: **24**	G: **20** (175), **24** (3) A: **20** (176), **57** (22) **24** (11, 12)
Pineus floccus (Patch) P: Picea rubens S: Pinus strobus	N. America	G: **36** (p. 44), **110** A: **110**	G: A:

Family Eriosomatidae (=Pemphigidae)

Subfam. Eriosomatinae

Colopha compressa (Koch) P: Ulmus (U. laevis, U. effusa) S: Carex (roots)	Europe, Turkey	G: **19** (p. 224), **21** (7323) **84** (1991) A:	G: A:
Colopha moriokaensis (Monzen) P: Zelkova serrata S:	Japan	G: A:	G: **103** A:

1	2	3	4
Colopha ulmicola (Fitch) = Byrsocrypta ulmicola Fitch P: Ulmus americana	N. & C. Europe, N. America	G: 20 (p. 293)	G: 20 (126), 36 (186) 56 (40)
S: Leersia, Aira, Agrostis, Panicum, Triticum		A: 56 (p. 348), 78 (p. 347)	A: 78 (417)
Eriosoma (Schizoneura) lanuginosum (Htg.) = Schizoneura lanuginosa (Htg.) P: Ulmus (U. carpinifolia, U. scabra, U. campestris)	England, N. & C. Europe, N. Africa, Iran, Syria, India, Pakistan, New Zealand	G: 19 (p. 224), 21 (7322), 59 (578), 71 (p. 103), 84 (1986), 105 (p. 283)	G: 19 (51.3), 21 (424), 71 (118), 84 (228) 105 (169)
S: Pyrus mamorensis, P. communis		A: 19 (p. 227), 26 (p. 336) 105 (p. 282–8)	A: 26 (96)
Eriosoma rileyi Thomas = Schizoneura rileyi Thos. P: Ulmus	N. America	G: 36 (p. 228) A:	G: 36 (184) A:
S:			
Kaltenbachiella ulmifusa (Walsh & Riley) = Gobaishia ulmifusa Walsh & Riley P: Ulmus fulva	N. America	G: 20 (p. 293), 56 (p. 359) A: 78 (p. 354)	G: 20 (127), 56 (45), A: 78 (424)
S:			
Kaltenbachiella pallida (Haliday) = K. ulmi Licht = Byrsocrypta pallida Hal. = Gobaishia pallida Haliday P: Ulmus (U. carpinifolia, U. scabra, U. campestris, U. montana)	Europe, C. Asia, Middle East, N. Africa (roots)	G: 19 (167; p. 224), 20 (p. 293), 21 (7315)	G: 19 (51.2), 20 (124b), 21 (422), 84 (227) 105 (185)

	Distribution		
S: Mentha, Nepata (Artemisia?)		A: 41 (p. 488–9, 105 (p. 305)	A: 105 (104)

Tetraneura (Tetraneurella) akinire Sasaki

	Distribution		
P: Ulmus (U. campestris, U. parvifolia)	USSR, Hungary, Yugoslavia, Japan	G:	G:
S:		A: 55 (p. 36–41)	A: 55 (52, 53)

Tetraneura caerulescens (Passerini)
=Pemphigus caerulescens Pass.
=T. rubra Licht.
=T. aegyptiaca Theob.
=Byrsocrypta caerulescens Pass.

	Distribution		
P: Ulmus campestris	Mediterranean, Iran N. Africa, N. Europe	G: 19 (168–9, p. 224) 20 (p. 289), 59 (577)	G: 59 (305)
S: Panicum, Saccarum, Setaria		A: 55 (p. 45–49)	A: 20 (123), 55 (3, 26, 41, 54)

Tetraneura (Tetraneurella) javensis van der Goot

	Distribution		
P: Ulmus wallichiana	Java, Pakistan, India	G:	G:
S: Saccharum		A: 42 (260–2), 55 (p. 49–52)	A: 55 (2, 13–15, 27, 49)

Tetraneura (Tetraneurella) nigriabdominalis (Sasaki)

	Distribution		
P: Ulmus davidiana	Japan, India, Indonesia	G:	G:
S: Oryza (+ other graminae)		A: 55 (p. 52)	A: 55 (28–9)

Tetraneura (Tetraneurella) polychaeta HRL

	Distribution		
P: Ulmus villosa	Pakistan	G:	G:
S:		A: 55 (p. 64)	A: 55 (36, 44, 57–8)

Tetraneura radicicola Strand

	Distribution		
P: Ulmus davidiana	Japan, India, Malaya	G:	G:
S:		A: 55 (p. 68)	A: 55 (20, 60)

Tetraneura ulmi L.
=T. ulmisacculi Patch
=Byrsocrypta ulmi (L.)
=T. gallarum (Gmel.)

1	2	3	4
=Colopha ulmisacculi (Patch) = T. ulmifoliae Baker			
P: Ulmus (U. campestris, U. carpinifolia)	Palaearctic, introduced in N. America; Middle East, North Africa, (roots)	G: 19 (p. 224), 20 (p. 290), 21 (7324), 57 (2048), 59 (576), 84 (1990)	G: 20 (124), 21 (425), 36 (187), 59 (306–9), 66 (71D.), 84 (230–1), 105 (181)
S: Ribes, Triticum, Avena, Agropyron, and other Gramineae		A: 19 (p. 227), 41 (484–7), 55 (p. 75), 78 (p. 348), 105 (p. 293)	A. 20 (125), 55 (7, 35, 45, 55) 78 (418), 105 (176–80)
Tetraneura yezoensis Matsumura			
S: Pennisetum	Australia, Japan	G: A: 55 (p. 81)	G: A: 55 (37)
Subfamily Melaphinae			
Melaphis rhois (Fitch)			
P: Rhus glabra	N. America	G: 20 (p. 322)	G:
S: mosses?		A:	A:
Schlechtendalia chinensis Bell			
P: Rhus (R. semialata, R. japonica)	Japan, Korea, Formosa China	G: 59 (1774)	G: 20 (153), 59 (1017–25)
S: mosses?		A:	A:
Subfamily Hormaphidinae			
Astegopteryx styracicola Takahashi			
P: Styrax suberifolia	Taiwan	G: 6, 8	G: 6, 8
S: Bamboo?		A: 6	A: 6
Astegopteryx styracophila Karsch			
P: Styrax (S. serratum, S. benzoin)	India, Java, Sumatra	G: 59 (2580–1), 67 (p. 194)	G: 59 (1583–8)
S: Bamboo?		A:	A:

Ceratovacuna nekoashi (Sasaki)
=*Astegopteryx nekoashi* Sasaki

P: Styrax japonicum	Japan	G: **59** (2583)	G: **59** (1589–90)
S: Bamboo?		A:	A:

Hamamelistes spinosus Shimer

P: Hamamelis virginiana	N. America	G: **56** (p. 375)	G: **36** (205)
S: Betula fontinalis		A: **56** (in key), **78** (p. 378)	A: **56** (50), **78** (455)

Hormaphis hamamelidis Fitch

P: Hamamelis virginiana	N. America	G: **56** (p. 376)	G: **36** (206)
S: Betula		A: **56** (in key)	A:

Nipponaphis distyliicola Monzen

P: Distylium racemosum	Japan	G: **128**	G: **128**
S:		A:	A:

Monzenia globuli (Monzen)

P: Distylium racemosum	Japan	G: **128**	G: **128**
S:		A:	A:

Reticulaphis distyli (van der Goot)
=*Schizoneuraphis distyli* van der Goot

P: Distylium stellare	Java	G: **59** (1105)	G:
S: Fagaceae?		A: **42** (p. 247–9)	A:

Schizoneuraphis foliorum van der Goot

P: Distylium stellare	Java	G:	G:
S: Fagaceae?		A: **42** (p. 250–2)	A:

Schizoneuraphis gallorum van der Goot

P: Distylium stellare	Java	G: **59** (1104)	G: **59** (657)
S: Fagaceae?		A: **42** (p. 252–6)	A:

Subfamily Pemphiginae

Tribe Prociphilini

Thecabius affinis (Kaltenbach)
=Pemphigus affinis Kalt.

1	2	3	4
P: Populus (p. nigra v. italica, P. deltoides, P. pyramidalis) S: Ranunculus, Filago? Chenopodium? Soncus?	Palaearctic, N. Africa (roots)	G: 20 (p. 304), 21 (5115, 5117), 59 (210), 84 (1291), 105 (p. 223–4) A: 41 (p. 456–60), 105 (p. 220–3)	G: 19 (52.4), 20 (138), 105 (125) A: 105 (124)
Thecabius populiconduplifolius (Cowen) P: Populus (P. deltoides, P. balsamifera, P. sargentii, P. monilifera) S. Ranunculus	Canada	G: 51, 2 A: 78 (370)	G: 51 (3, 4) A: 68 (2), 78 (448)
Thecabius populimonilis (Riley) P: Populus, (P. angustifolia, P. balsamifera) S: Ranunculus	N. America, Canada	G: 46, 68, 104 (169) A: 78 (p. 371)	G: 46 (8) A: 68 (1), 78 (449)
Tribe Pachypappini			
Mordwilkoja vagabunda Walsh P: Populus (P. deltoides, P. sargentii) S:	N. America, Canada	G: 20 (p. 298), 46, 56 (p. 360), 60, 61 A: 78 (p. 358)	G: 20 (132), 36 (42), 46 (9, 10), 56 (46), 60, 61 A: 78 (430)
Pachypappa lactea Tullgr. P: Populus tremulae S:	N. Europe	G: 58 (6360), 84 (1293) A:	G: A:
Pachypappa marsupialis Koch P: Populus (P. nigra v. italica, P. maximowczii) S:	Europe, Japan	G: 20 (p. 296), 21 (5093) 57 (539) A:	G: 20 (130) A:
Pachypappa populi (L.) = P. grandis Tullgr. P: populus (P. alba, P. tremulae) S:	N. & C. Europe	G: 20 (p. 298), 21 (5095) A:	G: 20 (131) A:

Pachypappa sacculi Gillette
=Pemphigilachnus kaibabensis Knowlton
=Asiphum sacculi (Gill.)
P: Populus tremuloides
S:

N. America, Canada

G: 20 (p. 298), **51**
A:

G: **51** (1)
A:

Pachypappa varsoviensis (Mordv.)
=Pemphigus varsoviensis Mordv.
=Asiphum varsoviensis (Mordv.)
P: Populus

S:

N. Europe

G: **21** (5066), **57** (470) **84** (1262)
A:

G:
A:

Pachypappa vesicalis Koch
P: Populus (P. alba, P. bollcana)

S:

C. Europe, Italy

G: **21** (5065), **57** (469) **84** (1261)
A:

G:
A:

Tribe Pemphigini

Epipemphigus imaicus (Cholodk.)
=Pemphigus imaicus Cholodk.
P: Populus ciliata

S.

India

G: **59** (215), **67** (p. 241–2)
A:

G: **59** (139), **66** (VI.1) **67** (X.1)
A:

Epipemphigus niisimae HRL
P: Populus maximowiczii
S:

Japan

G: **11**
A:

G: **11**
A: 11

Pemphigus betae Doana
=P. balsamifera Williams
P: Populus (P. balsamifera, P. angustifolia)
S: Beta, Chenopodium, Achillea, Agropyron

N. America, Canada

G: **46, 47, 49, 104** (164)

A: 20 (p. 309), **49**, **78** (p. 359)

G: **46** (1, 14, 17), **118** (2)

A: **68** (7), **78** (431)

Pemphigus borealis Tullgr.

1	2	3	4
Pemphigus bursarius (L.) =Pemphiginus bursarius = P. lactuarius Pass. =Aphis bursarius L. P: Populus (P. laurifolia, P. nigra v. italica, P. pyramidalis) S: Lactuca, Taxacum, Bidens	Sweden, Russia, Britain, Iran?	G: **19** (p. 317), **20** (p. 310), **21** (5053), **84** (1269) A: **105** (258–60)	G: **20** (144), **29** (1), **105** (152) A: **29** (4), **105** (151)
P: Populus (P. nigra v. italica)	Europe, N. America N. Africa, Middle East, New Zealand (roots)	G: **19** (p. 225, 174), **20** (p. 308) **21** (5073), **59** (201, 206) **71** (p. 97), **84** (1275) 30 A: **26** (p. 341), **41** (p. 464–7), **78** (p. 360), **105** (p. 245–51)	G: **2** (4, 5), **21** (184), **51** (2), **57** (135), **58** (1379–80), **66** (71 a, b), **84** (76, 77), **105** (144–5) A: **20** (142), **26** (97–8), **78** (432), **105** (139–43)
S: Beta, Chenopodium, Euphorbia, Daucus, Crepis, Cichorium, Taraxacum, Rumex			
Pemphigus dorocola Matsumura P: Populus maximowiczii S:	Japan	G: A: **7**	G: **7, 10** A: **7**
Pemphigus immunis Buckton =P. lichtensteini Tullgr. = P. globulosus Theob. P: Populus (P. nigra v. italica, P. pyramidalis, P. euphratica) S:	N. & C. Europe, N. Africa, Middle East, C. Asia	G: **19** (p. 224, 176), **20** (p. 310), **21** (5054), **67** (p. 243), **71** (p. 99), **84** (1270) A:	G: **19** (52.1), **71** (112, 113) A:
Pemphigus junctisensoriatus Maxson P: Populus (P. sargentii, P. deltoides) S:	N. America, Canada	G: **2** A: **78** (p. 361)	G: **2** (11), **78** (433) A: **78** (433)
Pemphigus monophagus Maxson P: Populus (P. angustifolia, P. balsamifera, P. deltoides) S: none (monoecious)	N. America, Canada	G: **2** A: **78** (p. 361)	G: **2** (14) A: **78** (434)

Pemphigus mordvilkoi Cholodk.
  P: Populus ciliata
  S:
India
G: 67 (p. 242), 59 (213) A:
G: 59 (137) A:

Pemphigus napaeus Buckton
  P: Populus ciliata
  S:
India
G: 67 (p. 242) A:
G: A:

Pemphigus nortoni Maxson
  P: Populus sargentii
  S:
N. America, Canada
G: 46, 47 A: 78 (p. 362)
G: 46 (5, 20) A: 78 (435)

Pemphigus populi Courchet
= Pemphiginus populi (Courchet)
  P: Populus nigra
  S:
C. Europe, Mediterranean, C. Asia
G: 19 (172, p. 225), 20 (p. 311) 57 (537), 59 (208) A:
G: 20 (145b), 57 (141–2), 59 (140) A:

Pemphigus populicaulis Fitch
  P: Populus (P. deltoides, P. sargentii, P. fremonti, P. trichocarpa)
  S: Umbelliferae?
N. America, Canada
G: 2, 5 (p. 312), **46, 68,** 104 (165) A: 68, 78 (363)
G: 2 (9, 10), 20 (146a), 46 (3,15,16, 19), 68 (3) A: 68 (3), 78 (436)

Pemphigus populiglobuli Fitch
  P: Populus (P. balsamifera, P. angustifolia, P. deltoides)
  S:
N. America, Canada
G: 2, 20 (p. 312) **46, 68** A: 68, 78 (p. 363)
G: 20 (146b) 46 (2) 68 (4) A: 68 (4), 78 (437)

Pemphigus populinigrae (Schrank)
= P. filaginis (Fonsc.)
  P: Populus (P. nigra v. italica, **P. deltoides,** P. pyramidalis, P. monilifera)
  S: Gnaphalium, Filago
Europe, C. Asia, Middle East, Britain (roots)
G: 19 (175, p. 225), **20** (p. 307) 21 (5100), **59** (209) 84 (1281), **105** (p. 255) A: 41 (p. 469–73), **105** (p. 251–4)
G: 19 (52.3), **20** (140) 21 (181, 186), **66** (71c), 84 (80–1), **105** (149) A: **105** (146–8)

Pemphigus populiramulorum Riley
  P: Populus (P. balsamifera, P. sargentii)
  S:
Canada
G: 20 (p. 313), **46** A: 78 (p. 364)
G: 20 (146c), 46 (6) A: 68 (6), 78 (438)

1	2	3	4
Pemphigus populitransversus Riley P: Populus (P. deltoides, P. sargentii, P. monilifera, P. balsamifera) S: Cruciferae (Brassica; Lactuca)	N. America, Canada	G: 2, 20 (p. 312) 56 (p. 367), 104 (166) A: 56 (in key), 78 (p. 364)	G: 2 (6), 20 (143) 36 (1.6) 46 (4, 11–13, 18), 56 (48) A: 78 (439)
Pemphigus populivenae Fitch P: Populus (P. angustifolia, P. balsamifera, P. trichocarpa) S: Beta vulgaris	N. America, Canada	G: 46, 68, 20 (p. 312) A: 78 (p. 365)	G: 46 (7) A: 78 (440)
Pemphigus protospirae Lichtenstein P: Populus (P. nigra v. italica, P. pyramidalis) S: Umbelliferae?	Europe, Mediterranean, C. Asia	G: 19 (177), 20 (p. 307), 21 (5070), 84 (1278) A: 41 (p. 474–6)	G: 20 (139) A:
Pemphigus spirothecae Passerini P: Populus (P. nigra v. italica, P. pyramidalis, P. tremulae) S: none (monoecious)	N. & C. Europe N. Africa, Middle East, N. America (Canada)	G: 2, 20 (p. 306), 21 (5069) 57 (535), 59 (203), 71 (p. 101), 84 (1277) 105 (p. 257) A: 41 (p. 476), 105 (p. 256–7)	G: 2 (1–3), 19 (52.2), 21 (177), 57 (8–40), 71 (116–7), 84 (78, 79), 105 (150) A:
Pemphiginus vesicarius passerini =Pemphiginus vesicarius Pass. P: Populus (P. nigra, P. alba, P. suaveolens) S:	Mediterranean, N. Africa, C. & S. Europe,	G: 19 (173, p. 224), 20 (p. 311), 21 (5071), 57 (524), 59 (207) A:	G: 20 (145a), 57 (136) A:

Subfamily Fordinae

Tribe Fordini

[Fundatrix galls—see text]

=Pemphigus sp.
P: Pistacia (P. atlantica, P. terebinthus) Mediterranean, N. Africa G: 58 (7017), 59 (1735), 71 (p. 92), 112 G: 58 (1479–80), 59 (1014–5), 63 (12), 71 (97–100), 112 (6)

S: A: A:

Forda dactylidis Börner
=F. mordvilkoi Börner
P: Pistacia atlantica Crimea, Israel?
S: Graminae G: 19 (185) G:
A: A:

Forda formicaria Heyden
=Pemphigus semilunaria Pass.
P: Pistacia (P. palaestina; P. mutica, P. khinjouk; P. terebinthus) Mediterranean, roots also: N. America N. Europe G: 19 (184), 28 (p. 119) G: 28 (74), 63 (6)
S: Graminae (Bromus, Agropyron, Triticum, Festuca. Poa, Soncus) A: 28 (p. 118), 78 (p. 373) A: 28 (84, 87, 99, 102), 78 (450), 105 (90–2)

Forda hirsuta Mordv.
P: Pistacia vera, P. mutica, P. terebinthus G: 28 (p. 123) G: 28 (75)
S: A: 28 (p. 124) A:

Forda kaussarii Davatchi & Remaudiere
P: Pistacia khinjouk Iran G: 28 (p. 120) G: 28 (72)
S: A: 28 (p. 119) A: 28 (90, 100)

Forda marginata Koch
=Pemphigus follicularius Pass.
=Forda follicularia Pass.
=F. olivacea Rohwer
=Pentaphis marginata Koch
P: Pistacia (P. palaestina; P. mutica, P. terebinthus) Middle East Mediterranean G: 19 (183) 59 (1734, 1740, 1748) G: 63 (7), 71 (101–3)
S: Graminae (Bromus, Elymus, Hordeum, Agropyron, Setaria) A: 78 (p. 373), 105 (186–7) A: 28 (89, 93, 98), 78 (451), 105 (104–5)

Forda riccobonii (Stephani)

1	2	3	4
=Pemphigus riccobonii Stephani P: Pistacia (P. atlantica, P. terebinthus, P. mutica) S: Graminae (Triticum)	Middle East, N. Africa, Canary Islands	G: **19** (186), **28** (p. 122), **59** (1749), **71** (100) A: **28** (p. 119)	G: **19** (23.2), **28** (73), **59** (1013), **63** (14), **71** (114–5), **113** (2) A: **28** (88)
Paracletus cimiciformis Heyden =Pemphigus pallidus Derbes =Pemphigus derbesi Licht. =Paracletus portschinskyi Mordv. P: Pistacia (P. palaestina, P. vera, P. terebinthus) S: Graminae (Hordeum, Triticum, Agropyron, Andropogon, Festuca, Erianthus)	Mediterranean, Middle East, roots also: C. Asia, Europe	G: **19** (190), **20** (p. 314) **28** (p. 116) A: **19** (p. 228), **20** (p. 314) **105** (p. 202–7)	G: **28** (68–9), **63** (8) A: **20** (147), **28** (85, 91, 92, 96) **105** (117–9)
Smynthurodes betae Westwood =Trifidaphis phaseoli (passerini) =T. radicicola Essig. P: Pistacia (P. atlantica; P. mutica, P. vera) S: Triticum, Hordeum, Lathyrus, Lycopersicum, Gossypium, Amaranthus, Euphorbia.	Mediterranean. roots also: N. & C. Europe, N. America, N. Africa, E. Africa, New Zealand	G: **19** (192), **28** (112–114), **31** (100–1) A: **19** (p. 228), **26** (p. 323) **28** (112–4), **78** (p. 375) **104** (163), **105** (207–212)	G: **19** (23.4), **28** (67), **63** (13), **113** (4) A: **26** (92), **28** (86, 95, 97, 101) **78** (453), **105** (120–1)
Tribe Baizongiini *Aploneura lentisci* (Passerini) =Tetraneura lentisci Pass. =Rhizobius graminis Buckt.			

57

Taxon / Hosts	Distribution	G (1)	A (1)	G (2)	A (2)
=Tychea silvestrii Mordv. P: Pistacia lentiscus S: Triticum, Hordeum, Poa, Agropyron, Bromus, Anthoxanthum	Mediterranean. roots also: Britain, N. &. C. Europe, New Zealand, Canary Islands	19 (196), 20 (p. 318) 28 (p. 135), 59 (1727)	19 (p. 229), 26 (p. 329), 105 (213–6)	20 (149), 28 (82), 59 (1008–9) 63 (17), 71 (106, 108)	19 (46.15), 26 (94), 28 (117, 121), 105 (122)
Asiphonella cynodonti Das =*Baizongiini, sp. P: Pistacia (P. khinjouk-Iran, P. palaestina-Israel, [rare]) S: Cynodon dactylon	Israel, Iran, roots: Egypt, E. Africa	19 (200), 28 (p. 98–99)	28 (p. 140–1)	28 (79a, b, 106), 63 (3)*	28 (114, 118, 125–6)
Baizongia pistaciae (L.) =Aphis pistaciae L. =Pemphigus cornicularius Pass. P: Pistacia (P. palaestina; P. terebinthus, P. khinjouk) S: Avena, Poa, Hordeum	Mediterranean, Middle East, roots also: C. Europe	19 (194), 20 (318) 28 (p. 134), 112 (p. 82–3) 19 (p. 229)		20 (148), 28 (105) 63 (1), 122	19 (46.17), 28 (110, 116)
Chaetogeoica foliodentata (Tao) P: Pistacia sinensis S:	Iran	28 (p. 132)		28 (78)	28 (103)
Geoica mimeuri (Gaumont) =Pemphigella mimeuri Gaumont P: Pistacia (P. atlantica, P. terebinthus) S:	Morocco	71 (p. 96) 28 (p. 130)		71 (109–10)	71 (111)
Geoica setulosa (Pass.) =Tychea setulosa Pass. P: Pistacia khinjouk S: Setaria, Oryza, Triticum	Iran, roots also: Britain		28 (p. 130), 105 (p. 200)		105 (112–3)
Geoica utricularia (Passerini) =Pemphigus utricularius Pass.					

58

1	2	3	4
P: Pistacia (P. palaestina, P. atlantica P. khinjouk) S: Graminae	Mediterranean, N. Africa, roots also: Europe, N. America	G: **19** (199), **20** (p. 320), **28** (p. 128), **71** (p. 94) **112** (p. 84–6) A: **19** (p. 230), **28** (p. 109) **78** (374)	G: **20** (151), **28** (77), **59** (1010–12), **63** (2, 9, 10), **71** (104, 105), **112** (5), **122** A: **78** (452)
Rectinasus buxtoni Theob. P: Pistacia (P. khinjouk, P. palaestina) S: Artemisia, Gossypium	Mediterranean, Iran, roots also: Europe	G: **28** (p. 125) A: **19** (p. 228), **28** (p. 106)	G: **28** (76), **63** (4) A:
Slavum esfandiari Davatchi & Remaudiere P: Pistacia mutica S: none (monoecious ?)	Iran	G: **28** (p. 139–40) A:	G: **28** (127) A: **28** (113)
Slavum lentiscoides Mordvilko P: Pistacia vera S: Agropyron	Iran, roots also: Europe	G: **20** (p. 319), **28** (p. 137) A:	G: **20** (150), **28** (81) A: **28** (109, 111, 119, 120)
Slavum mordvilkoi Kreutzberg P: Pistacia vera? S:	Turkey, Afghanistan, Iran	G: **28** (p. 138) A: **28** (p. 138)	G: A: **28** (122)
Slavum wertheimae H.R.L P: Pistacia (P. atlantica, P. khinjouk, P. mutica) S: none (monoecious)	Israel, Iran	G: **19** (197), **28** (p. 138) **113** A: **19** (p. 229), **54**	G: **19** (23.1), **28** (107), **54** (1), **63** (11), **113** (1), **122** A: **19** (46.16), **28** (108, 112, 123–4)

3. The Biology and Ecology of the Gall-forming Psylloidea (Homoptera)

I.D. Hodkinson

Introduction

The Psylloidea or jumping plant-lice comprise a group of about 2,000 species of Sternorrhynchous Homoptera, characterised by the form of the hind legs which are modified for jumping (Eastop, 1978). A recent classification, based on a study of both nymphal and adult characters, divides the psyllids into eight families, namely the Psyllidae, Triozidae, Aphalaridae, Spondyliaspidae, Carsidaridae, Calophyidae, Homotomidae and Phacopteronidae (White, 1980). The vast majority of species feed, by stylet insertion, on plants belonging to the Dicotyledones, although members of the small subfamily Liviinae feed on *Juncus* and *Carex* (Monocotyledones). Other odd species such as *Megatrioza palmicola* Crawf. and *Eutrioza opima* Log. feed on *Pritchardia* (Palmae) and *Pinus* (Coniferae) respectively.

Most species of psyllid are narrowly host specific. They normally feed on a narrow range of host plants, usually within a single plant genus and more rarely on more than one genus within a single plant family (Hodkinson, 1974). Polyphagous species such as the European *Trioza nigricornis* Först. or the South-East Asian *Paurocephala psylloptera* Crawf., which feed on plants in more than one family, are exceptional. Psyllids, like their relatives the aphids, are known to transmit a number of viral, bacterial and mycoplasmal diseases of plants including citrus greening disease, citrus leaf-mottle yellows, pear decline condition and fireblight of pears (Martinez and Wallace, 1967, McLean and Oberholzer, 1965, Schwarz et al., 1970, Jensen et al., 1974, Wilde et al., 1971, and Chen et al., 1973).

The life cycle of a typical psyllid consists of an egg stage followed by

59

five nymphal stages. The final instar nymph becomes quiescent and moults to produce the adult. Reproduction is normally sexual; males and females occurring in approximately equal numbers. However, a few species such as *Psylla myrtilli* Wag., *Psylla rara* Tut. and some *Glycaspis* species may be facultatively parthenogenetic under certain conditions and males can be absent from the population (Hodkinson, 1974, 1978; Moore, 1970).

Psyllid life cycles in temperate regions are synchronised with the seasons and there is usually a period of winter dormancy or inactivity. By contrast, generations of tropical species are usually continuous, with growth rates governed by the prevailing climate and host plant condition. Consequently, temperate species are usually restricted to at most four generations per year and many species have an annual life cycle. Tropical species, such as *Diaphorina citri* Kuw. may have as many as 10 generations per annum (Hodkinson, 1974).

Detailed feeding studies have been made on a number of psyllid species including *Strophingia ericae* (Curt.) on *Calluna vulgaris*, *Cardiaspina albitextura* Tay., *C. densitexta* Tay., *Creiis costatus* Frog., *Lasiopsylla rotundipennis* Frog. and *Glycaspis* sp. on *Eucalyptus* spp. and *Paratrioza cockerelli* Sulc on *Solanum tuberosum*. These studies indicate that psyllids feed primarily on the soluble contents of phloem tissue, either the sieve tubes or the sheathing parenchyma (Woodburn and Lewis, 1973, Hodkinson, 1973, Clark, 1962, Eyer, 1937). Different species may, however, select vascular bundles of different sizes and there is also evidence that for *C. densitexta* and *Glycaspis* sp. there is also some breakdown of the pallisade mesophyll layer of the leaf (Woodburn and Lewis, 1973, White, 1970). Some of the gall-forming *Pachypsylla* species also appear to feed primarily on mesophyll tissues (see later). A salivary sheath is normally secreted around the stylets during feeding (Walton, 1960, Woodburn and Lewis, 1973).

A number of psyllids are known to induce gall formation in their host plants and many galls have been described in the literature. J. and W. Docters van Leeuwen-Reijnvaan (1914) working in the tropical forests of Java found that about 5 per cent of all gall types collected were formed by psyllids. Unfortunately, many of these earlier gall descriptions, particularly those described from the old and new world tropics by such prolific authors as Tavares, Kieffer, Rubsaamen and J. and W. Docters van Leeuwen-Reijnvaan are not accompanied by taxonomic accounts of the accompanying psyllids, which often remain unknown to this day.

While earlier descriptions give some indication of the taxonomic range of plants galled by psyllids and illustrate variation in gall morphology, they are of strictly limited value. Houard (1908, 1909, 1922, 1923, 1933), Mani (1973), Hodkinson and White (1981) and Hodkinson (1983) should be consulted for a full list of references to these earlier papers. European psyllid galls are better documented and Buhr (1964) gives a useful recent bibliography.

This account concentrates on variations in the types of gall produced by psyllids. It examines the physiological basis of gall formation, the adaptations of psyllids for life in a gall and the adaptive significance of gall formation in the Psylloidea. It concludes with a consideration of the economic significance of the gall-forming species.

Specificity of site and shape of psyllid galls

Psyllids usually show a high degree of specificity with respect to the site of gall formation, whether it be on a leaf, flower stem or root. For example, a number of North American *Pachypsylla* species feed on hackberry (*Celtis*) and can be recognised by the shape and position of their galls. *Pachypsylla celtidis-gemma* Riley, *P. celtidis-internis* Mally and *P. pallida* Patch form bud or twig galls, *P. venusta* (O.S.) forms a purse-shaped gall on the leaf petiole whereas *P. celtidis-mamma* (Fletcher) and *P. celtidisvesicula* Crawf. form galls of different shape on the leaf lamina (Riley, 1883, 1890; Tuthill, 1943). Similarly, four species belonging to the family Triozidae form galls on *Metrosideros collina* in Hawaii (Nishida et al., 1980). *Trioza hawaiiensis* Crawf. makes pit-shaped galls on the stem. *Kuwayama minuta* Crawf. and two further *Trioza* species form either flat, cone shaped or pit galls on the leaf lamina.

Some psyllids are, however, less specific in their choice of feeding site. The Indian species *Megatrioza hirsuta* (Crawf.) feeding at high densities on *Terminalia* spp. forms leaf, flower, fruit and branch galls (Mani, 1935). The European *Trioza centranthi* (Vall.) forms both leaf and flower galls on *Centranthus ruber* and the Central African species *Phytolyma fusca* Alib. forms leaf, bud and shoot galls on *Chlorophora excelsa*, an important timber-producing tree (André, 1878; Sampo, 1975; Vosseler, 1906).

Within certain genera closely related species may produce galls that are quite different in appearance. The Japanese species *Togepsylla matsumurana* Kuw. forms pit galls on the leaves of its host *Lindera erythrocarpa* whereas the related species in Taiwan, *T. takahashii* Kuw., forms leaf roll galls on *Lindera communis* and *L. oldhami* (Takahashi, 1936; Miyatake, 1970). Alternatively, some species within a genus may form galls whereas others do not. In the *Metrosideros* example (above) there is a further *Kuwayama* species which is not a gall former. Similarly, most members of the Palaearctic genus *Psyllopsis* form roll leaf galls on *Fraxinus excelsior* but one species, *P. fraxinicola* (Först.), does not.

Life stages responsible for gall initiation

Feeding by the nymphal stages is usually responsible for gall initiation and development. Adult psyllids generally cause less visible damage to the plant. There are, however, exceptions: adults of certain pests such as the

North American *Paratrioza cockerelli* Sulc and the European *Trioza apicalis* Först, feeding on potato (*Solanum tuberosum*) and carrot (*Daucus carota*) respectively, may produce severe growth distortion of leaf and stem tissue. Adult *Trioza alacris* Flor feeding on bay (*Laurus nobilis*) produce slight leaf curling and eggs are laid in the rolled part of leaf. The hatching nymphs, however, are responsible for the characteristic roll leaf gall (Lizer, 1918; Miles, 1928).

Oviposition may also initiate gall formation. Eggs of the Holarctic *Craspedolepta nebulosa* (Zett) are laid in rows along the inferior leaf margin of *Epilobium angustifolium*, inducing a downward curling of the leaf edge (Sampo 1975). However, in this case the nymphs do not form galls.

The number of feeding nymphs associated with each gall structure is variable, dependent on the type of gall. Small pit galls and related structures on the leaf lamina usually contain a single nymph, whereas galls resulting from leaf curling or rolling usually contain several nymphs living together. Enclosed galls may contain a single nymph living in a simple chamber as in the Mexican *Trioza anceps* Tut. living on Avocado (*Persea americana*) or the Austro-Oriental species *Pauropsylla udei* Rüb. (≡ *montana*) feeding on *Ficus variegata* (Stahler, 1941; Uichanco, 1919). Alternatively, enclosed galls may be multi-chambered and contain several nymphs as in the Indian mango psyllid, *Apsylla cistellata* (Buckt.), which forms galls on the buds of *Mangifera indica* (Buckton, 1893). Newstead and Cummings (1913) record a remarkable gall made by an undescribed psyllid on a tree, thought to be *Tamarindus indica*, in Lebanon. The gall, eighteen centimetres long, shaped like a pod, contained "an immense quantity of mealy psyllids".

Morphology of psyllid galls

Psyllids are known to gall plants in a variety of ways. The effects of their feeding range from simple distortion of the plant to the formation of structurally complex galls. Leaves, petioles, shoots, stems, flowers, buds and roots are all known to be attacked but leaf galls are by far the most common. The biology of most psyllid species is unknown and it is difficult to arrive at a reliable estimate of the proportion of species which form galls. There is some indication, however, that gall formation is more frequent in some psyllid families than others (see later).

Leaf Galls
The various types of leaf gall appear to represent an intergrading series of increasing complexity. Feeding, in its simplest form, results in a distortion or twisting of young leaves. The European *Psylla buxi* (L.) feeds on the newly developing leaves around the shoot apex of *Buxus sempervirens*. The margins of the leaves begin to curl upwards and the leaf takes on a shallow cup-like appearance, with a regularly concave upper surface (Wilcke, 1941;

Sampo, 1975; Nguyen, 1965). Other species, such as *Paurocephala psyllop-tera* Crawf. feeding on *Ficus ulmifolia* in the Philippines, induce a down-ward curling of the leaf lamina (Uichanco, 1919, 1921). Docters van Leeuwen (1920) figures a leaf gall, formed by an undescribed psyllid on *Vernonia cinerea* in which the whole leaf lamina is crinkled and distorted. This appears to result from the mid-rib growing more slowly than the lamina.

The effects of psyllid feeding are not always so localised. Systemic damage, in which the symptoms are apparent some distance from the site of feeding, is found in several pest species. The European carrot psyllid, *Trioza apicalis* Först. causes severe curling of the leaves of *Daucus carota*, with the intensity of curling related to the level of infestation (Laska, 1974; Markkula and Laurema, 1971; Markkula et al., 1976; Rygg 1977). Similarly, *Trioza tremblayi* Wag. causes twisting and distortion of the leaves of onion (*Alium cepa*) in Italy (Tremblay, 1965a, b). The systemic effects of the North American potato psyllid *Paratrioza cockerelli* Sulc are particularly well documented. They include rolling, cupping and marginal yellowing of leaves, necrosis of tissues and abnormal stem elongation (Eyer, 1937; Pletsch, 1947). Systemic effects are not confined to herbaceous plant species. *Diclidophlebia eastopi* Vond. feeding on young *Triplochiton sclero-xylon* in Nigeria causes shoot dieback, even at very low population densi-ties (Osisanya, 1969).

So far, in the examples examined, the psyllids do not form an enclosing gall in which to live. However, many species do: such galls can result from folding or rolling of the leaf lamina or they can be more specialised struc-tures resulting from abnormal growth of the lamina or petiole. The simp-lest form is where the leaf folds in half along its long axis enveloping a space in which the nymphs live. Good examples of these galls are those formed on *Caesalpinia japonica* and *Gleditsia japonica* by *Euphalerus hiurai* Miy. and *E. robinae* (Shinji) in Japan and by *Gyropsylla spegazziniana* (Lizer) on *Ilex paraguariensis* in South America (Miyatake, 1973; Brethes, 1921). An analagous gall is formed on the segments of the palmate leaves of *Triplochiton scleroxylon* by *Diclidophlebia harrisoni* Osis. in Nigeria (Osisanyi, 1974).

A similar but slightly more advanced type of gall is formed by a rolling of the leaf margin to form an enclosed tube, often of irregular shape. Several psyllids form this type of gall. They include the European *Psyllop-sis fraxini* (L.) on *Fraxinus excelsior*, *Trioza alacris* Flor on *Laurus nobilis*, the Indian *Cecidopsylla schimae* Kieff. on *Schima wallichii* and the South and Central American species *Leuronota leguminicola* Crawf. and *Triozoida johnsonii* Crawf. on undetermined species of Leguminosae and Myrtaceae respectively (Hodkinson and Flint, 1971; Nguyen, 1970; Miles, 1928; Lizer, 1918; Houard, 1933; Sampo, 1975; Mathur, 1975). A modification of the roll leaf gall is the purse-shaped gall of the European *Trichochermes walkeri* (Först.) which is formed by a rolling and thickening on just part of the leaf margin of *Rhamnus catharticus* (Sampo, 1975).

Many psyllids form localised galls on the leaf lamina rather than along the leaf margin. The simplest takes the form of a shallow pit or concave depression, usually on the underside of the leaf. The nymph feeds within the pit and often the gall appears as a small discoloured bulge or blister on the upper surface of the leaf. Examples of pit gall forming species include the Japanese species *Trioza machilicola* Miy. on *Machilus thunbergii*, *Trioza cinnamomi* (Bos.) on *Cinnamomum japonica* plus *Trioza camphorae* Sas. on *Cinnamomum camphora*, and the South American species *Trioza ocoteae* Houard and *Calophya rotundipennis* White and Hod. on *Ocotea acutifolia* and *Protium* sp. respectively (Miyatake, 1968b, 1969; Sasaki, 1910; Lizer, 1919b; Lizer and Molle, 1945; White and Hodkinson, 1980). Hawaiian species belonging to the genus *Hevaheva* also form pit galls on *Pelea* spp. as do the Asian species *Ceropsylla minuta* Mathur, *Trioza pitiformis* Mathur and *Megatrioza pallida* Uich.; the first on *Shorea robusta* the last pair on *Mallotus philippinensis* (Zimmerman, 1948; Uichanco, 1919; Mathur 1975). Deeper open pit galls which appear as hemispherical or subcylindrical bumps on the opposite leaf surface are found in a few species such as the southern Palaearctic *Egeirotrioza* spp. feeding on *Populus* (Bergevin, 1926; Boselli, 1931).

A further level of complexity occurs when the plant tissue grows across the mouth of the pit to form a closed chamber containing a feeding nymph. Sealed galls of this type are usually initiated by nymphs feeding on the underside of the leaf: the galls generally develop on the upper leaf surface. However, some species, such as *Pachypsylla celtidis-mamma* (Fletch.) feed on the upper leaf surface and the plant tissue grows upward to enclose the nymph (Walton, 1960). The South American *Trioza ulei tenuicornis* (Rüb.) feeding on *Nectandra* sp. is unusual in that the gall develops primarily on the lower side of the leaf (Lima, 1942). Closed leaf galls may be subspherical or globose as in the Philippine species *Pauropsylla udei* Rüb. on *Ficus variegata* and two Indian species, *Trioza jambolanae* Crawf. and *Pauropsylla depressa* Crawf. living on *Syzygium cumini* and *Ficus glomerata* respectively (Uichanco, 1919, 1921; Mani, 1935; Mathur 1975). The galls of other species may be barrel shaped as in the Asian species *Pseudophacopteron tuberculata* (Crawf.) on *Alstonia scholaris* or elongate/sub-conical as in the New World species *Calophya gallifex* (Kieff. and Jörg.) on *Schinus dependens* and *Pachypsylla celtidis-mamma* (Fletch.) on *Celtis occidentalis* (Mani, 1935; Brethes, 1921; Riley, 1883, 1890). Sometimes the closed galls are attached to the leaf by a narrow constriction which marks the site of gall initiation and often they become partly lignified as for example in *Pauropsylla depressa* Crawf. One of the most interesting psyllid galls known is that formed by the Indian species *Phacopteron lentiginosum* Buck. on the terminal leaves of *Garuga pinnata* (Mani, 1935; Mathur, 1975). The lignified galls, up to 2 cm long and 1 cm diameter, resemble small nuts.

Sealed galls normally occur singly although in some species at high densities, such as *Phytolyma fusca* Alib. on *Chlorophora excelsa*, they may become grouped or fused together (Harris, 1936). It is normally possible, however, still to distinguish the individual galls.

Some psyllids form galls on the leaf petiole or the basal part of the midrib rather than on the lamina. These galls can vary markedly in shape. The North American *Pachypsylla venusta* (O.S.) forms a purse-shaped petiole gall on the leaves of *Celtis occidentalis* whereas the South American *Neolithus fasciatus* Scott makes spherical galls on *Sapium aucuparium* (Riley, 1883, 1890; Houard, 1933).

BUD GALLS

Bud galls may be formed when nymphs invade and feed on developing buds. Nymphs of the North American hackberry psyllid *Pachypsylla celtidisgemma* Riley initiate bud gall formation on the woody twigs of *Celtis occidentalis*. Buds are usually invaded by more than one nymph but each nymph occupies an isolated chamber within the developing gall. At first the gall is green and covered by a bud scale. Later, during the autumn, the fully developed gall turns brown, becomes lignified and assumes a globular form with a smooth external surface (Walton, 1960). The Indian *Apsylla cistellata* (Buck.) forms somewhat different bud galls on the apical shoots of *Mangifera indica*. The galls appear as green or yellow-green false-buds which resemble miniature fir cones. Each 'cone' contains a number of developing nymphs and several may develop on each shoot, leading to cessation of growth.

STEM GALLS

Examples of stem galling species are rare among the Psylloidea. However, several members of the genus *Pachypsylloides*, found throughout the arid regions of Soviet Central Asia, form galls on the vegetative shoots or woody twigs of *Calligonum* species. The galls, initially formed as an invagination of the stem tissue, develop into a variety of shapes including spherical, conical and purse-shaped types (Loginova, 1970). Mathur (1975) also illustrates the galls of an undescribed psyllid on the woody shoots of *Terminalia alata* var *tomentosa* from India.

FLOWER GALLS

Some psyllids may cause severe growth distortion by feeding on developing flowers but usually they do not form enclosing galls. Nymphs of the European *Trioza centranthi* (Vall.) feeding on *Centranthus* and *Valerianella* cause severe distortion of the flowers, which remain green and fail to develop their normal pink colouration (André, 1878; Sampo, 1975). Another European species, *Trioza rumicis* Löw, lives on the flowers of dock (*Rumex scutatus*). Feeding induces hypertrophy of the pistil and the normally small ovary develops into a pod-shaped structure of up to 2 cm long (Sampo, 1975).

Rosette Galls

One of the most unusual types of psyllid gall is the rosette gall formed by *Livia juncorum* (Latr.) on *Juncus* species throughout the west Palaearctic region. At high psyllid densities the whole above ground part of the plant assumes a rosette form and is barely recognisable. Eggs of *L. juncorum* are laid in batches within the tissue of the basal shoots and newly emerged nymphs migrate to the growing point and commence feeding. The normally linear rush-type leaves become puffy and swollen basally and the tips wither and curl. Often every leaf is affected, giving the plant its characteristic rosette form. Gall formation is accompanied by a colour change: the whole plant turns a deep reddish brown (Vervier, 1929; Heslop-Harrison, 1949).

Root Galls

There is only a single known example of psyllids inducing the formation of root galls. Nymphs of the Holarctic species *Craspedolepta subpunctata* (Först.) feed for part of their life cycle on the roots of *Epilobium angustifolium*. The gall comprises an irregular shaped ball of up to 34 coiled and bleached rootlets. Each gall is around 1 cm in diameter and may contain one to three developing nymphs (Lauterer and Baudys, 1968).

Morphological adaptations for gall formation

As a very broad generalisation nymphs of the families Triozidae (including the Pauropsyllini), Calophyidae and, to a slightly lesser extent, the Homotomidae are strongly flattened dorso-ventrally and the dorsal sclerites of the thorax are fused. The wing pad margins are often confluent with the body outline and the head, wing pad and abdomen margins are often fringed with dense short setae. The appendages such as the legs, labium and antennae are relatively short and do not extend far beyond the body margin. The antennae are usually mounted on the underside of the head. In the extreme case such as certain *Ceropsylla* species the nymph in dorsal view resembles a flat circular disc.

By contrast, nymphs of the families Psyllidae, Carsidaridae, Spondyliaspidae, Aphalaridae and Phacopteronidae are, while still flattened, more robust. The thoracic sclerites are less fused, the wing buds are more pronouncedly developed and the fringeing setae are less dense and often longer. The body appendages are generally longer and the antennae are mounted at the front or on top of the head.

It can be argued that the flattened nymphs as typified by the Triozidae are adapted for a sedentary existence lying flat against the surface of the leaf lamina or other suitable organ whereas the more robust types of nymph, as typified by the Psyllidae, are better adapted for walking over the surface of the plant. Thus it might be expected that the more flattened species are more likely to form galls on 'flat' plant structures such as the leaf lamina. As a generality this is true; most pit gall forming species belong to the families Triozidae and Calophyidae. The nymphs form a depression in the leaf in which the body fits while feeding: the flat dorsal surface forms an almost smooth cover over the pit. However, there are exceptions: pit galls are also formed by some species of *Pseudophacopteron* (Phacopteronidae) which have nymphs of the more robust type (Aulmann, 1911).

It appears logical to suggest that the more complex types of closed leaf gall have evolved from pit galls. If this were so it could be predicted that the species involved would be more likely to belong to the families Triozidae and Calophyidae. This is often the case but is by no means exclusively so: *Pachypsylla* species, for example, belong in the family Spondyliaspidae and *Phacopteron lentiginosum* belongs in the Phacopteronidae.

It is also logical to suggest that the more mobile, robust types of nymph are better adapted for a free-living existence feeding on shoots or flowers or for communal life in a large leaf-fold or leaf-roll gall. This again holds true as a generality but there are, for example, several Triozid species, such as *Trioza alacris* Flor and *Trichochermes walkeri* (Först.) which also form leaf-roll galls.

Table 1 summarises the frequency and types of gall formed by the different families of Psylloidea. It shows that the known gall-forming species are unevenly spread across the families. They are rare or absent in the Psyllidae and Carsidaridae but common in the Triozidae, Calophyidae and Phacopteronidae. This distribution, coupled with the variability in the types of gall produced within each family, suggests that the ability to induce gall formation must have evolved independently on numerous occasions, even within a single family. Nevertheless, nymphal morphology and the associated feeding behaviour patterns ensure that certain types of nymph are better pre-adapted to form particular kinds of gall.

Unusual morphological structures specifically associated with gall formation are rare in the Psylloidea. Nymphs of *Paurotriozana adaptata* Cald. in Hawaii form blister galls on the leaves of *Cryptocarya*. Each gall is up to 3 mm in diameter with a circular opening of 0.5 mm diameter. The nymph sits in the gall with its sclerotised dorsal surface pressed tightly into the opening, thus sealing it (Zimmerman, 1948). This is a simple gall closing mechanism but nymphs of the southern Palearctic, *Egeirotrioza ceardi* (Berg.) and *E. intermedia* Baeva, feeding on *Populus* spp. have evolved a much more elaborate mechanism. They feed at the bottom of deep hemispherical pit galls, the circular entrance of which is sealed by a remarkable outgrowth, arising at the base of the insect's abdomen (Boselli, 1931). This

Table 1. Types of gall formed by the different families of Psylloidea

Family	Size of family	Leaf fold, roll or purse gall	pit gall	enclosed leaf gall	bud gall	flower gall	shoot/stem gall	root gall	rosette gall
					Type of gall				
Psyllidae	large								
Triozidae (inc. Pauropsylla)	large	++	+++	++		+	+		
Aphalaridae	large	++	+			+	+	+	+
Spondyliaspidae	large	+		+	+				
Carsidaridae	small								
Homotomidae	small			+					
Calophyidae	small	+++				+			
Phacopteronidae	small	++		+					

+++indicates commonly, ++indicates frequently, +indicates rarely

structure, which is as long as the nymph itself, takes the form of a tube which is narrowly constricted basally and which broadens apically. The tube apex is formed into a circular, sclerotised plate, which serves to seal the gall entrance.

Physiological aspects of gall formation

Experiments using radioactive tracers have shown that psyllids inject saliva into the plant while feeding. This saliva, in species such as *Trioza apicalis* Först. and *Psylla pyricola* Först., can be translocated within the plant (Williams and Benson, 1966, Markkula and Laurema, 1971). There is little doubt that salivary injection is the primary stimulus for gall formation but the exact nature of the salivary component-inducing cecidogenesis is, however, unknown. Markkula and Laurema (1971) found no evidence for the presence of indoleacetic acid (IAA) or a related phytohormone in the saliva of *Trioza apicalis* although there was some evidence to suggest an imbalance of IAA levels in the galled leaves of the host plant. White (1966) records the presence of IAA-like auxins in the leaves of *Chlorophora* galled by *Phytolyma* sp. These were absent from ungalled leaves but were present in body extracts of both nymphal and adult psyllids. White's suggestion that the auxins are produced by the psyllid and injected into the plant is, however, unproven; the psyllids may simply accumulate ingested IAA.

Williams and Lindner (1965) tested the toxicity of various compounds, obtained from whole body macerates of *Psylla pyricola*, on a test plant *Phaseolus vulgaris* (the species normally feeds on cultivated pears, *Pyrus* spp., where it causes necrosis of phloem sieve tubes). They found four injurious compounds, a U.V. fluorescing phenol and three purine-like substances but were unable to demonstrate whether the compounds were actively injected with the saliva. Lewis and Walton (1964) claim to have demonstrated that a virus-like material, found in the body tissues of *Pachypsylla* species, was responsible for gall initiation on *Celtis*. They provide little convincing evidence, however, to support their contention.

Anatomical structure and development of galls

Detailed anatomical and developmental studies on psyllid galls are scarce. The most complete investigations involve the galls formed by *Pachypsylla* species on *Celtis* (Wells, 1916, 1920; Weiss 1921; Walton 1960, Lewis and Walton, 1964). Wells summarises and illustrates the stages of development of the enclosed 'mammiform' gall of *P. celtidis-mamma* (Fletch.) on the leaf lamina. Newly emerged nymphs feed on the upper surface of the leaf and initiate a number of changes. The stylets are inserted and a salivary sheath begins to form. The nymphs then appear to retain this initial position throughout gall development. Feeding induces hypertrophy of the epidermal and adjacent mesophyll cells on the opposite side of the leaf and hyperplasia is obvious in the cells adjacent to the stylet apex. Chloroplasts in the zone of tissue beneath the nymph degenerate and the cell nuclei increase in size. These changes produce a downward evagination of the leaf, in which the nymph sits. Cells immediately surrounding the nymph then grow upwards to produce the characteristic cone-shaped cover. Multinucleate cells appear in the tissues forming the floor of the nymphal feeding chamber but most nuclei soon disintegrate.

Walton (1960) describes similar changes in the tissues of *Celtis* buds, invaded by *Pachypsylla celtidis-gemma* Riley, which involve cell hypertrophy and hyperplasia. By contrast with *P. celtidis-mamma*, the nymphs repeatedly withdraw and re-insert their stylets. Differences in feeding pattern between species probably plays a significant role in the way in which the shape of the gall develops.

Similar detailed studies of gall structure and development have been made on the South American species *Holotrioza duvauae* (Scott), *Calophya gallifex* (Kieff. and Jörg.) and *Tainarys schini* Breth., all on *Schinus polygamus* and *Trioza ocoteae* Houard on *Ocotea acutifolia* (Lizer and Molle, 1945).

Advantages of gall formation to the psyllid

A fundamental question which must be posed is why do psyllids form galls? What is their adaptive significance? Several potential advantages may accrue to psyllids living in a gall. Feeding psyllids derive their nutrition from the plant sap. The quality and quantity of available nutrients will determine their rates of growth and reproduction. Most free living, non gall forming, species are highly selective in their choice of feeding site. Nymphs feed on actively growing tissues such as young shoots, expanding leaves or flowers where the supply of high quality soluble nitrogenous compounds, particularly amino acids, is maximal. For example, the palaearctic *Psylla peregrina* Först. aggregates on the terminal shoots and flower clusters of *Crataegus monogyna* and nymphs change their feeding position as the plant grows. Shoot and flower tissues contain the highest levels of soluble nitrogenous compounds (Sutton, 1982). For several other free-living species such as *Trioza erytreae* D.G. and *Diaphorina citri* Kuw. on *Citrus* spp., *Acizza russellae* W & M on *Acacia karroo*, *Glycaspis* spp. on *Eucalyptus blakelyi* and various Alaskan *Psylla* species on *Salix*, psyllid performance appears to be related to the availability of actively growing, nutrient rich, tissues (Catling, 1971; Pande, 1972; Webb and Moran, 1978; Journet, 1980; Hodkinson et al., 1979).

By contrast, mature tissues, particularly leaves before senescence, usually contain lower levels of soluble nitrogen and represent a much poorer food resource (Hodkinson and Hughes, 1982). However, many psyllids feed for at least part of their development on mature leaves and it seems that gall formation is one mechanism by which they can ameliorate their food supply. It appears to involve a breakdown or modification of plant tissue to produce a nutrient sink which offers an enhanced supply of soluble nutrients to the insect. Thus gall formation can be seen as an adaptation for the exploitation of nutritionally inferior plant tissues. There are, however, exceptions: not all gall-forming species exploit mature plant tissue. The African species of *Phytolyma* living on *Chlorophora* feed on young growth in preference to older leaves (White, 1966, 1967).

Galls may also serve to protect psyllids from climatic extremes and natural enemies. Nymphs appear particularly susceptible to desiccation. To overcome this problem the free-living species either cover themselves with waxy threads or secretions or alternatively seek out protected microhabitats such as partially open leaf axils. Many Australian species of Spondyliaspidae construct a shell-like cover or lerp of dried honeydew on the leaf surface, under which they live. An interesting link between lerp and gall formation occurs in the closely related genera *Pachypsylla* and *Celtisaspis* on *Celtis*. The North American *Pachypsylla* species form galls but the Asian *Celtisaspis* species form either galls or lerps (Miyatake, 1968, 1980; Yang and Li, 1982). This could suggest an evolutionary sequence in which gall formation superseded lerp construction as a mechanism of humidity control. At the other extreme, nymphs of *Livia juncorum* (Latr.), living in rosette galls on *Juncus*, appear able to survive lengthy periods of inundation by water (Heslop-Harrison, 1949).

Some species, particularly those living in closed galls, derive further benefit in that periods of suspended development can be spent in a protected environment, thereby avoiding unfavourable external conditions. For example, *Pachypsylla celtidis-gemma* Riley spends up to eight months as a fifth instar nymph within the bud gall. This coincides with the winter period and adults emerge to continue the life cycle in the following summer. Protection against natural enemies is another potential advantage for gall-forming species. The evidence here, however, is less convincing. Psyllids living in closed galls are probably isolated from generalist predators but they still appear to suffer heavy mortality from specialised parasitoids, particularly the Hymenoptera (Jensen, 1957). *Phytolyma* species for example are attacked by wasps of the genera *Aprostocetus* (Eulophidae) and *Psyllaephagus* (Encyrtidae) (White, 1966). The former uses its ovipositor to deposit eggs within the developing gall (Alibert, 1947). *Pachypsylla celtidis-gemma* is similarly heavily parasitised by *Psyllaephagus pachypsyllae* (Walton, 1960).

Nymphs living in more open galls appear to be subject to normal levels of parasitism and predation. Many natural enemies display patterns of host-finding behaviour which seek out aggregations: galls often contain such aggregations. Hodkinson and Flint (1971) examined the predators found in the leaf roll gall of *Psyllopsis fraxini* in Britain. Predators included two species of *Anthocoris* (Cimicidae), *Psallus flavellus* Stich. (Miridae), *Melanostoma mellinum* L. (Syrphidae) and an unidentified lacewing larva (Chrysopidae). Almost a quarter of the 305 galls examined contained at least one predator. As most predators can move between galls this high level of predation suggests that these galls offer little protection to the nymphs inside.

Galls as a means of dispersal

Crawford (1918) noted that on the Hawaiian Islands a significantly high proportion of the psyllid fauna formed galls and that all belonged to the important gall-making family the Triozidae. He suggested that these geographically remote islands were originally colonised by a few gall-forming Triozids which diversified to give rise to the distinctive endemic fauna. He went on to postulate that psyllids in galls could be transported more successfully by strong winds and ocean currents than their free-living counterparts.

Evidence to test this idea is contradictory. The most widely distributed psyllids found throughout the scattered South Pacific islands and the Malay Archipelago are *Leptynoptera sulfurea* Crawf. on *Calophyllum inophyllum*, *Megatrioza vitiensis* Kirk. on *Eugenia malaccensis* plus *Syzygium* species and *Mesohomotoma hibisci* (Frogg.) on *Hibiscus tiliaceus*. The first two species are gall formers but *M. hibisci* is not. Two of the host plants, *C. inophyllum* and *H. tiliaceus*, are strand-line species which are easily dispersed by ocean currents. Nevertheless, *M. hibisci* is most likely to be wind dispersed and there is little reason to suppose that the other species cannot equally well be dispersed in a similar manner.

Colonisation of new island habitats, however, can be very rapid. Following the catastrophic destruction of vegetation by the explosion on the island of Krakatau in 1883, gall-forming psyllids became re-established on plants such as *Homalanthus populneus* and *Thespezia populnea* within 36 years (Docters van Leeuwen, 1920).

Economic significance of gall-forming psyllids

Most of the major psyllid pests are free-living species, such as *Trioza erytreae* D.G., *Diaphorina citri* Kuw., *Psylla pyricola* Först. and *P. mali* Schmid., which usually feed on actively growing tissues. However, as we have noted, some free-living species, such as *Paratrioza cockerelli* Sulc, *Trioza apicalis* Först. and *T. tremblayi* Wag. can cause severe growth distortion on crops like potato, carrots and onions. A few true gall-forming species are, nevertheless, of local economic importance, attacking agricultural crops, forest trees and ornamental plants.

The South and Central American species *Trioza anceps* Tut. and *T. perseae* Tut. are pest of Avocado and the Indian *Apsylla cistellata* (Buck.) causes damage to mango (Armenta, 1973; Tuthill, 1959; Mathur 1975). The bay psyllid *Trioza alacris* Flor, which has been introduced into both North and South America, attacks both cultivated bay and related ornamental species (Crawford, 1912; Miles, 1928; Lizer, 1918; Essig, 1917; Weiss, 1917; Weiss and Dickerson, 1918). Some trees, harvested for their natural products, such as cinnamon and camphor, carry large populations of pit gall-forming *Trioza* species (Boselli, 1930; Miyatake, 1969; Sorin, 1959). The South American species *Calophya schini* Tut. and *Gyropsylla spegazziniana* (Lizer) feed on the mastic producing *Schinus molle* and the Paraguay tea (*Ilex paraguariensis*) (Tuthill, 1959; Lizer 1919a).

Gall-forming psyllids can also be serious pests of forest trees. West African species of *Phytolyma* cause severe damage to newly established plantations of the important timber-producing iroko (*Chlorophora*) (Vosseler 1906; Harris, 1936; White, 1966, 1967). Other species, such as the Indian *Trioza fletcheri minor* Crawf. on *Terminalia arjuna*, a shade tree, may be sufficiently abundant to necessitate chemical control (Bindra and Varma, 1969).

A few psyllid species are of minor economic importance as pests of ornamental shrubs and trees. For example, *Psylla buxi* (L.) distorts the apical leaves of ornamental box plants, *Psyllopsis* species may, at high population densities, damage the foliage of ash trees and *Pachypsylla* species can be extremely abundant on *Celtis*, where their galls disfigure the leaves and shoots (Weiss and Dickerson, 1917; Nguyen, 1965; Wilcke, 1941; Loginova, 1954; Tuthill, 1943; Thompson 1962).

REFERENCES

1. Alibert, H. 1947. *Phytolyma lata* Scott var. *fusca* var. N., Psyllidae vivant sur iroko (*Chlorophora excelsa*). l'*Agronomie Tropicale, Nogent.* 2 : 165–166.

2. Andre, E. 1878. Memoire pour servir a l'histoire de la *Tiroza centranthi* Vallot. *Ann. Soc. Ent. Fr.* 8 : 77–86.

3. Anonymous, 1893. Mango Psylla. *Indian. Mus. Notes.* 3(1) : 13–14.

4. Armenta, H.M.C. 1973. Biologic data on *Trioza anceps* Hom. Psyl. which produces the leaf gall of the avocado, with observations in 4 avocado producing regions of Mexico. *Fol. Ent. Mex.* 25: 33.

5. Aulmann, G. 1911. Schadlinge an Kulturpflanzen aus Deutschen Kolonien. Mitteilungen aus dem Zoologischen Museum in Berlin. 5 : 261–273.

6. Bergevin, E. De 1926. Description d'une nouvelle espece de *Trioza* (Hemiptere, Psyllidae) produisant une galle sur *Populus euphratica* Oliv. var *Bonnetiana* Dod. dans le Sud orano-marocain. *Bull. Soc. d'hist. Nat. l'Afrique du Nord.* 149–153.

7. Bindra, O.S., Varma, G.C. 1969. Studies on the chemical control of *Trioza fletcheri minor* Crawford (Psyllidae: Hemiptera), a pest of *Terminalia arjuna*. *Ind. Forester.* 95 : 263–266.

8. Boselli, F.B. 1930. Studii sugli Psyllidi (Homoptera: Psyllidae o Chermidae) VI. Psyllidi di Formosa raccolti dal Dr. R. Takahashi. *Boll. Lab. Zool. Gen. e. agraria della Facolta agraria in Portici.* 24 : 175–210.

9. Boselli, F.B. 1931. Studii sugli Psyllidi (Homoptera: Psyllidae o Chermidae) X. Istituzione di un nuovo genere e descrizione di *Egeirotrioza ceardi* (De Bergevin) *euphratica* n. var., Triozina galligena su *Populus euphratica* in Mesopotamia. *Boll. Lab. Zool. gen. e agraria della Focolta agraria in Portici.* 24 : 267–278.

10. Brethes, J. 1921. Un nuevo Psyllidae de la Republica Argentina (*Gyropsylla ilicicola* Brethes). *Revista de la Facultad de agronomia* (Universidad nacional de La Plata). 14 : 82–89.

11. Buckton, G.B. 1893. The mango shoot psylla. *Psylla cistellata* n. sp. *Indian Mus. Notes* 3(2) : 91–92.

12. Buhr, H. 1964. *Bestimmungstabellen der Gallen (Zoo-und Phytocecidien) an Pflanzen Mittel und Nordeuropas.* Jena: Fischer.

13. Catling, H.D. 1971. The biononomics of the South African citrus psylla *Trioza erytreae* (Del Guercio) (Homoptera: Psyllidae). 5. The influence of host plant quality. *Jour. ent. Soc. Sth. Afr.* 33 : 341–348.

14. Chen, M.H., Miyakawa, T., Matsui, C. 1973. Citrus Likubin pathogens in salivary glands of *Diaphorina citri*. *Phytopath.* 63 : 194–195.

15. Clark, L.R. 1962. The general biology of *Cardiaspina albitextura* (Psyllidae) and its abundance in relation to weather and parasitism. *Aust. Jour. Zool.* 10 : 537–586.

16. Crawford, D.L. 1912. A new insect pest (*Trioza alacris* Flor.), *Monthly Bulletin of the California Commission of Horticulture.* 1 : 86–87.

17. Crawford, D.L. 1918. The jumping plant-lice (Family Psyllidae) of the Hawaiian Islands. *Proc. Hawaiian Ent. Soc.* 3 : 430–457.

18. Docters van Leeuwen-Reijnvann, W. & J. 1914. Einige gallen aus Java. Siebenter beitrag. *Bull. Jard. Bot. de Buitenzorg.* 16 : 1–68.

19. Docters van Leeuwen, W. 1920. The galls of Krakatau and Verlaten Eiland (Desert Island) in 1919. *Ann. Jard. Bot. de Buitenzorg.* 31 : 57–82.

20. Eastop, V.F. 1978. Diversity of the Sternorrhyncha within major climatic zones. In *Diversity of Insect Faunas*, ed. Mound, L.A. & Waloff, N. Oxford: Blackwell Scientific Publications.

21. Essig, E.O. 1917. The tomato and laurel psyllids. *Jour. Econ. Ent.* 10 : 433–444.

22. Eyer, J.R. 1937. Physiology of psyllid yellows of potato. *Jour. Econ. Ent.* 30 : 891–898.

23. Harris, W.F. 1936. Notes on two injurious psyllids and their control. *East. Afr. Agric. J.* 1 : 498–500.

24. Heslop-Harrison, G. 1949. The subfamily Liviinae of the Homopterous family Psyllidae. *Ann. & Mag. Nat. Hist.* 2(12) : 241–270.

25. Hodkinson, I.D. 1973. The biology of *Strophingia ericae* (Curtis) (Homoptera: Psylloidea) with notes on its primary parasite *Tetrastichus actis* (Walker) (Hym., Eulophidae). *Norsk. ent. Tidjsk.* 20 : 237–243.

26. Hodkinson, I.D. 1974. The biology of the Psylloidea (Homoptera): a review. *Bull. ent. Res.* 64 : 325–339.

27. Hodkinson, I.D. 1978. The psyllids (Homoptera: Psylloidea) of Alaska. *Syst. Ent.*, 3 : 333–360.

28. Hodkinson, I.D. 1983. The psyllids (Homoptera: Psylloidea) of the Austro-oriental, Pacific and Hawaiian Zoogeographical realms: an annotated check list. *J. Nat. Hist.* 17 : 341–377.

29 Hodkinson, I.D., Flint, P.W.H. 1971. Some predators from the galls of *Psyllopsis fraxini* (L.) (Hem. Psyllidae). *Ent. Monthly Mag.* 107 : 11–12.

30. Hodkinson, I.D., Jensen, T.S., MacLean, S.F. 1979. The distribution, abundance and host plant relationships of *Salix*-feeding psyllids (Homoptera: Psylloidea) in arctic Alaska. *Ecol. Ent.* 4 : 119–132.

31. Hodkinson, I.D., Hughes, M.K. 1982. *Insect Herbivory*, London: Chapman & Hall.

32. Hodkinson, I.D., White, I.M. 1981. The Neotropical Psylloidea (Homoptera: Insecta): an annotated check list. *Jour. Nat. Hist.* 15 : 491–523.

33. Houard. C. 1908. *Les Zoocécidies des Plantes d'Europe*, I, Paris: Librairie Scientifique Jules Hermann.

34. Houard, C. 1909. *Les Zoocécidies des Plantes d'Europe*, II, Paris: Librairie Scientifique Jules Hermann.

35. Houard, C. 1922. *Les Zoocécides des Plantes d'Afrique d'Asie et d'Oceanie*. Paris: Librairie Scientifique Jules Hermann.

36. Houard, C. 1923. *Les Zoocecidies des Plantes d'Afrique d'Asie* Vol. 2. Paris: Librairie Scientifique Jules Hermann.

37. Houard, C. 1933. *Les Zoocecidies des Plantes de l'Amerique du Sud et de l'Amerique Centrale*. Paris: Librairie Scientifique Hermann et Cie.

38. Journet, A.R.P. 1980. Intraspecific variation in food plant favourability to phytophagous insects, Psyllids on *Eucalyptus blakelyi* M. *Ecol. Ent.* 5 : 249–261.

39. Jensen, D.D. 1957. Parasites of the Psyllidae. *Hilgardia* 27 : 71–99.

40. Jensen, D.D., Griggs, W.H., Gonzales, C.Q., Schneider, H. 1964. Pear decline virus transmission by pear psylla. *Phytopath.* 54 : 1346–1351.

41. Laska, P. 1974. Studie uber den Mohrenblattfoh (*Trioza apicalis* Forst.) (Triozidae, Homoptera). *Acta Scient. natur. Acad. Scient. Bohem.* 8 : 1–44.

42. Lauterer, P., Baudys, E. 1968. Description of a new gall on *Chamaenerion angustifolium* (L.) Scop. produced by the larva of *Craspedolepta subpunctata* (Forst) with notes on the bionomics of this psyllid. *Casopis Moravskeho Musea.* 53 : 243–248.

43. Lewis, I.F., Walton, L. 1964. Gall formation on leaves of *Celtis occidentalis* L. resulting from material injected by *Pachypsylla* sp. *Trans. Amer. Microscop Soc.* 83 : 62–78.

44. Lima, A.M. Da C. 1942. *Insectos do Brazil*; Homopteros. 3, Imprenso National, Rio de Janeiro. 327pp.

45. Lizer, C. 1918. Sobre la presencia en Argentina de un psilido exotico (*Trioza al-*

acris F.). *Ann. Zool. aplicada.* 5 : 16–21.

46. Lizer, C. 1919a. Description d'une nouvelle espece de psyllide cecidogene, de l'Amerique Meridionale (*Paurocephala spegazziniana* n. sp.). *Marcellia.* 16 : 103–107.

47. Lizer, C. 1919b. Sobre una neuva hemipterocecidia Argentina. *Reunion nacional de la Sociedad Argentina de ciencias naturales.* 1 : 383–388.

48. Lizer, C.A Molle, C.C. 1945. Estructura anatomica de filocecidias Neotropicas. *Lilloa.* 11 : 153–187.

49. Loginova, M.M. 1954. On the biology of leafhoppers of the genus *Psyllopsis* Low in the Stalingrad region. *Trudy Zool. Inst. Leningrad.* 15 : 35–53 (In Russian).

50. Loginova, M.M. 1970. New psyllids (Homoptera, Psylloidea) from Soviet Central Asia. *Entomol. Obozr.* 49 : 601–623 (In Russian).

51. Mani, M.S. 1935. Notes on some Indian gall forming Psyllidae (Homoptera). *Jour. Asiat. Soc. Bengal* (Sci.). 1 : 99–109.

52. Mani, M.S. 1964. *Ecology of Plant Galls.* The Hague: W. Junk. 434pp.

53. Mani, M.S. 1973. *Plant Galls of India.* New Delhi: Macmillan India. 354pp.

54. Markkula, M. Laurema, S. 1971. Phytotoxaemia caused by *Trioza apicalis* Forst. (Hom., Triozidae) on carrot. *Ann. Agric. Fenn., Ser Animalia Nocentia.* 10 : 181–184.

55. Marrkula, M., Laurema, S., Tiittanen, K. 1976. Systematic damage caused by *Trioza apicalis* on Carrot. In *The host-plant in relation to insect behaviour and reproduction.* Ed. Jermy, T. New York: Plenum Publishing Co.

56. Martinez, A.L. Wallace, J.M. 1967. Citrus leaf mottle yellows in the Philippines and transmission of the causal virus by a psyllid, *Diaphorina citri. Pt. Dis. Reptr.* 51 : 692–695.

57. Mathur, R.N. 1975. *Psyllidae of the Indian Subcontinent.* New Delhi, Indian Council of Agricultural Research.

58. McLean, A.P.D. Oberholzer, P.C.J. 1965. Citrus psylla, a vector of the greening disease of sweet orange. *Sth. Afr. J. Agr. Sci.* 8 : 297-298.

59. Miles, H.W., 1928. The bay psyllid, *Trioza alacris* Flor. *North Western Naturalist.* 3 : 8–14.

60. Miyatake, Y. 1968a. *Pachypsylla japonica* sp. nov, a remarkable lerp-forming psyllid from Japan, (Homoptera: Psyllidae). *Bull. Osaka Mus. Nat. Hist.* 21 : 5–12.

61. Miyatake, Y. 1968b. A new Japanese species of *Trioza* from *Machilus thunbergii*, with descriptions of the immature stages and notes on biology (Hemiptera: Psyllidae). *Trans. Shikoku Ent. Soc.* 10 : 1–10.

62. Miyatake, Y. 1969. On the life history and the immature stages of *Trioza cinnamomi* (Boselli), with the redescription of adult (Hemiptera: Psyllidae). *Bull. Osaka Mus. nat. Hist.* 22 : 19–30.

63. Miyatake, Y. 1970. Some Taxonomical and Biological Notes on *Togepsylla matsumurana* Kuwayama, Jr. (Hemiptera: Psyllidae). *Bull. Osaka Mus. nat. Hist.* 23 : 1–10.

64. Miyatake, Y. 1973. Notes on the genus *Euphalerus* of Japan with description of a new species (Homoptera: Psyllidae). *Bull. Osaka Mus. nat. Hist.* 27 : 23–28.

65. Miyatake, Y. 1980. Notes on the genus *Pachypsylla* of Japan with description of a new species (Homoptera: Psyllidae). *Bull. Osaka Mus. nat. Hist.* 33 : 61–70.

66. Moore, K.M. 1970. Observations on some Australian forest insects 24. Results from a study of the genus *Glycaspis* (Homoptera: Psyllidae). *Aust. Zool.* 15 : 343–376.

67. Newstead, R., Cummings, B.F. 1913. On a remarkable gall-producing psyllid from Syria. *Ann. & Mag. nat. Hist.* 9 : 206–208.

68. Nguyen, T.X. 1965. Observations sur la ponte preferentielle de *Psylla buxi* L. (Homopteres, Psyllides) sur les differentes varietes de buis. *Bull. Soc. d'Histoire nat. de Toulouse.* 100 : 299–311.

69. Nguyen, T.X. 1970. Recherches sur la morphologie et la biologie de *Psyllopsis fraxini* (Hom. Psyllidae). *Ann. Soc. ent. Fr.* 6 : 757–773.

70. Nishida, T., Haramoto, F.H., Nakahara, L.M. 1980. Altitudinal distribution of Endemic Psyllids (Homoptera: Psyllidae) in *Metrosideros* Ecosystem. *Proc. Hawaiian ent. Soc.* 23 : 255–262.

71. Osisanya, E.O. 1969. The effect of attack of *Diclidophlebia eastopi* (Vond.) (Homoptera: Psyllidae) on the survival of *Triplochiton scleroxylon* (K. Schum). *Nigerian Ent. Magazine* 2 : 19–25.

72. Osisanya, E.O. 1974. Aspects of the biology of *Diclidophlebia eastopi* Vondracek and *D. harrisoni* Osisanya (Hom., Psyllidae). *Bull. ent. Res.* 64 : 9–18.

73. Pande, Y.D. 1972. Seasonal fluctuations in the abundance and host preference of *Diaphorina citri* Kuw. in relation to certain species of citrus. *Ind. J. Agric. Res.* 6 : 51–54.

74. Pletsch, D.J. 1947. The potato psyllid *Paratrioza cockerelli*. Its biology and control. *Bull. Montana agric. Exptl. Sta.* no. 446 : 1–95.

75. Riley, C.V. 1883. Hackberry psyllid galls. *Can. Ent.* 15 : 157–159.

76. Riley, C.V. 1890. Hackberry psyllids. In *Forest and shade tree insects*. Ed. Packard, A.S. U.S. Entomological Commission. 5th report. pp. 614–622.

77. Rygg. T. 1977. Biological investigations on the carrot psyllid *Trioza apicalis* Forster (Homoptera, Triozidae). *Meldinger Norges Landbrukshogskole.* 56(3) : 1–20.

78. Sampo, A. 1975. Di alcuni Psillcocecidi nuovi o poco noti della Valle d'Aosta. *Rev. Valdot. d'Histoire nat. (Aosta).* 29 : 153–174.

79. Sasaki, C. 1910. On the Life History of *Trioza camphorae* n. sp. of Camphor Tree and its Injuries. *Jour. Coll. Agric. Imp. Univ. Tokyo.* 2 : 277–285.

80. Schwarz, R.E., McLean, A.P.D., Catling, H.D. 1970. The spread of greening disease by the citrus psylla in South Africa. *Phytophylactica.* 2 : 59–60.

81. Sorin, M. 1959. On the life history and immature stages of *Trioza camphorae* Sasaki (Psyllidae, Homoptera). *Kontyu.* 27 : 244–248.

82. Stahler, N. 1941. Psyllid galls on avocado. *Bull. Calif. Dept. Agric.* 30 : 217 pp.

83. Sutton, R.D. 1982. The ecology of three species of psyllid (Homoptera: Psylloidea) living on hawthorn (*Crataegus monogyna* Jacq.). Liverpool: Liverpool Polytechnic. Ph.D. Thesis.

84. Takahashi, R. 1936. Food habits and new habitats of Formosan Psyllidae, with notes on the peculiar food habits of Formosan insects. *Kontyu.* 10 : 291–296.

85. Thompson, H.E. 1962. Control of hackberry nipple gall maker with new organic insecticides. *Jour. econ. Ent.* 55 : 555–556.

86. Tremblay, E. 1965a. Risultati di prove di lotta contro la psilla della cipolla (*Trioza tremblayi* Wagner). *Ann. del. Fac. Sci. Agr. della Univ. Napoli.* (III) 30 : 15–27.

87. Tremblay, E. 1965b. Studio morfo-biologico sulla *Trioza tremblayi* Wagner (Hemiptera-Homoptera: Psyllidae). *Bull. del Lab. ent. Agr. 'Filippo Silvestri'.* 23 : 37–138.

88. Tuthill, L.D. 1943. The psyllids of America north of Mexico (Psyllidae : Homoptera) (Subfamilies Psyllinae and Triozinae). *Iowa State Coll. Jour. Sci.* 17 : 443–660.

89. Tuthill, L.D. 1959. Los Psyllidae del Peru Central (Insecta : Homoptera). *Rev. Peruana Ent. Agric.* 2 : 1–27.

90. Uichanco, L.B. 1919. A biological and systematic study of Philippine plant galls. *Phil. Jour. Sci.* 14 : 527–550.

91. Uichanco, L.B. 1921. New records and species of Psyllidae from the Philippine

Islands, with descriptions of some preadult stages and habits *Phil. Jour. Sci.* 18: 259–288.

92. Vervier, M.L. 1929. Contribution a l'etude de la cécidie de *Livia juncorum* Latr. (Hem. Psyllidae) sur *Juncus conglomeratus* L. *Bull. Soc. ent. Fr.* 19 : 77–80.

93. Vosseler, J. 1906. Eine psyllide als erzeugerin von gallen am Mwulebaum (*Chlorophora excelsa* (Welw.) Benth. et. Hook). *Zeit. Wiss. Insektenbiol.* 2 : 276–285 & 305–316.

94. Walton, B.C.J. 1960. The life cycle of the hackberry gall-former *Pachypsylla celtidis-gemma* (Homoptera : Psyllidae). *Ann. Ent. Soc. Amer.* 53 : 265–277.

95. Webb, J.W., Moran, V.C. 1978. The influence of the host plant on the population dynamics of *Acizzia russellae* (Homoptera: Psyllidae). *Ecol. Ent.* 3:313–321.

96. Weiss, H.B. 1917. The bay flea-louse, *Trioza alacris* Flor as a new pest in New Jersey. *Can. Ent.* 49 : 73–75.

97. Weiss, H.B. 1921. Notes on the life history of *Pachypsylla celtidis-gemma* Riley. *Can. Ent.* 53 : 19–21.

98. Weiss, H.B., Dickerson, E.L. 1917. *Psyllia buxi* Linn. in New Jersey (Homop.). *Ent. News* 28 : 40–41.

99. Weiss, H.B., Dickerson, E.L. 1918. Notes on *Trioza alacris* Flor in New Jersey, *Psyche.* 25 : 59–63.

100. Wells, B.W. 1916. The comparative morphology of the zoocecidia of *Celtis occidentalis. Ohio Jour. Sci.* 16 : 249–298.

101. Wells, B.W. 1920. Early stages in the development of certain *Pachypsylla* galls on *Celtis. Amer. J. Bot.* 7 : 275–285.

102. White, I.M. Hodkinson, I.D. 1980. New psyllids (Homoptera: Psylloidea) from the cocoa region of Bahia, Brazil. *Rev. brasil. ent.* 24 : 75–84.

103. White, I.M. 1980. Nymphal taxonomy and systematics of the Psylloidea. Liverpool Polytechnic: Unpublished Ph. D. Thesis.

104. White, M.G. 1966. The problem of the *Phytolyma* gall bug in the establishment of *Chlorophora*, University of Oxford, Institute paper of the Commonwealth Forestry Institute. 37 : 1–52.

105. White, M.G. 1967. Research in Nigeria on the Iroko gall bug (*Phytolyma* sp.). *Nig. For. Inform. Bull.* (N.S.) No. 18 : 1–72.

106. White, T.C.R. 1970. The nymphal stage of *Cardiaspina densitexta* (Homoptera : Psyllidae) on leaves of *Eucalyptus fasiculosa. Aust. Jour. Zool.* 18 : 273–293.

107. Wilcke, J. 1941. Biologie en morphologie van *Psylla buxi Tidjschr. o. Plantenz.* 47 : 41–89.

108. Wilde, W.H.A., Carpenter, J., Liberty, J., Tunnicliffe, J. 1971. *Psylla Pyricola* (Hemiptera : Psyllidae) vector relationships with *Erwinia amylovora. Can. Ent.* 103 : 1175–1178.

109. Williams, M.W., Benson, N.R. 1966. Transfer of C_{14} components from *Psylla pyricola* Foer to pear seedlings. *Jour. Ins. Phys.* 12 : 251–254.

110. Williams, M.W., Lindner, R.C. 1965. Biochemical components of pear psylla and their relative toxicity to excised bean plants. *Jour. Ins. Phys.* 11 : 41–52.

111. Woodburn, T.L., Lewis, E.E. 1973. A comparative histological study of the effects of feeding by nymphs of four psyllid species on the leaves of Eucalypts. *Jour. Aust. Ent. Soc.* 12 : 134–138.

112. Yang, C.K., Li, F. 1982. Descriptions of the new genus *Celtisaspis* and five new species of China (Homoptera : Psyllidae) *Entomotaxonomia* 4 : 183–198.

113. Zimmerman, E.C. 1948. Superfamily Psylloidea. *Insects of Hawaii* 5 : 12–38.

4. Gall-forming Coccoidea

John W. Beardsley, Jr.

Introduction

The Coccoidea, often referred to collectively as the scale insects, is the largest and most diverse superfamily within the homopterous suborder Sternorrhyncha. Classification at the family level within this group is rather unsettled, and the number of families recognized by different coccid systematists during the past half century or so has ranged mostly between 12 and 24. For the purpose of this chapter a fairly conservative classification has been adopted (Table 1) based upon family concepts of Balachowsky (1948), Ferris (1957), and Borchsenius (1958). Gall-producing species are known in eight of the 16 family taxa listed.

Without exception, female Coccoidea are neotenic; lacking all vestiges of the wings and external genitalia which normally are present in adult insects. In contrast, adult males are generally much smaller than their female counterparts, usually winged, with well developed copulatory organs, but, lacking mouthparts, are very short lived. Males undergo a complex metamorphosis which is comparable to that of the holometabolous orders, and, in all species which have been carefully studied, there is at least one more instar in the male developmental cycle than in the female. In all Coccoidea the dispersal phase of the life cycle is the newly hatched first instar larva, commonly termed the "crawler" stage. The crawlers are often dispersed by air currents, which permit the insects to spread from plant to plant, and perhaps over distances of up to several kilometers or more (Beardsley and Gonzalez, 1975).

In most gallicolus Coccoidea, gall formation is apparently initiated individually by the newly settled first-stage larvae. The colonial gall-inhabiting mealybugs, such as *Pseudotrionymus* spp., may be exceptional in this regard. Once gall formation has been initiated, gallicolus females usually

79

Table 1. Families of Coccoidea[1]

Section	Family	Genera with gall-forming species
Margaroidi	Margarodidae*	*Matsucoccus, Araucariacoccus*
	Ortheziidae	none
	Phenacoleachiidae	none
Coccoidi	Pseudococcidae*	*Phylococcus, Grewiacoccus,* etc.
	Dactylopiidae	none
	Eriococcidae*	*Apiomorpha, Gallacoccus,* etc.
	Kermesidae	*Fullbrightia, Olliffiella*
	Coccidae	*Cissococcus*
	Lacciferidae	none
	Aclerdidae	none
	Asterolecaniidae*	*Asterolecanium, Frenchia,* etc.
	Lecanodiaspididae	*Gallinococcus, Lecanodiaspis*
	Stictococcidae	none
	Conchaspidae	none
Diaspidoidi	Halimococcidae	none
	Diaspididae	*Cryptophyllaspis, Maskellia,* etc.

[1]The family level classification presented here is a fairly conservative one. The families marked with asterisks have been divided into two or more families by some authors. The section Margaroidi is generally considered to be the most primitive; the Diaspidoidi is the most specialized.

spend their entire lives within the gall. In contrast, mature males must leave their galls and seek receptive females in order to mate. In those forms in which males and females produce separate galls, the galls of the two sexes usually are quite dissimilar. Male galls are generally smaller, with relatively large apical openings when mature which permit egress of the adult. In some species of gallicolus Diaspididae, males apparently do not form galls and are similar in appearance to males of non-gallicolus species [e.g. *Parachionaspis galliformens* (Green)]. Female galls generally are much larger than those of conspecific males. In many species which inhabit enclosive galls, the galls of females develop a small apical opening as the coccid approaches maturity, through which mating occurs and the first stage larvae of the next generation escape. Females of gallicolus species often possess specialized structures which serve to block the gall aperture and limit access by predators and parasitoids. In forms which produce honeydew, this liquid excrement is normally voided through the gall aperture.

In most, if not all, gallicolus coccids, gall formation is probably initiated and maintained largely by salivary secretions. Like other Homoptera, all Coccoidea feed by means of fine elongate stylets which, in most species, are inserted into vascular tissues of the host. The initiation of feeding by the newly settled crawler stage involves the injection of salivary secretions which, very probably, contain gall-inducing substances. Parr (1939) injected young pine twigs with extracts of salivary glands of *Matsucoccus gallicolus* Morrison and produced symptoms of gall formation similar to those associated with the living insects. Similar results were obtained by the same author using salivary gland extracts of *Asterolecanium variolosum* Ratzeburg, which forms pit galls on twigs of oaks (Parr 1940). Possible mechanisms of gall initiation other than salivary secretions apparently have not been investigated in coccids. Because the newly hatched first-stage larvae disperse before settling, secretions associated with maternal oviposition play no role in gall formation in this group. It is possible that anal secretions or secretions from epidermal glands may function in gall formation in some coccids.

Galls produced by species of Coccoidea range from irregular leaf and stem deformations and shallow pits in leaf or stem tissues to complex enclosive structures. Species which produce the less specialized kinds of deformations, such as witches brooms and shallow pits, are often polyphagous or oligophagous. In such species the occurrence or non-occurrence of gall formation may depend upon the species of host infested. For example, the polyphagous mealybug *Nipaecoccus vastator* (Maskell) produces simple leaf deformations on many, but not all, hosts. Certain species of the genus *Asterolecanium* (e.g. *A pustulans* Cockerell) can produce conspicuous pit galls on hosts such as oleander (Fig. 1), but the same species may produce no galls on some other plants (e.g. mango).

Taxonomy, distribution and hosts of gall-forming Coccoidea

The geographic distributions and hosts of the known gall-forming Coccoidea are summarized in Table 2. Obviously, the ability to produce plant galls has arisen independently in many different evolutionary lines within the superfamily. The majority of the known gallicolus species, however, belong to a relatively small number of family level taxa. Gall formation apparently is rare among the margaroid coccids, the group which contains the least specialized members of the superfamily. The greatest numbers of gallicolus species occur in the Eriococcidae and Asterolecaniidae which are generally considered to be of an intermediate degree of specialization. The Diaspididae, the largest and most specialized family of Coccoidea, also contains a number of gall-producing species, but these comprise a relatively small fraction of the family.

The margaroid coccids (Margarodidae, Ortheziidae and Phenacoleachiidae) are characterized by the possession of abdominal spiracles (absent in all other coccids), relatively unspecialized chromosomal systems in most groups, and adult males usually with compound eyes (greatly reduced in other coccids). This group, which contains around 350 described species, is believed to be more closely related to the Aphidoidea than are other, more specialized, groups within the superfamily. Only two species of margaroids

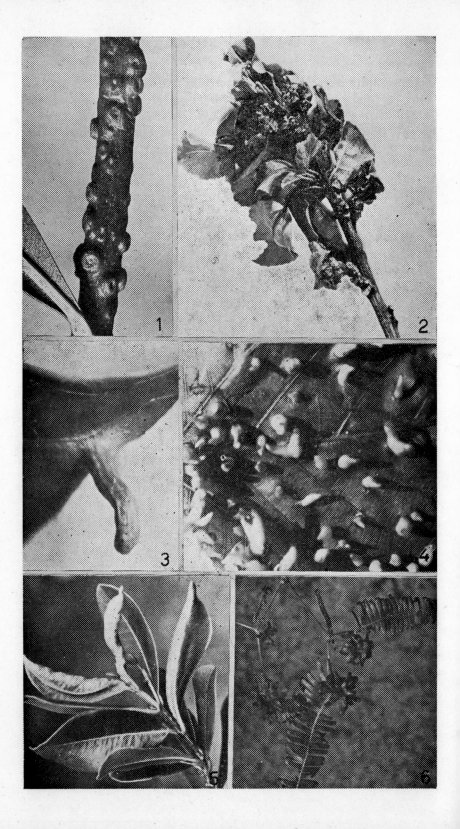

are known which produce distinct galls. The genus *Matsucoccus*, an essentially Holarctic group confined to hosts of the genus *Pinus*, includes one known gall-forming species, *M. gallicolus* Morrison, which is found primarily on *P. rigida* in eastern North America. This species produces galls in the form of deep, well-like depressions accompanied by the swelling of the surrounding tissues, on host twigs (Parr 1939). *Araucariococcus queenslandicus* Brimblecombe (1960), an Australian species possibly allied to *Matsucoccus*, reportedly develops within large, fleshy galls on twigs and young branches of *Araucaria* pines. It is likely that similar galls may be produced by other twig and bark-infesting margaroid scales elsewhere, but none have been documented.

The cosmopolitan Pseudococcidae (mealybugs) is the second largest family within the Coccoidea, containing around 1200 described species. Several polyphagous mealybugs (e.g. *Nipaecoccus vastator* and *Maconellicoccus hirsutus* (Green) (Fig. 2), are known which cause irregular plant deformations characterized by internode shortening and asymetrical growth, on some of their hosts.

Obligate gall formation seems to be relatively uncommon among mealybugs, except in Hawaii where several such species are known. Hawaiian gallicolus mealybugs range from species with relative unspecialized morphology, such as *Pseudococcus gallicola* Ehrhorn and several related forms which produce leaf galls on *Santalum* spp. (Fig. 3), to highly specialized types such as *Phylococcus oahuensis* (Ehrhorn), which forms cylindrical leaf galls on *Urera* (Fig. 4). The leaf-pocket galls of the *Pseudococcus gallicola* complex, and the curled leaf-edge galls of *Pseudotrionymus* spp. (Fig. 5) each characteristically contain several to many mealybugs, frequently of both sexes and often of several different developmental stages. In contrast, galls of the morphologically specialized mealybugs (e.g. *Phylococcus oahuensis*) contain either a single female or one or more males (unpublished observations). Three species of the endemic Hawaiian genus *Nesopedronia* form

---

Figure 1. Pit galls of *Asterolecanium pustulans* Cockerell (Asterolecaniidae) on twigs of *Nerium oleander* (Apocynaceae); Hawaii.

Figure 2. Deformed terminal growth of *Hibiscus rosa-sinensis* (Malvaceae) caused by *Maconellicoccus hirsutus* (Green) (Pseudococcidae); Hawaii.

Figure 3. Leaf-pocket gall of *Pseudococcus gallicola* Ehrhorn (Pseudococcidae) on *Santalum freycinetianum* (Santalaceae); Hawaii.

Figure 4. Galls of *Phylococcus oahuensis* (Ehrhorn) (Pseudococcidae) on leaf of *Urera sandwicensis* (Urticaceae); Hawaii.

Figure 5. Curled leaf-edge galls on *Eugenia sandwicensis* (Myrtaceae) caused by *Pseudotrionymus multiductus* (Beardsley) (Pseudococcidae); Hawaii.

Figure 6. Rosette deformation of *Dicranopteris linearis* (Gleicheniaceae) associated with *Nesopedronia hawaiiensis* (Ferris) (Pseudococcidae); Hawaii.

small open-ended cavity galls in aborted pinnule tips of *Dicranopteris* ferns. A fourth species, *N. hawaiiensis* (Ferris), produces a peculiar rosette deformation on *Dicranopteris* fronds (Fig. 6) (Beardsley 1957). These appear to be the only coccid galls yet reported from pteridophyte hosts. *Ohiacoccus cryptus* Beardsley (1971) forms fleshy open top pit galls at the bases of young leaves of a glabrous form of its host, *Metrosideros polymorpha*, but the same mealybug produces little or no gall formation on tomentose varieties of the same polymorphic host species (unpublished observations).

There is strong morphological evidence that the gall-forming habit has arisen independently in at least five different evolutionary lines of Hawaiian mealybugs. Additionally, there are several Hawaiian mealybug species, mostly undescribed, but including *Gallulacoccus tenorioi* Beardsley (1971a), which are morphologically specialized to inhabit abandoned psyllid galls on leaves of *Metrosideros* (unpublished observations).

Outside of Hawaii, obligately gallicolus mealybugs have been reported only from South Africa where *Grewiacoccus gregalis* Brain forms pitcher-shaped galls on leaf margins of *Grewia occidentalis* (Brain 1918). However, it seems likely that many additional gallicolus mealybugs await discovery.

The great complex of genera which comprises the family Eriococcidae, in the broad sense of Hoy (1963), contains the majority of all known gall-forming Coccoidea. The Eriococcidae, a family containing over 400 described species wordwide, have attained their greatest numbers and diversity on the Australian continent, and the gall-forming Eriococcidae of Australia form the largest assemblage of gallicolus coccids known.

The majority of gall-forming eriococcids are highly specialized forms contained in genera which are composed entirely of gallicolus species. Some less specialized gall-formers are known, including several species of the cosmopolitan genus *Eriococcus*. In the Australian genus *Lachnodius* approximately two-thirds of the known species develop in open-top pit galls (Fig. 13); the remainder apparently are not gallicolus. In general, eriococcid species tend to be highly host specific and none of the gall-producing forms are known to occur on more than one host genus.

The largest and most conspicuous of all coccid galls are formed on Australian *Eucalyptus* species by Eriococcids of the genera *Apiomorpha*, *Ascelis* and *Opistoscelis*. Many of these galls were described and illustrated by Froggatt (1921). Mature females of some *Apiomorpha* and *Ascelis* species are among the largest of all known Coccoidea. Fully developed females of several *Apiomorpha* species range between half and three-quarters of an inch in length, and the mature female of *Ascelis pomiformis* (Froggatt) is stated to attain a length of one inch. *Apiomorpha*, with approximately forty known species, is the largest genus of strictly gallicolus coccids known. Females of this group commonly form woody, globular or cylindrical twig galls, half to one-and-a-half inches or more in length, each with a small

85

Table 2. Summary of known gall-forming Coccoidea

Family	Genus	Known Gall-forming species	Host Family	Host Genus	Distribution of Gall-formers
1	2	3	4	5	6
MARGARODIDAE	Araucariococcus*	queenslandicus Brimblecombe	Araucariaceae	Araucaria	Australia
	Matsucoccus	gallicolus Morrison	Pinaceae	Pinus	North America
PSEUDOCOCCIDAE	Grewiacoccus*	gregalis Brain	Tiliaceae	Grewia	South Africa
	Nesopedronia	4 spp.	Gelecheniaceae	Dicranopteris	Hawaii
	Ohiacoccus*	cryptus Beardsley	Myrtaceae	Metrosideros	Hawaii
	Phylococcus*	ohauensis (Ehrhorn)	Urticaceae	Urera	Hawaii
	Pseudococcus	gallicola Ehrhorn: complex	Santalaceae	Santalum	Hawaii
	Pseudotrionymus*	2 spp.	Myrtaceae	Eugenia	Hawaii
ERIOCOCCIDAE	Aculeococcus*	morrisoni Lepage	unidentified	unidentified	Brazil
	Apiomorpha*	ca 40 spp.	Myrtaceae	Eucalyptus	Australia
	Ascelis*	5 spp.	Myrtaceae	Eucalyptus	Australia
	Atriplicia*	gallicola Cockerell	Chenopodiaceae	Atriplex	North America
	Beesonia	dipterocarpi Green	Dipterocarpaceae	Dipterocarpus	Burma
	Beesonia	undescribed species	Dipterocarpaceae	Shorea	Singapore
	Calycicoccus*	merwei Brain	Icacinaceae	Apodytes	South Africa
	Capulinia	crateraformans Hempel	Myrtaceae	Eugenia	Brazil
	Carpochloroides	mexicanus Ferris	Myrtaceae	Eugenia	Mexico
	Casuarinaloma*	leaii (Fuller)	Casuarinaceae	Casuarina	Australia
	Cylindrococcus*	4 spp.	Casuarinaceae	Casuarina	Australia
	Eriococcus	abditus Hoy	Myrtaceae	Metrosideros	New Zealand
	Eriococcus	arcanus Hoy	Podocarpaceae	Phyllocladus	New Zealand

1	2	3	4	5	6
	Eriococcus	elytranthae Hoy	Loranthaceae	Elytranthe	New Zealand
	Eriococcus	fossor (Maskell)	Oleaceae	Gymnalia	New Zealand
	Eriococcus	montanus Hoy	Compositae	Celmisia	New Zealand
	Eriococcus	orbiculus (Matesova)	Tamaricaceae	Tamarix	Kazakhstan, U.S.S.R.
	Floracoccus*	elevans (Maskell)	Myrtaceae	Eucalyptus	Australia
	Gallacoccus*	2 spp.	Dipterocarpaceae	Shorea	Singapore
	Lachnodius	14 spp.	Myrtaceae	Eucalyptus	Australia
	Macracanthopyga*	verganianus Lizery Trelles	Myrtaceae	Campomanesia	Argentina
	Madarococcus*	cunicularius Hoy	Fagaceae	Nothofagus	New Zealand
	Neotectococcus*	lenticularis Hempel	unidentified	unidentified	Brazil
	Opisthoscelis*	ca. 15 spp.	Myrtaceae	Eucalyptus	Australia
	Pseudotectococcus*	anonae Hempel	Anonaceae	Anona	Brazil
	Sphaerococcopsis*	4 spp.	Myrtaceae	Eucalyptus	Australia
	Stegococcus*	oleariae Hoy	Compositae	Olearia	New Zealand
	Tectococcus*	ovatus Hempel	Myrtaceae	Psidium	Brazil
	undescribed genus	2 spp.	Fagaceae	Nothofagus	Australia
KERMESIDAE	Fullbrightia*	gallicola Ferris	Fagaceae	Quercus	China
	Olliffiella*	cristicola Cockerell	Fagaceae	Quercus	North America
	Reynvaania*	gallicola Reyne	Fagaceae	Quercus	Indonesia
COCCIDAE	Cissococcus*	fulleri Cockerell	Vitaceae	Cissus	South Africa
ASTEROLECANIIDAE	Amorphococcus*	acaciae Brain	Leguminosae	Acacia	South Africa
	Amorphococcus	mesuae Green	Guttiferae	Mesua	Sri Lanka
	Amorphococcus	undescribed species	Guttiferae	Calophyllum	Singapore
	Asterolecanium	many spp.	various	various	cosmopolitan
	Callococcus	leptospermi (Maskell)	Myrtaceae	Leptospermum	Australia
	Cerococcus	gallicolus Mamet	Euphorbiaceae	Euphorbia	Madagascar

		Family	Host	Distribution
*Eremococcus**	*pirogallis* (Maskell)	Myrtaceae	*Leptospermum*	Australia
*Frenchia**	2 spp.	Casuarinaceae	*Casuarina*	Australia
Frenchia	*banksiae* Lambdin & Kosztarab	Proteaceae	*Banksia*	Australia
LECANODIASPIDIDAE				
*Gallinococcus**	*ferrisi* Lambdin & Kosztarab	Myrtaceae	*Leptospermum*	Australia
Lecanodiaspis	several spp.	various	various	cosmopolitan
*Stictacanthus**	*azadirachtae* (Green)	Meliaceae	*Azadirachta*	Sri Lanka, India
Abgrallaspis	*liriodendri* Miller & Howard	Magnoliaceae	*Liriodendron*	North America
DIASPIDIDAE				
*Adiscoforinia**	*secreta* (Green)	Tiliaceae	*Grewia*	Sri Lanka
*Anastomoderma**	*palauensis* Beardsley	Verbenaceae	*Premna*	Palau Is.
Aonidia	*loranthi* Green	Loranthaceae	*Loranthus*	Sri Lanka
Aspidiotus	*putearius* Green	Acanthaceae	*Strobilanthes*	Sri Lanka
*Cryptophyllaspis**	*bornmuelleri* (Rubsaamen)	Globulariaceae	*Globularia*	Canary Is.
Cryptophyllaspis	*rubsaameni* Cockerell	Euphorbiaceae	*Codiaeum*	Bismark Archipelago
Cryptophyllaspis	2 spp.	Tiliaceae	*Grewia*	Sri Lanka
Diaspidiotus	*liquidambaris* (Kotinsky)	Hamamelidaceae	*Liquidambar*	North America
Emmereziaspis	*galliformens* (Grandpré and Charmoy)	Rubiaceae	*Gaertnera*	Mauritius
*Maskellia**	*globosa* Fuller	Myrtaceae	*Eucalyptus*	Australia
*Mauritiaspis**	*malloti* Mamet	Euphorbiaceae	*Mallotus*	Mauritius
Mauritiaspis	*mimusopis* Mamet	Sapotaceae	*Mimusops*	Mauritius
*Nudachaspis**	*fodiens* (Green)	Loranthaceae	*Loranthus*	Sri Lanka
*Parachionaspis**	*galliformens* (Green)	Rubiaceae	*Hedyotis*	Sri Lanka

1	2	3	4	5	6
UNPLACED	"Sphaerococcus"	cantentulatus Froggatt	Leguminosae	Acacia	Australia
	"Sphaerococcus"	ferrugineus Froggatt	Myrtaceae	Melaleuca	Australia
	"Sphaerococcus"	froggatti Maskell	Myrtaceae	Melaleuca	Australia
	"Sphaerococcus"	morrisoni Fuller	Myrtaceae	Melaleuca	Australia
	"Sphaerococcus"	pustulans Green	Myrtaceae	Eucalyptus	Australia
	"Sphaerococcus"	rugosus Maskell	Myrtaceae	Leptospermum	Australia
	"Sphaerococcus"	socialis Maskell	Myrtaceae	Melaleuca	Australia
	"Sphaerococcus"	tepperi Fuller	Myrtaceae	Melaleuca	Australia
	"Sphaerococcus"	turbinata Froggatt	Myrtaceae	Melaleuca	Australia

[1]Genera marked with an asterisk are composed entirely of gall-forming species.

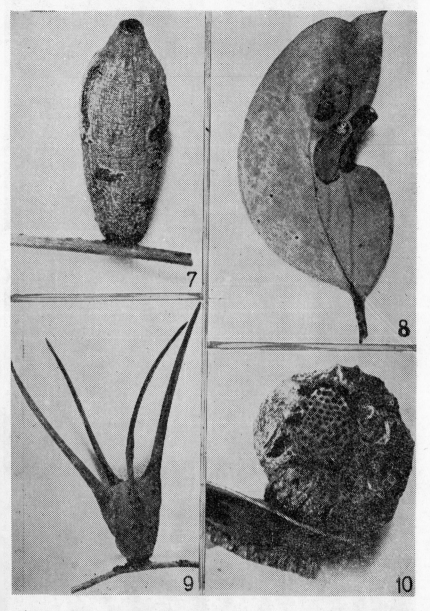

Figure 7. Gall of *Apiomorpha conica* (Froggatt) (Eriococcidae) on twig of *Eucalyptus* sp. (Myrtaceae); Australia.

Figure 8. Female galls of *Apiomorpha spinifer* (Froggatt) (Eriococcidae) on leaves of *Eucalyptus* sp. (Myrtaceae); Australia.

Figure 9. Female gall of *Apiomorpha munita* (Schrader) (Eriococcidae) on twig of *Eucalyptus* sp. (Myrtaceae); Australia.

Figure 10. Compound male gall of *Apiomorpha pharetrata* (Schrader) (Eriococceidae) arising from female gall on leaf of *Eucalyptus* sp. (Myrtaceae) Australia.

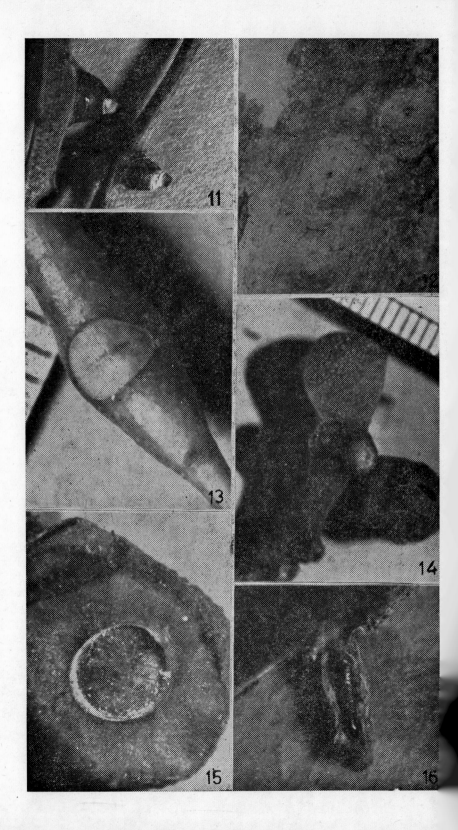

apical opening (Fig. 8). The interior of the gall forms a large, smooth-walled chamber which the insect occupies. In other *Apiomorpha* species the galls are formed attached to leaves (Fig. 8) or from aborted flower buds. In some species female galls are adorned with rib-like ridges (*A. helmsii* Fuller) or elongate horn-like projections [*A. munita* (Schrader), Fig. 9]. In *A. duplex* (Schrader) the large, sessile, female galls form elongate, rigid, four-sided structures averaging one-and-a-half to two-and-a-half inches in length from base to apical orifice. The orifice forms a narrow slit between two flattened, curled, leaf-like horns which extend outward for up to seven or eight inches from the body of the gall (Froggatt 1921). In a few *Apiomorpha* (e.g. *A. hilli* Froggatt) the galls develop as solid, elongate cones which taper to a blunt point. When mature, the galls of *A. hilli* are up to three inches in length. The apical portion of the cone then dehisces leaving a truncate, cylindrical structure about one inch long which contains the female coccid. The insect is protected by a flat septum below the rim of the gall which contains a small central orifice (Froggatt 1921, 1931).

Galls of *Apiomorpha* males are much smaller than those of females, and are generally tubular in shape. Male galls may be entirely separate from those of females, or they may arise from the surface of the female galls. Such satellite male galls may develop singly or in compound clusters. In species such as *A. pharetrata* (Schrader), female galls typically bear a disc or cup-like appendage which is composed of a large number of coalesced male galls (Fig. 10). When mature, male galls develop a large apical aperture through which the winged insect emerges. Adult males of this genus characteristically possess an elongate abdomen and genitalia which are presumed to facilitate mating through the female gall aperture.

Species of the genus *Opisthoscelis* inhabit galls which are generally smaller than those of *Apiomorpha* species. Female *Opisthoscelis* galls, which

Figure 11. Female galls of *Opisthoscelis maskelli* Froggatt (Eriococcidae) on twigs of *Eucalyptus* sp. (Myrtaceae); Australia.

Figure 12. Galls of *Sphaerococcopsis simplicior* (Maskel) (Eriococcidae) on bark of *Eucalyptus camaldulensis* (Myrtaceae); Australia.

Figure 13. Female of *Lachnodius lectularius* Maskell (Eriococcidae) in pit gall on twig of *Eucalyptus* sp. (Myrtaceae); Australia.

Figure 14. Galls of undescribed *Lachnodius* sp. (Eriococcidae) formed by aborted flower buds of *Eucalyptus baxteri* (Myrtaceae); Australia.

Figure 15. Gall of undescribed *Lachnodius* sp. (Eriococcidae) formed by aborted flower bud of *Eucalyptus baxteri* (Myrtaceae); gall dissected to show position of insect; Australia.

Figure 16. Leaf-pocket gall of undescribed genus and species (Eriococcidae); broken open to show position of insect; on *Nothofagus moorei* (Fagaceae); Australia.

occur singly or in small groups on *Eucalyptus* leaves or twigs, usually are about half an inch or less in diameter. *Opisthoscelis* galls are spherical to conical or spine-like in form (Fig. 11). For example, the leaf galls of *O. subrotunda* Schrader are nearly spherical, with a small opening on the side of the leaf opposite the main body of the gall. *O. maskelli* Froggatt and *O. spinosa* Froggatt produce thorn-like galls on twigs with the opening at or near the apex of the thorn. Females of *O. mammularis* Froggatt form groups of irregular, truncate, conical galls on host leaves. Like *Apiomorpha*, *Opisthoscelis* species are sexually dimorphic, and male galls are elongate tubular structures open at the apex when mature. Adult females of *Opistoscelis* spp. have greatly enlarged and elongated hind legs which may function in mating, although this has not been documented.

Females of *Ascelis schraderi* Froggatt produce circular, blister-like galls, about half an inch in diameter with a small central aperture, on leaves of *Eucalyptus corymbosa* (Froggatt 1921). Blister-like galls which commonly occur on *Eucalyptus* bark in Australia are caused principally by species of the genus *Sphaerococcopsis* (Fig. 12) and by a few other aberrant eriococcids such as *Floracoccus elevans* (Maskell) (Beardsley 1974, 1974a). In *Sphaerococcopsis*, males desert their galls at the end of the second instar and form pupal cocoons under loose bark, but the females remain permanently within their galls. Like *Opisthoscelis* females, those of *Sphaerococcopsis* have enlarged hind legs.

Gallicolus species of the genus *Lachnodius* inhabit open-top pit-like depressions in leaves or swollen twigs, or enclosive galls formed from aborted flower buds, on *Eucalyptus* spp. (Figs. 13–15). Unlike most obligate gall formers, females of *Lachnodius* leave their galls when they attain sexual maturity and migrate to the trunks and branches of host trees. There, eggs are deposited in long, strip-like aggregations. Males form smaller galls which they leave at the end of the second stage, and form pupal cocoons under loose bark. Beardsley (1974a) has shown that the genera *Lachnodius* and *Sphaerococcopsis* are closely related and, more recently (Beardsley 1982), placed the gall-forming species *Pseudopsylla hirsutus* Froggatt (1921a) in the former. At present there are four described species placed in *Lachnodius*, all of which are gallicolus. Ten of 17 known undescribed species which appear to belong in this genus are believed to be gallicolus (unpublished data).

Endemic Australian gall-forming Eriococcidae which occur on nonmyrtaceous hosts include two genera associated with *Casuarina* spp. and an apparently undescribed genus found on *Nothofagus* spp. The latter contains two undescribed species, one forming shallow pocket galls on leaves of *N. cunninghami* in Victoria and Tasmania, the other forming deep, tubular, pocket galls on leave of *N. moorei* in New South Wales (Fig. 16) (unpublished data). Another eriococcid, *Madarococcus cunicularius* Hoy (1962), forms leaf galls on two *Nothofagus* species in New Zealand.

On Australian *Casuarina*, the genus *Cylindrococcus* contains four species which produce foliaceous twig galls that resemble small cones or seed capsules (Froggatt 1921). The gall of *C. amplior* Maskell (Fig. 18) is surrounded at the base by a series of leaf-like bracts which partially enclose the conical inner gall that contains the female insect. The genus *Casuarinaloma* (Froggatt 1933) with one included species, *C. leaii* (Fuller), forms subspherical twig galls with strongly fluted sides (Froggatt 1921a). Hoy (1963) tentatively assigned this aberrant genus to the Eriococcidae.

Outside Australia gallicolus Eriococcidae are apparently less common, but are widely distributed; particularly in the tropics and subtropics. It seems likely that many additional gall-forming eriococcids remain to be discovered. For example, during a careful study of plant galls in a small forested area of Singapore, Anthony (1974) discovered seven previously unreported gallicolus coccid species. Four of these were Eriococcidae, including representatives of two new genera. The galls formed by two species of one genus, *Gallacoccus* (Beardsley 1971), were studied in detail (Anthony 1974a, 1977). Both species were found only on *Shorea curtisii* where each produces a distinctive type of epiphyllous gall from developing buds. The gall of one species, *G. secundus* Beardsley, is a solid enclosive structure while that of the other; *G. anthonyae* Beardsley, takes the form of a small, foliaceous cone which may contain one or several specimens of the coccid. Mature females of *Gallacoccus* are sac-like unsclerotized creatures with vestigeal appendages.

The unusual genus *Beesonia* (Green 1926) which contains two described species, one gailicolus, was placed tentatively within the Eriococcidae by Hoy (1963). However, Ferris (1950) considered this genus to constitute a distinct family level taxon, the Beesoniidae. The type species, *B. dipterocarpi* Green, was described from galls on *Dipterocarpus* collected in Burma. These morphologically aberrant coccids live deeply embedded in host twigs where they form irregularly shaped woody galls. The second species, *B. quercicola* Ferris, which was taken from the bark of oaks near Canton, China, apparently is non-gallicolus. Hoy (1963) has suggested that these two species may not be truly congeneric. Anthony (1977) has reported an apparently undescribed species of *Beesonia* from twig galls of *Shorea curtisii* in Singapore. In *Beesonia*, the mature female insects remain within the sclerotized exuviae of the previous stage. The pre-adult female stages are comprised chiefly of a much enlarged head region. The thorax and abdomen are relatively small and the thoracic spiracles are displaced to near the posterior extremity of the body.

The gallicolus Eriococcidae of New Zealand include *Madarococcus cunicularus* on *Nothofagus*, previously mentioned, as well as *Stegococcus oleariae* Hoy and five species of the widespread genus *Eriococcus*. All form leaf galls of various sorts on their host plants (Table 2) (Hoy 1962). Another *Eriococcus*, *E. orbiculus* (Matesova), is reported to form galls which appear as open-top swellings at the bases of twigs on *Tamarix* spp. in Kazakhstan,

94

Figure 17. Woody leaf galls of *Olliffiella cristicola* Cockerell (Kermesidae) on *Quercus emoryi* (Fagaceae); Arizona, U.S.A.

Figure 18. Foliaceous gall of *Cylindrococcus amplior* Maskell (Eriococcidae); bisected to show position of insect; on *Casuarina* sp. (Casuarinaceae); Australia.

Figure 19. Galls of *Frenchia casuarinae* Maskell (Asterolecaniidae) on twig of *Casuarina* sp. (Casuarinaceae); Australia.

Figure 20. Twig gall of *Maskellia globosa* Fuller (Diaspididae) on *Eucalyptus* sp. (Myrtaceae); Australia.

USSR (Matesova 1960). In Western North America the genus *Atriplicia*, which is closely allied to *Eriococcus*, contains a single species, *A. gallicola* Cockerell and Rohwer, forming galls on the leaves of *Atriplex canescens*. The galls are produced by a swelling of the leaf base and its subsequent folding, leaving a slit-like opening at the top (Ferris 1955).

Gall-forming eriococcid species probably are fairly numerous in tropical America. However, the coccid fauna of that vast region is, at present, very incompletely known, and as yet only a few Neotropical gall-producing forms have been described. These comprise several unique genera the relationships of which are as yet obscure. The following species are known.

1) *Aculeococcus morrisoni* Lepage (1941), which forms slender conical leaf galls on an unidentified Brazilian plant.

2) *Capulinia crateraformans* Hempel (1900) forms small, crater-shaped galls on bark of limbs and twigs of *Eugenia jambotica* in Brazil.

3) *Carpochloroides mexicanus* Ferris (1957), which produces twig galls on *Eugenia acapulcensis* in Oaxaca, Mexico. Galls, which are stated to involve the petioles of several small leaves, are "almost globular" and attain lengths of around two centimetres.

4) *Macrocanthapyga verganianus* Lizer y Trelles (1955) forms galls on branches of *Campomanesia* sp. in Brazil.

5) *Neotectococcus lectularius* Hempel (1937) makes small, lenticular galls on foliage of an unspecified Brazilian host plant.

6) *Pseudotectococcus anonae* Hempel (1934) produces small leaf galls, about 3 mm high and 1.5 mm diameter at base, on cultivated *Anona* in Brazil. Female galls are conical in shape; male galls are cylindrical. Galls protrude from the upper surface, and open onto the lower surface.

7) *Tectococcus ovatus* Hempel (1901) forms small, spherical leaf galls on *Psidium variable* in Brazil. The galls are about 8 mm in diameter and each protrudes from both upper and lower surfaces of the host leaf, with the gall aperture on the lower surface.

The only gall-forming eriococcid reported from the Ethiopian zoogeographical realm is *Calycicoccus merwii* Brain (1918) from South Africa, which forms leaf galls resembling small, laterally compressed cones, or thorns, on *Apodytes*. Lepage (1941) suggested a relationship between this species and *Aculeococcus morrisoni* from Brazil; however, the morphological similarities between the two are likely due to evolutionary convergence.

The family Kermesidae is a small group of around 60 described species which are exclusively associated with hosts of the genus *Quercus* and a few other closely related Fagaceae. This family, which some authors include within the Eriococcidae (e.g. Hoy 1963), is of largely Holarctic distribution. Most of the species are placed in the non-gallicolus genus *Kermes*. However, three gall-forming species which appear related to that genus have been described, one from North America, one from Asia and one from Indonesia. *Olliffiella cristicola* Cockerell (1896) forms distinctive thorn-like woody pocket galls on the undersides of leaves of *Querçus emoryi* in south-western North America (Fig. 17). These galls open on the upper leaf surface through an elongate slit parallel to the leaf midrib (Ferris 1919). *Fulbrightia gallicola* Ferris, known from Yunan Province, China on *Quercus delavayi*, produces galls from aborted twigs which resemble small foliaceous cones. The area shared by the bases of several of these cones becomes swollen, and the insect is found buried among the bases of the cones (Ferris 1950). *Reynvaania gallicola* Reyne produces bud galls on *Quercus lineata* in Java. The terminal buds of infested twigs develop into red-brown plushy balls, each 7–12 mm in diameter, from which one or more leaves sometimes arise. Within the gall is an irregular central cavity which houses the insect (Reyne 1954).

The Coccidae (sens. str.) is one of the more common and widespread families of Coccoidea, containing between eight and nine hundred named species. The only confirmed gall-forming member of this large group is the South African species *Cissococcus fulleri* Cockerell (1902), which forms large, globular, pear-shaped or urn-shaped galls on the stems, tendrils and leaf stalks of *Cissus cuneifolia*. The caudal end of the coccid is highly modified to form a sclerotized operculum for the gall aperture (Brain 1918).

The Asterolecaniidae is a moderately large family of worldwide distribution which contains around 150 described species. The majority of these are placed in the cosmopolitan genus *Asterolecanium* and are commonly referred to as "pit scales" as many of the species produce pit-like depressions with swollen margins at their site of attachment on the host (fig. 1). These pits are a type of gall formation resulting from lignification of the host cortex beneath the insect, and elongation of the cortical cells surrounding the lignified area (Habib 1943). Pit formation by *A. variolosum* Ratzenburg has been described in detail by Parr (1940). This type of gall also is produced by some species of the widespread asterolecaniid genus *Cerococcus*. For example, Mamet (1959) described *C. gallicolus* from stem galls on *Euphorbia stenoclada* in Madagascar.

Several additional genera which contain gallicolus species have been placed in the Asterolecaniidae. Some of these are highly aberrant forms, and their placement in this family appears, in some cases, to be based upon relatively weak morphological evidence. The genus *Amorphococcus*, referred to the Asterolecaniidae by Lambdin and Kosztarab (1973), contains two described species, *A. acaciae* Brain (1918), which forms flat, circular, blister-like galls on stems of *Acacia* sp. in South Africa, and *A. mesuae* Green, which inhabits enclosed, rounded to conical, woody twig galls on *Mesua ferrea* in Ceylon (Green 1909). Anthony (1974) has described an unusual bivalved gall formed by an undescribed *Amorphococcus* species on *Calophyllum inophylloide* in Singapore. The galls, which resemble dehescent seed capsules, are formed in clusters from aborted leaves. Each gall is about one centimeter in diameter and contains a large central cavity occupied by the insect.

The Australian genus *Frenchia*, recently reviewed by Lambdin and Kosztarab (1981), contains three described species, all obligate gall-formers. Two of these occur on species of *Casuarina*; the third on *Banksia*. *F. casuarinae* Maskell females produce distinctive twig galls which bear a cylindrical, peg-like protrusion arising from a central gall cavity (Fig. 19). The elongate attenuated abdomen of the mature insect extends outward into the projecting peg. Females of *F. semiocculata* Maskell live under host bark and produce irregular swellings, each with a minute opening, in infested *Casuarina* twigs (Froggatt 1921a). Males of this species make shallow, conical twig galls with relatively broad apical openings through which the adult insects escape. *F. banksiae* Lambdin and Kosztarab, which is found on *Banksia serrata* in coastal southeastern Australia, forms globular leaf galls.

Two additional gall-forming Asterolecaniidae are found on *Leptospermum* in Australia. *Eremococcus pirogallis* (Maskell) makes pear-shaped woody galls on stems *L. flavescens* (Ferris 1919, Froggatt 1921a). *Callococcus leptospermi* (Maskell) forms irregular galls in the form of elongate blister-like swellings on twigs of *L. laevigatum*. These swellings occur on either side of the insect, which occupies a cavity, with a narrow, slit-like opening, in the wood between them. Both of these asterolecaniids were described originally in the catch-all genus *Sphaerococcus* (Froggatt 1921a), the type species of which is a non-gallicolus mealybug.

The family Lecanodiaspididae is allied to the Asterolecaniidae and formerly was included as a subfamily of the latter. The family contains around 60 described species placed in eight genera (Lambdin and Kosztarab 1973). The type genus, *Lecanodiaspis*, includes several species (e.g. *L. acaciae* Maskell) which produce pit galls similar to those formed by many *Asterolecanium* species (Howell and Kosztarab 1972). Anthony (1974) described and figured pit galls formed by an undetermined *Lecanodiaspis* sp. on twigs of *Xylopia malayana* (Anonaceae) in Singapore. The mature female insect occupies a large pit surrounded by a fusiform swelling approximately a centimeter in length, and up to 6 mm wide. Under the growth of cortical tissues which participate in gall morphogenesis, the bark is fissured. Other gall-forming lecanodiaspidids include *Stictacanthus azadirachtae* (Green), which produces woody pit galls on *Azadirachtia indica* in Sri Lanka, and *Gallinococcus ferrisi* Lambdin and Kosztarab (1973), which forms enclosed stem galls, consisting of swellings with small, pore-like apertures, on *Leptosperum laevigatum* in Australia.

The Diaspididae (armored scale insects) is the largest and most highly specialized family of Coccoidea. There are presently recognized about 350 genera containing approximately 1800 species worldwide (Borchsenius 1966). Typically, diaspidids develop within a thin, detached, shield-like external covering which the insect produces. Relatively few armored scales have been reported to produce plant galls.

The formation of shallow pit galls on host twigs and leaves has been reported for a few diaspidids. For example, *Anastomoderma palauensis* Beardsley (1966) produces pit galls in callous growth on twigs of *Premna integrifolia* in the Palau Islands. The formation of such relatively inconspicuous galls by Diaspididae may be more common, particularly among tropical species, than the paucity of published references suggests. However, in the temperate regions of the world, where the diaspidid faunas generally have been reasonably well studied, very few species have been associated with plant galls. In North America, for example, *Diaspidiotus liquidambaris* (Kotinsky) is one of two diaspidid species for which a gallicolus habit is well documented. This insect produces small, open-top pit galls on the leaves of *Liquidambar styraciflua* (Kotinsky 1903, Ferris 1938). A second North American gallicolus diaspidid *Abgrallaspis liriodendri*, was recently described by Miller and Howard (1981). Females of this species occupy open-top pit galls on leaves of *Liriodendron tulipifera*. The galls protrude from the upper surface of host leaves and the gall opening on the lower surface is closed by the scale covering of the insect.

In his monumental work the Coccidae of Ceylon, E.E. Green described and illustrated several distinctive gallicolus Diaspididae. The relative wealth of such forms which are known from Sri Lanka probably reflects relatively thorough collecting of that island by Green and others, and the paucity of gallicolus diaspidids known from other regions of the tropics may be due largely to the lack of such thorough field work. The following gall-forming diaspidids are known from Sri Lanka:

1) *Aonidia loranthi* Green (1896). Females of this scale occupy small cavities in the centre of circular swellings on stems of *Loranthus* sp.

2) *Aspidiotus putearius* Green (1896). Females occupy pit-like depressions on the under surfaces of *Strobilanthes viscosus* leaves. The external scale forms an operculum over the opening of the pit. The upper surfaces of the leaves bear small discoloured prominences over the site of each pit.

3) *Cryptophyllaspis occulatus* (Green) (1896). Females of this species form minute, rounded galls on the upper surface of the leaves of *Grewia orientalis*. The insects actually settle on the lower surfaces of the host leaves, first producing a pit which deepens as the insect grows and later bulges out on the upper surface until the mature diaspidid is largely enclosed within the pouch-like gall. The external scale covering produced by the insect serves to close the gall aperture. Normally there is one insect per gall; occasionally two. A related species, *Cryptophyllaspis elongata* (Green), produces elongate pouch-like galls on the lower surface of a *Grewia* sp. in Sri Lanka (Green 1905).

4) *Adiscofiorinia secreta* (Green) (1896). Females of this diaspidid form deep, pouch-like galls on the upper surface of leaves of *Grewia occidentalis*. The insect assumes a vertical position within the gall with the cephalic end uppermost and the posterior end toward the gall opening on the lower surface of the leaf. A scale covering is formed, but the adult female insect

is enclosed within the second stage exuviae. Males of this species occupy shallow pits in the lower leaf surface.

5) *Nudachaspis fodiens* (Green) (1899). Females of this species form small, blister-like galls on the leaves and twigs of *Loranthus* sp. The mature insect is completely imbedded within the gall tissue and no external scale covering is formed. The position of the insect is marked externally by a wart-like swelling with a central depression and perforation which is closed by the exuviae of the first-stage insect. Males of this species apparently are of the normal diaspidid type and do not produce galls.

6) *Parachionaspis galliformens* (Green) (1899). Females of this diaspidid form small pustule galls on stems and leaf midribs of *Hedyotis lasertiana*. The insect occupies a hollow cell within the gall and the external scale is reduced to a thin film of secretory matter which lines the gall cavity. Males of this species are not gallicolus.

The genus *Cryptophyllaspis*, which presently contains four species, all gallicolus, was treated as a synonym of the large cosmopolitan genus *Aspidiotus* by Ferris (1941). However, Borchsenius (1966) considered *Cryptophyllaspis* as a valid genus, and included, in addition to the two Green species from Sri Lanka, *C. bornmuelleri* (Rubsaamen) from the Canary Islands and Maderia, and *C. rubsaameni* Cockerell, from the Bismark Archipelago. The former forms small conical galls on the leaves of *Globularia* and the latter inhabits subcylindrical galls, about 2 mm in length, on leaves of *Codiaeum*.

Mamet (1939) described the gallicolus genus *Mauritiaspis*, containing two species, from the island of Mauritius. Females of *M. malloti* Mamet are stated to form galls "—represented by a globose or truncated cylindrical protuberance with an irregular lobed rim; situated on the undersurface of the leaves of *Mallotus integrifolius*." The galls, which are about 2 to 3 mm in diameter, and about 2 mm high, open onto the upper surfaces of the leaves and the female insects occur singly within the central cavity of the gall. Males of this species apparently are not gallicolus, and male scales are found on the under surfaces of the leaves. Ferris (1941) has figured the gall of this species. A second species, *M. mimusopis* Mamet, produces small globular galls on leaves of *Mimusops petiolans*. *Emmereziaspis galliformens* (Grandpré and Charmoy), which produces galls described as small swellings, on leaves of *Gaertnera* sp., is also known only from Mauritius (Mamet 1941.)

The highly aberrant Australian genus *Maskellia* Fuller (1897) contains a single described species, *M. globosa* Fuller, and is presently included with the Diaspididae on the basis of overall morphology. Females of *M. globosa* develop completely enclosed within fleshy galls produced from aborted young twig terminal on *Eucalyptus* spp. (Fig. 20). The dark purple, globular, flask-shaped female insects are found firmly imbedded within the irregularly shaped galls. No vestige of the typical diaspidid external scale covering is produced. The males of this species are stated to produce galls in the form of horn-shaped pockets projecting from the upper surface, and opening on the lower surface, of host leaves (Frogatt 1915).

In addition to the galls produced by the apparently obligate gallicolus diaspidids listed in Table 2, there are a few published reports of other sorts of plant deformations associated with armored scale insects. For example, Anthony (1974) reported a witches' broom deformation of *Lithocarpus sundaicus* (Fagaceae) associated with an undetermined diaspidid in Singapore. She stated that the scale crawlers settle in the axils of young leaves of lateral buds and induce internode shortening, cessation of growth on the primary axis, and formation of secondary and tertiary shoots by axillary buds. Probably, deformations of this sort due to armoured scale infestation are not uncommon.

The remarkable Australian fauna of gallicolus Coccoidea contains several named species the family relationships of which are obscure. These species were placed by their describers in the genus *Sphaerococcus*, the type species of which is a mealybug (Pseudococcidae). However, none of these gallicolus forms appear to be pseudococcids, and their true taxonomic relationships are as yet undetermined, as is indicated in Table 2. Most of these probably have affinities with the Eriococcidae or the Asterolecaniidae, but critical study of the type material of each is required before they can be properly placed.

The galls produced by several of these unplaced species were described briefly by Froggatt (1921a). *Sphaerococcus rugosus* Maskell, for example, is stated to produce "remarkable dark green subglobular galls, stalked at the base, swelling out on the sides with the apex cone shaped" on leaves and twigs of an undetermined species of *Leptospermum* in Western Australia. *Sphaerococcus socialis* Maskell forms "globular galls of imbricated scales (like little pine cones) of a grayish green colour, varying in size from pin's head to half an inch in diameter." Of the nine *Sphaerococcus* species listed in Table 2, six are associated with *Melaleuca* spp.

Host plant relationships of gallicolus coccids

Species of Coccoidea which produce irregular deformations of host plant tissues (witches' brooms, etc.), such as *Nipaecoccus vastator*, or shallow pits, such as *Asterolecanium pustulans*, are often fairly polyphagous, while those which produce regularly structured enclosive galls are usually quite host specific. Virtually all species of obligate gall-forming Coccoidea known are restricted to a single host species or to a few closely related species in a single host genus. Furthermore, in most genera of obligate gall-formers all the included species are confined to members of a single genus of hosts.

Irregular plant deformations, such as bunchy tops or witches' brooms, are produced by coccids in several families, most notably by certain mealybugs and a few armoured scales. Plant deformations of this sort presumably enhance survival of the insects which produce them by providing some degree of shelter from natural enemies and other adverse environmental factors, and possibly also by altering host physiology to favour coccid nutrition. The cassava mealybug, *Phenacoccus manihoti* Matile-Ferrero, produces a severe bunchy-top deformation on its preferred host, cassava, but not on other euphorbiaceous hosts on which it has occasionally been found. Apparently, under natural conditions, this species does not occur on plants other than cassava, except where severe outbreaks force overflow populations to utilize less suitable hosts in the immediate vicinity of cassava plants (unpublished observations).

The great wealth of gallicolus coccids which occurs on the Australian continent is confined chiefly to plants of the family Myrtaceae (80 spp. on *Eucalyptus*, 6 spp. on *Melaleuca*, 4 spp. on *Leptospermum*). The genus *Eucalyptus* is host to more gall-forming Coccoidea than any other group of plants, and it is likely that many more gallicolus coccids await discovery on Australian *Eucalyptus*. Coccids are known to form galls on members of the Myrtaceae elsewhere, in New Zealand, Hawaii and tropical America. All together, nearly 100 species of obligate gall-producing Coccoidea confined to myrtaceous hosts have been discovered, and many more probably exist.

The Fagaceae, a family attacked by gall-forming insects of many kinds, are hosts of several obligate gallicolus coccids, but the number of these (six species in five genera listed in Table 2, plus several species of *Asterolecanium* which form pit galls), is far less than that which is found on hosts of the family Myrtaceae. Other plant families which are hosts for several species of gallicolus coccids include the Casuarinaceae, Dipterocarpaceae, Euphorbiaceae Guttiferae, Loranthaceae, Rubiaceae, and Tiliaceae. The overwhelming majority of coccid galls occur on dicotyledonous angiosperms. The only reported coccid galls on pteridophytes are those of the mealybug genus *Nesopedronia* in Hawaii, and only three species of gallicolus coccids are known from gymnosperms, although it seems likely that more exist. No coccid galls have been reported from any of the Monocotyledoneae.

Morphological specializations associated with gall-formation in Coccoidea

Facultative gall-forming coccids, such as *Nipaecoccus vastator* and *Asterolecanium pustulans*, exhibit no obvious morphological adaptation for gallicolus life. Similarly, many species which seem to be obligate gall producers (e.g. *Pseudococcus gallicola, Eriococcus orbiculus, Diaspidiotus liquidambarus*) possess no distinctive structural adaptations which would set them apart from their non-gallicolus congeners. On the other hand, among species which belong to strictly gallicolus genera, a wide range of specialized mor-

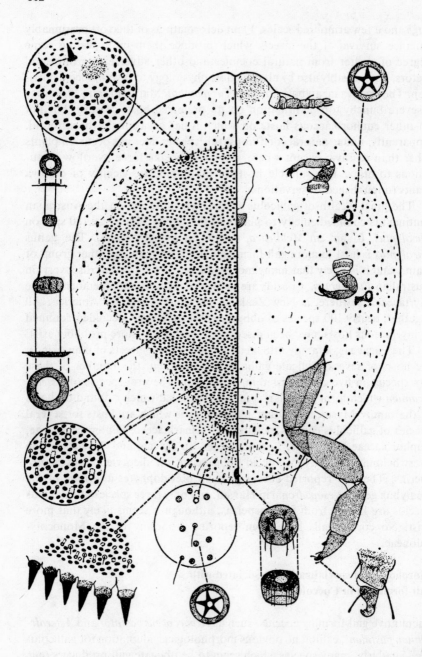

21

Figure 21. *Sphaerococcopsis platynotum* Beardsley (Eriococcidae); taxonomic drawing of mature female illustrating morphological specialization for gallicolus life; e.g. dorsal sclerotized area which functions as an operculum in the gall aperture; enlarged hind legs which may function to facilitate mating, etc.

phological adaptations for gallicolus existence are seen (Fig. 21). Examples of the sorts of adaptive specializations found among gallicolus coccids are:

1) Modification of the general body form to ovoid or spherical (e.g. *Amorphococcus, Carpochloroides, Eremococcus, Gallacoccus, Maskellia, Olliffiella, Opisthoscelis*) or pyriform (e.g. *Aculeococcus, Apiomorpha, Frenchia, Macracanthapyga, Phylococcus*). Species of these genera occur within completely enclosive galls or galls with small apical apertures.

2) Reduction or loss of appendages (legs and antennae) in species which belong to families in which these are normally fully developed (e.g. *Apiomorpha, Ascelis, Beesonia, Carpochloroides, Gallacoccus, Phylococcus*).

3) Specialization of the hind legs of adult females into elongated, or otherwise enlarged, structures [e.g. *Opisthoscelis, Sphaerococcopsis*) (fig. 21)]. It is thought that such modified hind legs may somehow facilitate mating by guiding or grasping the male abdomen when it is inserted into the aperture of the female gall. Such behaviour has not been documented, however.

4) Shortening of the feeding stylets (e.g. *Apiomorpha, Cylindrococcus, Frenchia, Gallacoccus, Opisthoscelis, Sphaerococcopsis*). The feeding stylets of most Coccoidae are very long, hairlike structures, sometimes several times as long as the main body of the insect. The extreme length of the stylets reflects the necessity for these insects to penetrate deeply into the tissues of the host in order to tap the conductive cells of the phloem. Species which inhabit enclosive galls apparently feed upon nutritive tissues in the walls of their galls. In these, the stylets have been reduced to relatively very short (usually much shorter than the length of the body), sometimes thickened structures.

5) Development of specialized sclerotic opercular structures which serve to plug the gall orifice. These are formed in several ways, i.e.

A) The posterior abdominal segments are modified to form sclerotized plates (e.g. *Cissococcus, Phylococcus*) or sclerotic plugs (e.g. *Ascelis*, some *Apiomorpha*).

B) The abdomen bears special enlarged spines associated with sclerotized dermal areas situated on the posterior portion (e.g. *Apiomorpha, Frenchia, Nesopedronia*).

C) Part or all of the dorsal surface becomes highly sclerotized and forms the operculum (e.g. *Floracoccus, Sphaerococcopsis* (Fig. 21), unnamed genus on Australian *Nothofagus*).

6) Loss or modification of external coverings which are typically present in related, non-gallicolus forms (e.g. *Amorphococcus, Maskellia*).

This chapter was prepared after review of available literature and is based upon that review, plus the author's personal unpublished observations. An attempt was made, in Table 2, to list all known obligate, gall-forming Coccoidea. However, it is quite likely that some gallicolus species have been overlooked.

104

REFERENCES

1. Anthony, M. 1974. Contribution to the knowledge ot Cecidia of Singapore. *Gardens Bull.* Singapore 27: 17–65.
2. Anthony, M. 1974a. Cecidogenese comparee de deux galles de coccides (*Gallacoccus anthonyae* Beardsley et *Gallacoccus secundus* Beardsley) developpees sur le *Shorea Curtisii* Dyer ex King (Dipterocarpaceae). *Marcellia* 38: 99–144.
3. Anthony, M. 1977. Morphological and anatomical comparison of normal and cecidial shoots in *Shorea Curtisii* Dyer ex King. *Marcellia* 40: 181–192.
4. Balachowsky, A.S. 1948. Les cochenilles de France d' Europe, du nord de l' Afrique, et du bassin Méditerranéen. IV. —Monographie des Coccoidea; Classification— Diaspidinae (Premiere Partie). Actualités Sci et Indus., *Entomol. Appl.* 1054: 243–394.
5. Beardsley, J.W. 1957. The genus *Pedronia* Green in Hawaii, with descriptions of new species (Homoptera: Pseudococcidae). *Proc. Hawaiian Entomol. Soc.* 16: 218–231.
6. Beardsley, J.W. 1966. The Coccoidea of Micronesia (Homoptera). Insects of Micronesia (B.P. Bishop Museum). 6: 377–562.
7. Beardsley, J.W. 1971. A new genus of gall-inhabiting Eriococcidae from Singapore (Homoptera: Coccoidea). *Proc. Hawaiian Entomol. Soc.* 21: 31–39.
8. Beardsley, J.W. 1971a. New genera and species of Hawaiian Pseudococcidae (Homoptera). *Proc. Hawaiian Entomol. Soc.* 21: 41–58.
9. Beardsley, J.W. 1974. A new genus of Coccoidea from Australian *Eucalyptus* (Homoptera). *Proc. Hawaiian Entomol. Soc.* 21: 325–328.
10. Beardsley, J.W. 1974a. A review of the genus *Sphaerococcopsis* Cockerell, with descriptions of two new species (Homoptera: Coccoidea). *Proc. Hawaiian Entomol. Soc.* 21: 329–342.
11. Beardsley, J.W. 1982. On the taxonomy of the genus *Pseudopsylla* Froggatt, with a redescription of the type species (Homoptera: Coccoidea). *Proc. Hawaiian Entomol. Soc.* 24: 31–35.
12. Beardsley, J.W., and R. Gonzalez, 1975. The biology and ecology of armored scales. *Ann. Review Entomol.* 20: 47–73.
13. Borchsenius, N.S. 1958. The evolution and phylogenetic relationships of the Coccoidea. (Insecta Homoptera), (in Russian, English summary). *Zool. Zhur.* 37: 765–780.
14. Borchsenius, N.S. 1966. A Catalog of the Armored Scale Insects of the World. 449 pp. Acad. Nauk SSSR, Zool. Inst. Leningrad.
15. Brain, C.K. 1918. The Coccidae of South Africa—II. *Bul. Entomol. Res.* 9. 107–139.
16. Brimblecombe, A.R. 1960. Studies of the Coccoidea. 11. New genera and species of Monophlebidae. *Queensland Jour. Agr. Sci.* 17: 183–193.
17. Cockerell, T.D.A. 1896. A gall-making coccid in America. *Science* (n.s.) 4: 299–300.
18. Cockerell, T.D.A. 1902. A new gall-making coccid. *Canad. Entomol.* 34: 75.
19. Ferris, G.F. 1919. A contribution to the knowledge of the Coccidae of the southwestern United States. Stanford Univ. Pubs., Univ. Ser., 68 pp.
20. Ferris, G.F. 1919a. Notes on the Coccidae—IV (Hemiptera). *Canad. Entomol.* 51: 249–253.
21. Ferris, G.F. 1938. Atlas of the Scale Insects of North America. Series II, no 223. Stanford Univ. Press.
22. Ferris, G.F. 1941. Contributions to the knowledge of the Coccoidea (Homoptera) X. *Microentomology* 6: 18.
23. Ferris, G.F. 1941a. The genus *Aspidiotus* (Homoptera; Coccoidea; Diaspididae). *Microentomology* 6: 33–69.

24. Ferris, G.F. 1950. Report upon a collection of scale insects from China. II. (Insecta: Homoptera). *Microentomology* 15: 69–97.

25. Ferris, G.F. 1955. Atlas of Scale Insects of North America, v. 7, The Families Aclerdidae, Asterolecaniidae, Conchaspididae, Dactylopiidae and Lacciferidae. iii+233 pp. Stanford Univ. Press.

26. Ferris, G.F. 1957. Notes on some little known genera of the Coccoidea (Homoptera). *Microentomology* 22: 59–79.

27. Ferris, G.F. 1957a. A review of the family Eriococcidae (Insecta: Coccoidea). *Microentomology* 22: 81–89.

28. Froggatt, W.W. 1915. A descriptive catalogue of the scale insects ("Coccidae") of Australia. Part I. Dept. Agric. N.S. Wales, Sci. Bul. 14, 64 pp.

29. Froggatt, W.W. 1921. A descriptive catalogue of the scale insects ("Coccidae") of Australia. Part II. Dept. Agric. N.S. Wales Sci. Bul. 18, 159 pp.

30. Froggatt, W.W. 1921a. A descriptive catalogue of the scale insects ("Coccidae") of Australia. Part III. Dept. Agric. N.S. Wales Sci. Bul. 19, 43 pp.

31. Froggatt, W.W. 1931. A classification of the gall-making coccids of the genus *Apiomorpha*. Proc. Linnean Soc. N.S. Wales 56: 431–454.

32. Froggatt, W.W. 1933. The Coccidae of the Casuarinas. *Proc. Linnean Soc.* N.S. Wales 58: 363–374.

33. Fuller, C. 1897. A gall-making diaspid. Agr. Gaz. N.S. Wales 8: 579–580.

34. Green, E.E. 1896. The Coccidae of Ceylon. Part I. xi+103 pp. Dulau Co., London.

35. Green, E.E. 1899. The Coccidae of Ceylon. Part II. pp. xiii–xli, 105–169. Dulau Co., London.

36. Green, E.E. 1905. Supplement arynotes on the Coccidae of Ceylon. *Bombay Nat. His. Soc. Jour.* 16: 340–357.

37. Green, E.E. 1909. The Coccidae of Ceylon—IV. pp. 250–344. Dulau Co., London.

38. Green, E.E. 1926. On some new genera and species of Coccidae. *Bul. Entomol. Res.* 17: 55–65.

39. Habib, A. 1943. The biology and bionomics of *Asterolecanium pustulans* Ckll. *Bul. Entomol. Soc. Egypt* 27: 87–111.

40. Hempel, A. 1900. Descriptions of three new species of Coccidae from Brazil. *Canad. Entomol.* 32: 3–7.

41. Hempel, A. 1901. Descriptions of Brazilian Coccidae. *Ann. and Mag. Nat. Hist.* (ser. 7) 7: 110–125.

42. Hempel, A. 1934. Descripcao de tres especies novas, tres generos novas e uma subfamilia nova de coccideos (Hemiptera: Homoptera). *Rev. de Entomol.* 4: 139–147.

43. Hempel, A. 1937. Novas especies de coccideos (Homoptera) do Brazil. *Arch. Inst. Biol.* (Sao Paulo) 8: 5–36.

44. Hoy, J.M. 1962. Eriococcidae (Homoptera: Coccoidea) of New Zealand. *New Zeal. Dept. Sci. and Indus. Res. Bul.* 146, 219 pp.

45. Hoy, J.M. 1963. A catalogue of the Eriococcidae (Homoptera: Coccoidea). *New Zeal. Dept. Sci. and Indus. Res. Bul.* 150, 260 pp.

46. Kotinsky, J. 1903. The first North American leaf-gall diaspine. *Proc. Entomol. Soc. Wash.* 5: 149–150.

47. Lambdin, P.L. and M. Kosztarab, 1973. Studies on the morphology and systematics of scale insects. No. 5, a revision of the seven genera related to *Lecanodiaspis*. *Virginia Polytechnic Inst. and State Univ. Res. Div. Bul.* 83, 110 pp.

48. Lambdin, P.L. and M. Kosztarab, 1981. Revision of the genus *Frenchia* with description of a new species (Homoptera: Coccoidea: Asterolecaniidae). *Proc. Entomol. Soc. Wash.* 83: 105–128.

49. Lepage, H.S. 1941. Descricao de um novo genero e nova espécie de coccideo produtor de galhas (Homoptera: Coccoidea). *Arq. Inst. Biol.* (Sao Paulo) 12: 141–145.

50. Lizer y Trelles, C.A. 1955. Description d'une nouvelle et bizarre cochenille de la region neotropicale (Hom. Eriococcidae). *Bul. Soc. Entomol. de France* 60: 37–38.

51. Mamet, R. 1939. Some new genera and species of Coccidae (Hemipt. Homopt.) from Mauritius. *Trans. Royal Entomol. Soc. London* 89: 579–589.

52. Mamet, R. 1941. On some Coccidae (Hemipt. Homopt.) described from Mauritius by de Charmoy. *Bul. Mauritius Inst.* 2: 23–37.

53. Mamet, R. 1959. Notes on Coccoidea of Madagascar 4. *Inst. Sci. de Madagascar* Mem. 11 (Ser. E. Entomol.) 369–479.

54. Maskell, W.M. 1898. Further coccid notes, with descriptions of new species and discussion of points of interest. *Trans. New Zealand Inst.* 30: 219–252.

55. Matesova, G.I. 1960. New species of soft scales Fam. Pseudococcidae (Homoptera, Coccoidea) of the Kazakhstan fauna (in Russian). *Acad. Nauk Kazakhskoi SSR Inst. Zool.* (Entomol.) Trudy 11: 205–217.

56. Miller, D.R., and F.W. Howard, 1981. A new species of *Abgrallaspis* (Homoptera: Coccoidea: Diaspididae) from Louisiana. *Ann. Entomol. Soc, America* 74: 164–166.

57. Parr, T. 1939. *Matsucoccus* sp., a a scale insect injurious to certain pines in the northeast (Hemiptera-Homoptera). *Jour. Econ. Entomol.* 32: 624–630.

58. Parr, T. 1940. *Asterolecanium variolosum* Ratzeburg, a gall-forming coccid, and its effects upon the host trees. *Yale Univ. School Forestry Bul.* 46, 49 pp.

59. Reyne, A. 1954. *Reynvaania gallicola*; a new eriococcid causing galls on *Quercus lineata* Bl. *Tijdschr. v. Entomol.* 97: 233–241.

5. Biology of Gall Thrips (Thysanoptera: Insecta)

A. Raman and T.N. Ananthakrishnan

Introduction

Thysanoptera as an essentially phytophagous group display in common with other gall makers, a remarkable potential to exploit their hosts by inducing specialized feeding and breeding facility, viz., the gall. Thrips galls tend to be always positive, directional responses, envisaging disharmonic growth effects, with the polarity gradient of the host-plant differentiation sequence controlled by the feeding action thrips. Irrespective of the nature of the 'form' in thrips-induced galls, gall formation is determined by the time of initiation and the population of thrips within the gall (Ananthakrishnan, 1981).

Although the precise physiological mechanism that controls gall-induction is not known, taxonomic treatments of gall-thrips highlight an advanced degree of monophagism. It may be true that this kind of an absolute host-specificity evident in cecidogenous thrips will result in a limited gene-flow between populations (Ananthakrishnan, 1979). The restriction of gall-forming trait mostly to the species belonging to the subfamily Phlaeothripinae, the abundance of galls in tropical and subtropical areas reflecting the diversity both in the floral and faunal components, and the restriction of galls to arboreal species, and gall induction on the foliage appear significant, and support this contention. From the viewpoint of adaptive specialization, both functional and morphological (Ananthakrishnan, 1979), the subfamily Phlaeothripinae is outstanding, with around 300 thrips species inducing galls on nearly 200 angiosperm species of about 60 natural orders. While the adaptation of the gall-thrips to particular hosts and their ability to induce galls appear significant on the one hand, the reciprocal action of the host plants by forming different types of galls involving different grades of morphogenetic phenomena on the other is indeed striking (Fig. 1, Fig. 2).

More than 5,000 species under two suborders, the Terebrantia and the Tubulifera are recognized. The tendency to form well-defined galls is more evident within the Tubulifera than the Terebrantia where, comparatively, very few species belonging to Thripidae are known to be involved in gall formation. Information on the biosystematics of this interesting group of insects which shows diverse types of feeding habits are reviewed extensively and consolidated in the works of Heming (1978), Ananthakrishnan (1979), and Mound et al. (1980).

Galls induced by thrips have been known since the turn of the present century. From the time of Rübsaamen (1902) reporting thrips galls on the leaves of *Galium verum* (Rubiaceae) and *Stellaria media* Wight (Caryophyllaceae) from Europe, much work has progressed through years. Notable contributions to the biology, systematics and bioecology of gall-thrips are those of Karny (1911–1913, 1923), Ramakrishna (1928), Bagnall (1928, 1929), Moulton (1927), Takahashi (1934), Costa Lima (1935, 1937, 1938), Baudys and Krotochvil (1941), and Wahlgren (1945). In recent years Priesner (1949, 1953, 1968), Mound (1970, 1971, 1971a), and Ananthakrishnan (1954–1983) have provided extensive information on the biology, taxonomy and bionomics of gall-thrips.

The thrips-induced galls differ from most of the other insect galls in a fundamental respect viz., the incidence of a large number of individuals in a mature gall with the number of thrips sometimes ranging from 200 to 5,000 in various stages of development as in the pouch galls of *Calycopteris floribundus* Lam (Combretaceae), *Terminalia chebula* Retz. (Combretaceae), etc. The gall community, besides the primary gall-making species, includes a number of secondary species, some of which are inquilines or predators. The population build-up of the cecidozoan is directly proportional to the developmental stage and the nature of the gall. Most thrips-galls contain one or more secondary species, though often they are relatively fewer than the gall-maker. In spite of feeding on the gall-tissue, these inquiline-organisms do not contribute to any modification of the morphology of the gall. It has been suggested that the inquilinous species have specialized secondarily for a life in the galls of other thrips, and in this process some of the inquilines have acquired a capacity to predate on gall thrips as evident in *Androthrips flavipes* Schmutz. By virtue of being associated with many thrips galls on diverse plants, inquilines such as *Mesandrothrips inquilinus* (Priesner) and *Androthrips melastomae* Zimmermann are known to be polyphagous. Some inquilines like *Androthrips ramachandarai* Karny, *Aeglothrips denticulus* Anan., *Mesothrips melinocnemis* Karny, and *Corycidothrips inquilinus* Anan., are absolutely specific to particular thrips-induced galls. Occasionally more than two organisms are involved in mature thrips galls, and such situations are recognized as 'company galls' (Ananthakrishnan, 1979, 1983a). Thrips occur as inquilines not only in thrips-induced galls, but also in the galls of other insects; *Haplothrips artiplicis* Priesner in the galls induced by *Asphondylia conglomerata* (Cecidomyiidae: Diptera), *Mallothrips indicus* Ramk. in the galls induced by *Trioza jambolanae* (Psyllidae: Homoptera), and *Dolichothrips inquilinus* Anan., in psyllid-induced galls are some of the common examples.

Distributional patterns of gall-thrips

An analysis of the distribution of cecidogenous thrips reveals their preponderance in tropical and subtropical regions of the world, viz., the Oriental, Australian, Neotropical, and Ethiopian regions; they are particularly abundant in the Indian (Ananthakrishnan, 1978), Malaysian (Anthony, 1974), Philippine (Docters van Leeuwen, 1929; Karny, 1923, 1928; Uichanco, 1919), Australian (Moulton, 1927, 1929, 1935, 1942; Mound, 1970, 1971), and Polynesian (Bagnall, 1928) subregions. While the number of species known from the entire Holarctic region is negligible, a few gall-inducing or gall-inhabiting thrips (about 7 per cent of the known species as on today) are on record from the Palaearctic region (Priesner, 1926–1928, 1960), although a majority of these are Terebrantia with elements like *Anaphothrips* and *Taeniothrips* of the family Thripidae associated with species of *Pinus*, *Salix*, *Krantia*, *Stellaria*, *Eurya*, to cite a few. In spite of a highly variegated and complex vegetational pattern, information relating to gall-thrips in the Neotropical and Ethiopian regions is meagre, although the possibility of the incidence of many gall-thrips, especially in view of the tropical climate and equatorial forest patterns that generally favour the cecidogenous trait among thrips, cannot be ignored.

The gall-thrips, as known from the Indian subcontinent, and more particularly from the peninsular India, mostly belonging to *Liothrips – Liophlaeothrips – Gynaikothrips*, and *Mesothrips* complexes (Ananthakrishnan and Muraleedharan 1974) associated with different Angiosperms appear interesting. Many other genera like *Arrhenothrips* and *Crotonothrips* inducing characteristic gall-types on the leaves of many native elements are very typical of this region. Though considered irrelevant phylogenetically (Mound, 1971a), the botanical affinity of gall-forming thrips presents a very interesting picture with specific gall-thrips genus being tied to a specific Angiosperm host (Ananthakrishnan, 1981, 1983, 1983a) suggesting an interesting relationship.

Nearly 85 per cent of the known species of gall-thrips occur in the 'Peninsular Indian – Indo-Malayan – Indonesian – North East Australian' belt with a few elements distributed sparsely and randomly in the neighbourhood. In spite of the variable influence of highly diverse climatic and

Figure 1. A representative set of thrips galls.

A.—Leaf fold galls of *Ficus benjamina* induced by *Gynaikothrips uzeli*
B.—Leaf roll galls of *Vitis lanceolaria* induced by *Liothrips viticola*
C.—Leaf roll galls of *Maytenus senegalensis* induced by *Alocothrips hadrocerus*
D.—Ceratoneon pouch galls on the leaflets of *Schefflera racemosa* induced by *Liothrips ramakrishnae* and *Liothrips associatus*
E.—Axillary bud galls of *Calycopteris floribundus* induced by *Austrothrips cochinchinensis*

Figure 2. Interspecific variability in the morphological patterns as in *Casearia tomentosa* leaf galls induced by *Gynaikothrips flaviantennatus*

A.—A young leaf showing feeding thrips; B.—Partial gall; C.—Complete gall

edaphic factors, and oceanic barriers, the high percentage of gall-thrips along this belt presents a very unique and puzzling situation, though as on date a possible explanation is not negotiable.

On the whole, gall-inducing thrips elements appear restricted to the region extending between the Tropic of Cancer and Tropic of Capricorn, and in a general way, from the 40°N to 40°S of the Earth. In this context, the occurrence of a few thrips in the Palaearctic region needs reconsideration as to whether they are true-gall formers, since most of them are terebrantians, and the morphology of the galls does not seem to conform to directed growth responses as under typical galling.

Trends in gall induction

PREREQUISITES AND THE GALL SYSTEM

Differing significantly from the galls of homopteran and dipteran insects (Mani, 1964), the galls of Thysanoptera develop as a result of the cumulative feeding effect of a population, although during the initial stages of cecidogenesis very few individuals occur. This is of special significance since population studies on gall thrips (Raman et al., 1978; Varadarasan, 1979; Ananthakrishnan, 1983a) have shown that their abundance as well as population build-up can be correlated with the availability of promordial leaves. Despite their occurrence throughout the year, gall-thrips like *Arrhenothrips ramakrishnae* Hood, *Gynaikothrips flaviantennatus* Moulton, and *Schedothrips orientalis* Anan., exhibit population peaks only when the hosts produce new shoots, and incidentally young foliage. Similarly, *Teuchothrips longus* Schmutz, *Crotonothrips dantahasta* Ramk., and *Thilakothrips babuli* Ramk., though restricted only to certain periods of the year in southern India, show identical trends of population abundance when young leaves and buds are available.

Gall-thrips prefer young plant tissue, and in many cases that of leaves, on which they feed and induce galls. Availability of tender tissue appears to be an important prerequisite, since gall induction attempted on mature, differentiated leaves in the laboratory failed to develop well-organised galls with their characteristic morphology. The continued presence of the gall-maker is yet another important feature that controls the development of the gall. Mechanical removal of thrips within the susceptible range of the host-leaf development leads to very little deformity, so that development is normal in the temporarily affected part (Docters van Leeuwen, 1956; Varadarasan, 1979).

In Nature, as with all the other gall-makers (Mani, 1964), the period of emergence of the cecidogenous thrips synchronizes with the susceptible range of host-development for gall-induction. Not all adults emerge at exactly the same time, but at spread-over periods of host-susceptibility. Generally gall-thrips complete a generation within 30 days, and in many cases the adult emergence coincides with the production of new host-shoots. These, indeed, suggest that adaptation of life-cycle patterns of cecidogenous thrips to the developmental biology of the host-organs.

Abundance of gall-thrips not only depends on continuous food supply in terms of young leaf tissue, but also on many biotic factors such as parasites, predators, and predator-inquilines which exercise a definite role in controlling thrips populations (Ananthakrishnan and Swaminathan, 1977; Ananthakrishnan and Varadarasan, 1977; Raman et al., 1978; Varadarasan, 1979). The latter is greatly facilitated by the 'simple' morphology of the galls, which are invariably 'open' gall-systems communicating with the exterior, thus aiding in the migration as well as in the establishment of various other arthropods which have very often considerable influence on the population of gall-maker.

Galls are induced by the feeding stimulus which results in the disruption of a few cells, and this contrasts well with the minute feeding punctures made by other gall-arthropods (Rey, 1979; Rohfritsch, 1966; Westphal, 1977). The initial damage caused by the continued feeding of thrips induces the leaf tissues to proliferate and contribute to the development of the gall. The role of primary lesions appears important since the initial damage which is in fact an enormous injury damaging two to three cells of the superficial layers, possibly prepares and stimulates the plant tissue to develop into gall tissue. Isolated lesions are ineffective, particularly in thrips galls, as continued feeding stimulus by numerous thrips is essential. Gall development is promoted by the proliferation of the mesophyll cells, probably initiated by the wound-stimulus. The patterns of divisions restricted to specific areas contribute to the epiphyllous or hypophyllous rolling of folding (Raman, 1981). But it is interesting to note the reaction processes of different types of tissues (palisade or spongy) occurring in more or less identical ways, although under different developmental requirements (Ananthakrishnan, 1983a).

Proliferation of the primordial mesophyll commences within 48–72 hours of initial injury. Galls of Thysanoptera, frequently induced on leaves, develop when the leaf is young and that has completed the marginal meristematic activity, but is still to undergo laminar expansion (Raman, 1981; Ananthakrishnan, 1981). The number of layers in gall-mesophyll increases either two-fold or three-fold, which again underlines the idea that galls are initiated during the late stage of ontogeny of leaves; perhaps this is largely responsible for the 'simplicity' in external form of thrips-galls (Ananthakrishnan, 1980/81, 1981).

Coupled with various other adaptations and tissue modifications to suit the requirements of the feeding thrips populations, thrips-galls show remarkable tissue specialization by forming a zone of nutrition which acts as a 'metabolic sink' concentrated with abundant food reserves. Feeding by thrips affects the outermost epidermal layers and a few successive mesophyll layers, invariably up to the vascular traces. This in turn results in the organization of a zone of nutritive cells characterized by their specialized and significantly modified subcellular morphology of the organelles well-reflecting their functional adaptations (Raman and Ananthakrishnan, 1983b). Lip-like growth meeting overhead and enclosing gall-thrips, though very rare, is known among the galls (Froggatt, 1927; Mound, 1970) induced by *Kladothrips rugosus* Froggatt on the phyllodes of *Acacia pendula* A. Cunn. ex. G. Don. (Mimosaceae).

PREFERENCE AND SUSCEPTIBILITY

Gall-forming species exist in abundance only during the seasons when young leaves are present on their respective host-plants. During other seasons, gall-maker populations are low or the insects undergo diapause. *Thilakothrips babuli* forming rosette galls on *Acacia leucophloea* Willd. (Mimosaceae) occur in abundance in mid-summer* (May-June) and early monsoon periods (September) only when tender foliage/inflorescences are available (Varadarasan and Ananthakrishnan, 1981; Raman and Ananthakrishnan, 1983). *Teuchothrips longus* and *Crotonothrips dantahasta* inducing leaf galls on *Pavetta hispidula* Hiern. (Rubiaceae) and *Memecylon lushingtonii* Gamble (Melastomaceae) occur from late summer (August) to early spring (February) and from late winter (December) to late summer (June), undergo akinesis during other periods (Varadarasan and Ananthakrishnan, 1982). They also possess abundant fat bodies which are gradually used-up particularly when host-plants are not available, suggesting that during the nonavailability of host-resources, gall-thrips undergo diapause for want of suitable food. Young leaves are available almost throughout the year in *Mimusops elengi* Linn. (Sapotaceae), *Ventilago maderaspatana* Gaertn. (Rhamnaceae), and *Casearia tomentosa* W. & A. (Samydaceae), susceptible to galling by *Arrhenothrips ramakrishnae*, *Schedothrips orientalis*, and *Gynaikothrips flaviantennatus* respectively. All these gall-thrips complete their life-cycles within a month, and the adult emergence from II pupae coincides with new shoot formation. Field as well as laboratory observations show that *Arrhenothrips ramakrishnae*, *Thilakothrips babuli* and *Crotonothrips dantahasta* complete gall-formation in four to five days, *Teuchothrips longus* in three to four days, and *Gynaikothrips flaviantennatus*, and *Schedothrips orientalis* in two to three days on their respective host-plants. The emerging adults usually spend two to three days in the old galls and migrate to young leaves. Their period of existence in the old galls in addition to the time required for the completion of gall-formation almost coincides with the preoviposition period thus establishing a well-defined synchrony between its life-cycle pattern and gall-induction. From

*References to seasons are in relation to conditions in southern India.

the day of expansion and opening-out of the leaf blades, gall-thrips begin to inhabit the leaves. As the oviposition period of gall-thrips is spread over a few days it is unlikely that all the adults emerge on the same day and infest simultaneously the host-plants (Varadarasan, 1979). A sequence in their emergence is obvious, and accordingly they attack the foliage in different stages of its development. The preoviposition period is always greater than the period required for gall-formation; further, the leaves remain susceptible to galling for sufficient number of days so that the emerging adults can successfully infest the foliage.

GALL-FORMATION SEQUENCE

Usually only one gravid female, rarely two, occupies a leaf and induces the gall as in *Mimusops elengi* where galls develop within four to five days of inhabitation by thrips along the midrib. In other instances, for example, in *Ventilago maderaspatana*, *Casearia tomentosa*, and *Memecylon edule* galls, two to 10 thrips occupy one leaf along the upper side margins, while in *Pavetta hispidula*, two to 10 thrips attack the lower side margins. With the feeding impact of thrips, reaction of leaf tissues causes the leaf margins to roll or fold, and eventually enclose the insects. Depending on the number of thrips infesting the leaves, the rolling involves one laminar half or both, as evident in the galls of *Casearia tomentosa* (Raman *et al.*, 1978) and *Pavetta hispidula* (Varadarasan, 1979). However, the rolling can be effective and complete involving both laminar halves as in *Ventilago maderasptana* and *Memecylon edule*, even when the number of infesting thrips is as low as two or three. In *Acacia leucophloea*, thrips infest the emerging buds which on maturation involve the entire rachi, and the gall-formation is gradual. Either a part of the entire rachi or sometimes more than one is involved in gall-formation (Raman and Ananthakrishnan, 1983).

From the time of opening of the leaf blade, the young leaves of various host plants are susceptible to gall formation up to a certain number of days (Table 1), and if thrips infest the leaves within this range, they develop into galls. Depending on the stage of infestation within the susceptible range of the host-plants the size and dimensions of the galls vary. If the infestation is at the beginning of the opening time of the leaf blades, the galls are comparatively small. Therefore, in addition to the number of thrips infesting, the nature of the gall depends on the stage of host-leaf development. Though the shape and size of a gall mainly depend on the stage at which the cecidozoa attack the young leaf and also the number of thrips attacking it (Krishnamurthy et al., 1975; Raman et al., 1978; Ananthakrishnan, 1980/81, 1981, 1983), the numbers of individuals in a mature gall do not depend on the size of the gall. In fact, a mature but small gall, is inhabited by a large number of individuals owing to their extreme compact condition which probably prevents the predatory organisms effectively.

Table 1. Susceptibility period and the time required for completion of gall-formation in different host plants and the preoviposition period of the respective gall-thrips

Gall Thrips (Host Plant)	Days required for completion of gall-formation	Preoviposition period (in days)	Number of days the leaf remaining susceptible to gall-formation
1. *Arrhenothrips ramakrishnae* (*Mimusops elengi*)	4–5	7–9	6–8
2. *Sehedothrips orientalis* (*Ventilago madaraspatana*)	2–3	6–8	7–9
3. *Gynaikothrips flaviantennatus* (*Casearia tomentosa*)	2–3	4–6	6–8
4. *Teuchothrips longus* (*Pavetta hispidula*)	3–4	5–8	5–7
5. *Crotonothrips dantahasta* (*Memecylon edule*)	4–5	8–10	7–9
6. *Thilakothrips babuli* (*Acacia leucophloea*)	4–5	4–6	4–6

(After Varadarasan, S. 1979)

Biology of gall thrips

VARIATIONS IN THE ADULT AND LARVAL MORPHS

A wide range of variability within species limits is evident among the gall-inhabiting thrips. These variations, generally occur in both sexes involving alary polymorphism as well as sex-limited diversity including the incidence of major and minor females, and oedymerous and gynaecoid males. Based on the variations in wing size, macropterous, brachypterous, and apterous types are recognized. The cecidicolous thrips are invariably macropterous, and they seldom show striking alary polymorphism. *Oncothrips tepperi*, known from Australia, inducing galls on *Acacia aneura* Muell. (Mimosaceae), *Byctothrips ayyari* Anan. inducing galls on a species of *Memecylon*, and *Thilakothrips babuli* inducing galls on *Acacia leucophloea* known from the Indian subcontinent, are some of the best examples showing remarkable alary polymorphism, and examples similar to these are not known so far. Macropterous forms possess well-developed eyes, ocelli, sense cones and sigmoid setae on the abdominal tergites, while the truly apterous forms have the eyes reduced to a few ommatidia, and show the absence of ocelli, shorter sense cones and a considerably reduced pterothorax.

Sex-limited intraspecific diversity is well-evident in many gall species, through more often confined to the females, with major and minor females as in *Mallothrips, Arrhenothrips, Byctothrips, Nesothrips*, etc. However, the degree of oedymerism and gynaecoidism in males is restricted. Ananthakrishnan (1978) has outlined the major trends in variability patterns in gall thrips populations. Size differences are equally conspicuous among the extreme morphs, the longest being twice as long as the smallest, accompanied by recognizable differences in body shape as in *Gryptothrips mantis* Karny and *Katothrips tityrus* (Girault) (Mound, 1971). In *Oncothrips tepperi*, the antennal segments are shorter and broader in the macropterous forms; the head is shorter with lateral margins weakly concave in the micropterae. In the micropterae, the pronotum is much longer and wider dorsally and with a median carina. Likewise, variations occur in the foretibial length and thickness in the micropterae and macropterae. Alate and brachypterous females are common in *Thilakothrips babuli* as also smaller males exhibit both alate and apterous conditions. *Euoplothrips*, *Warithrips* and *Thaumatothrips* are some of the interesting genera in which the forefemora of the major females are conspicuously armed with strong teeth or tubercles. Variations in endomorphic features relate to differences in lengths of ovarioles and basal oocytes, and in the widths of basal oocytes, lateral oviducts and the common oviduct occur among the major, normal, and minor females as in *Arrhenothrips ramakrishnae*.

SEX RATIO AND MATING

Females are always known to outnumber the males among all populations in the ratio of 3:1 to 4:1. Even in species where both sexes are produced in approximately equal numbers, the females apparently predominate because they often live longer than males. The sex ratios differ during different seasons owing to the fluctuations of the population.

Adult thrips usually mate within two or three days after the last pupal moult. After their emergence from the final pupal moult, males and females of gall-thrips remain lethargic for some time, and later congregate within galls. Owing to the precocious development, males mature fast and attempt to copulate soon after emergence. But the females become receptive and accept males only after the first or second day of their emergence. Besides the normal mating behaviour of mounting along the sides, they also tend to mount from the head-end of the females exhibiting a 180° rotation, before the tips of their genitalia are locked in copula. The mating behaviour of the inquilinous *Androthrips flavipes* is very much different from that of the gall-formers. The males grasping the posterior abdominal segments of the females for mounting are dragged to some distance in this process, before the copulation is finally effected. The adults remain in copula for one to three minutes. Copulation is accomplished within the old gall and gravid females then migrate to young leaves. Because of their confined habitat, both polyandry and polygamy are common, but once copulated females always try to mate in vain. Major females prefer gynaecoid males, though mating with oedymerous and normal males does take place with initial avoidance. The modifications in the phenotypic expressions as evident in the polymorphs of gall-thrips appear to be genetically controlled as evident in the experimental investigations of Varadarasan and Ananthakrishnan (1981).

OVIPOSITION AND FECUNDITY

Normally, one female (sometimes along with a male) or rarely two females inhabit a gall and the female oviposits generally in clusters with ovipositional behaviour differing from species to species. Females of *Arrhenothrips ramakrishnae* lay their eggs in one or two clusters, while those of *Gynaikothrips flaviantennatus*, *Schedothrips orientalis*, *Teuchothrips longus*, *Crotonothrips dantahasta*, and *Thilakothrips babuli*, lay their eggs in many smaller groups. *Gynaikothrips flaviantennatus* and *Teuchothrips longus*, in particular, lay eggs scattered all over the surface of the leaf that is destined to develop into a gall. *Thilakothrips babuli*, inducing rosette galls on *Acacia leucophloea*, oviposits along the inner basal surfaces of the leaflets. The predatory inquiline, *Androthrips flavipes* generally deposit their eggs in groups near or between the egg clusters of cecidogenous thrips. The total number of eggs laid by the gall-thrips varies in different species. The preoviposition and oviposition periods and the fecundity of some of the gall thrips of southern India are tabulated (Table 2).

Table 2. Preoviposition, oviposition periods, and fecundity (in days) of Gall thrips

Species	Preoviposition period	Oviposition period	Fecundity
Arrhenothrips ramakrishnae	7–9	7–13	17–66
Schedothrips orientalis	6–8	6–11	17–45
Gynaikothrips flaviantennatus	4–6	7–21	12–57
Teuchothrips longus	5–8	6–13	20–55
Crotonothrips dantahasta	8–10	6–10	12–28
Thilakothrips babuli	4–6	6–8	25–34
Androthrips flavipes	3–4	7–9	13–37

LIFE CYCLE PATTERNS

Gall-thrips generally complete their life cycles in about 30 days. In species like *Arrhenothrips ramakrishnae*, *Schedothrips orientalis*, and *Gynaikothrips flaviantennatus*, that occur throughout the year, there are 10–12 generations in a year. In *Teuchothrips longus* and *Crotonothrips dantahasta*, that occur abundantly only for seven months, seven to eight generations occur in a year. Populations of *Thilakothrips babuli* are seasonal with a span of 40–45 days in the rosette galls, and 20–25 days in the inflorescence galls. In the former, *Thilakothrips* completes two to three overlapping generations, while in the latter there are one to two overlapping generations. There is a considerable reduction in the duration of the life cycle of *Thilakothrips* in the inflorescence galls compared to the duration, in the rosette galls. The incubation period and the duration of various stages of the gall thrips are given in Table 3.

Table 3. Comparison of the duration (in days) of developmental stages of the predatory inquiline and gall-making species

Species	Incubation	I Larva	II Larva	Pre-pupa	I Pupa	II Pupa	Total
Predatory inquiline :							
Androthrips flavipes	4–6	3–4	2–3	1	2–3	1–2	13–16
Gall forming species:							
Arrhenothrips ramakrishnae	5–6	3–4	4–5	1	2–3	1–2	16–21
Schedothrips orientalis	5–8	2–3	4–5	1–2	1–2	2–4	14–21
Gynaikothrips flaviantennatus	7–9	4–5	4–5	1–2	1–2	2–4	19–27
Crotonothrips dantahasta	6–8	3–4	3–4	1	3–4	3–4	19–25
Teuchothrips longus	4–7	5–6	3–6	1–2	1–2	3–4	17–26
Thilakothrips babuli							
Rosette gall	5–6	3–4	2–3	1	2	2–3	15–19
Floret gall	4–5	3–4	2–3	1	2	2–3	14–17

(After Varadarasan, S. and Ananthakrishnan, T.N. 1981)

The life cycle of gall thrips has two feeding larval stages and three non-feeding pupal stages. The duration of the former is comparatively longer since this stage should build up sufficient food reserve for the subsequent development of the pupal stages. The duration of the postembryonic development of different gall-forming thrips and inquiline thrips varies, the predatory inquiline completing its life cycle much faster than the gall-forming thrips. The inquiline inhabits the galls only after the cecidozoa have established their population, feeds on the gall-thrips, and completes the life cycle earlier than the cecidozoa. A similar adaptation is seen in *Thilakothrips babuli* which completes the life cycle faster in the floret galls than in the rosette galls. This acceleration is only to facilitate the completion of its life cycle before the inflorescences of *Acacia leucophloea* become dry.

Ecology of gall thrips

INTERACTIONS WITH BIOTIC AND ABIOTIC FACTORS

The primary ecological demands for the survival and maintenance of a gall-maker which is a specialist plant-feeder depend on two major aspects, the first being the interaction and continued availability of the host, and the second being the influence of the associated arthropods. The duration of life cycles and fecundity, longevity and mortality rates of gall-thrips are affected by biotic as well as abiotic factors. Climatic factors mostly influence the populations of phytophagous insects by indirectly affecting the host-plants. Besides these, predation and parasitism also bring about effective changes in gall-thrips populations. In all the cecidogenous thrips, gall formation is an important prerequisite for the development and oviposition, which functions as an effective microhabitat. Normally gall-forming thrips complete their life-cycles in about 30 days, and this coincides with the regular monthly production of young leaves on host plants where the emergence of adult thrips synchronizes with the formation of young leaf buds (Varadarasan and Ananthakrishnan, 1982).

Some of the gall thrips like *Arrhenothrips ramakrishnae*, *Schedothrips orientalis*, and *Gynaikothrips flaviantennatus* occur throughout the year when the young leaves are also produced in large numbers on their respective host plants. Production of young leaves in *Mimusops elengi* occurs almost throughout the year, and the gall-maker, *Arrhenothrips ramakrishnae*, occurs in considerable number for nearly 180–240 days. On the other hand, in *Ventilago maderaspatana*, leaf production is evident for a period of 180–240 days in a year with thrips being abundant for 80–180 days. In *Gynaikothrips flaviantennatus* the peak population abundance varies from 120 to 270 days during which period their host plant *Casearia tomentosa*, abounds in numerous tender shoots and young leaves. Rarely, low populations of these three gall-making thrips occur, in spite of the normal leaf production by the host plants, and various other biotic factors appear responsible for this. The absence of *Crotonothrips dantahasta* and *Teuchothrips longus* for 150 days of the year coincides well with the absence of new shoots on their hosts. The few stragglers found during this lean period have been seen to have larger amounts of fat bodies in them which suggests that they are in a stage of 'akinesis' or temporary diapause. When such thrips are provided with young leaves which are normally absent in that season, they feed and readily induce the gall, and reproduce normally, increasing in number (Varadarasan and Ananthakrishnan, 1982). Unlike all the other earlier mentioned gall thrips, the populations of *Thilakothrips babuli* are very much restricted to a short period of three months, two months in rosette galls and one in floret galls, with one to two peaks in each habitat. With the maturation and drying up of the rosette gall, the thrips populations dwindle, leaving behind very few apterous adults. These adults remain within the dry galls and diapause. When *Acacia leucophloea* puts forth new floral sprouts, the apterous adults leave the diapausing site and migrate to the flowers to form floret galls where all polymorphic forms live. What is of significance is the strategy adopted by the thrips species with a shorter life cycle on the inflorescence gall (12–14 days) as compared to that on the bud gall (20–28 days), in view of the comparatively shorter life span of the inflorescence than the axillary short buds. Both in the rosette and floret galls, individuals of *Thilakothrips babuli* are predated by *Androthrips flavipes*, considerably reducing the populations of the gall-maker.

The role of intraspecific competition in population regulation within thrips-galls appears to be a significant adaptation. In this connection, the reproductive strategies as well as those involved in population regulation within galls, appear to be important parameters, with each species exhibiting very characteristic egg laying patterns varying from small groups of 4–16 to as many as 60-100, as for instance in *Arrhenothrips ramakrishnae*. The basic cause for this intraspecific variation appears to be the impact of polymorphism, the reproductive behaviour of the major and the minor females as well as the oedymerous and gynaecoid males in particular, which considerably influence fecundity. Such polymorphism also results in patterns of mating leading to recognisable variation in fecundity.

The favourable temperature range of 24-34°C, relative humidity of 55-88 per cent and rainfall of up to 120 mm per month seem to favour the populations of gall-thrips. With the decline in temperature (26-29°C), and increase in relative humidity (75-90 per cent) thrips population dwindles. This agrees with the favourable range of temperature and relative humidity reported for the non-cecidogenous thrips which prefer special microhabitats like the ears of panicles, or between grass leaf sheaths (Cederholm, 1963). The main visible effect on gall thrips population is by heavy rainfall. Though young leaves (the chief limiting factor) are available in plenty, the cecidozoa do not withstand heavy rainfall, which possibly subjects the gall thrips populations to choking within the gall, and results in heavy mortality. Moderate rainfall does not affect gall-thrips populations, and the ability of thrips to survive temporary submersion in water is well known (Morison, 1957; Lewis, 1955; Titschack, 1969).

Besides the availability of their respective host tissues, the presence of the predatory inquiline – *Androthrips flavipes*, the anthocorid bug – *Montandoniola moraguesi* (Puton) (Anthocoridae: Heteroptera), and the internal eulophid parasite – *Tetrastichus thripophonus* Waterson (Enlophidae: Hymenoptera) brings about considerable fluctuations in populations of the gall-forming thrips; the effect of these biotic factors on the gall-thrips population is of great significance. A study of the biology of the predators and the parasites appears essential for the establishment of the interrelationships between these organisms and the prey populations, and this would facilitate the understanding of the prey-predator-parasite interactions.

Predatory inquilines are not uncommon in thrips-induced galls; *Androthrips flavipes* seems to be the most effective species as it is associated with at least 15 thrips galls of the Indian subcontinent (Ananthakrishnan, 1978). This predatory-thrips always inhabit mature galls that are abundant in eggs and immature stages of the gall-makers. The slender body is well adept for swift movements in contrast to gall-makers. They find their prey by random searching, recognize in close vicinity or upon contact. On locating a suitable prey, they usually stop long enough to empty the contents of the eggs of immature stages. Within the gall, the prey and the predator are confined to a limited space, providing *Androthrips* enough chances to find their hosts thigmostatically. Individuals of *Androthrips flavipes* are not very ferocious in their predatory habit, and the larvae and the adult gall-thrips easily deter them by a violent shaking of the abdomen, but a hungry predator is always able to overcome such a resistance. *A. flavipes* occurs throughout the year generally exhibiting two peaks of abundance in early monsoon (September), winter (December) and early spring (February), late spring (April), yet the degree of incidence is variable: lower when higher numbers of *Montandoniola moraguesi* occur. The populations never outnumber those of the host, the predator-prey ratio being 1 : 20 (Ananthakrishnan, 1972). The first intruder into thrips-galls is invariably *Androthrips flavipes* which chooses to occupy mature galls having a full complement of eggs and immature stages of these gall-makers, and they prefer to lay the eggs dispersed along the eggs of the gall-maker thus providing a ready food-source to the young ones hatching out of eggs. They start laying eggs soon after their entry into the gall, consume about 1/10 of both the eggs and the larvae of the gall-maker; the very few adults in the gall are not preferred because of their high resistance. The incubation period of this inquiline is also short, and the emerging larvae find the food sources in close proximity. This fast rate of consumption soon depletes the food even before the inquiline reaches the non-feeding pupal stages. When deprived of the normal food, as a result of starvation, these inquilines resort to cannibalism. But, in spite of the regular supply of normal food, rarely, *Androthrips flavipes* resorts to cannibalism due to overcrowding, and the rate of mortality of immature stages is directly proportional to the total number of eggs laid. Notwithstanding their precocious development and fecundity similar to that of the gall-maker, the predatory inquiline fails to outnumber the populations of the gall-maker, the cannibalistic behaviour being a major factor in checking their numbers.

Information on the bionomics of the predatory anthocorid bug, *Montandoniola morguesi*, is provided by Muraleedharan and Ananthakrishnan (1971). The predatory anthocorid occurs in significant numbers only in specific galls, for example, those inhabited by *Gynaikothrips flaviantennatus* and *Teuchothrips longus*. It consumes three to four adult thrips/day up to 81–118 thrips during the entire life cycle (Muraleedharan and Ananthakrishnan, 1971). The incidence of this bug drastically brings down gall-thrips populations.

A high incidence of *Tetrastichus thripophonus* has been noticed in the cecidogenous *Schedothrips orientalis* (Ananthakrishnan and Swaminathan, 1977) and as traces in *Arrhenothrips ramakrishnae* and *Thilakothrips babuli* (Ananthakrishnan and Varadarasan, 1977). In these gall thrips, the larvae that are close to the base or extremities of the galls are prone to greater parasitic infections. *Tetrastichus* effectively brings down the populations of *Schedothrips orientalis* and as much as 20 per cent parasitization is known especially when *Schedothrips* populations are high. Although this parasite does not inhabit all the thrips galls of southern India, induced parasitization in the laboratory was effective in that the adults of *Tetrastichus thripophonus* readily oviposited and completed their life cycles (Varadarasan, 1979). Interestingly, this eulophid parasite also parasitizes *Androthrips* both under natural and laboratory conditions.

SUCCESSIONAL PATTERNS IN THRIPS GALL SYSTEMS

Though 'patchy' (Washburn and Cornell, 1981), the gall systems are subject to high disturbance frequencies, involving colonizations and extinctions, being part of a natural biological assemblage. The gall as a purposeful microhabitat of a specific and special type of the environmental requirement offers abundant scope for the successful inhabitation and maintenance of a number of organisms, permitting immigration, reproduction and survivorship. Thysanoptera-induced galls, though simple morphologically, provide sufficient basis for the development of a unique community. They display a complex food-web involving a characteristic succession pattern and the interaction of various arthropod populations.

After the gall-induction is achieved, many other primary consumers like coccids and microlepidopteran larvae get associated within the gall system. Once the gall-forming thrips populations are established, several secondary organisms like the predatory-inquilinous thrips, predatory bugs, pseudoscorpions, ants, psocids, and hymenopteran parasites enter this habitat in a regular sequence, leading to a complex succession that is found to vary in different types of thrips-galls; this depends on the compactness of rolling or folding. Hence, the ecological succession patterns occurring in thrips galls can be categorized into three distinct types: (a) the less compact and open types, (b) more compact and open types, and (c) closed types.

In the less compact types like the leaf fold galls of *Mimusops elengi*, leaf roll galls of *Casearia tomentosa* and *Pavetta hispidula* formed by *Arrhenothrips ramakrishnae, Gynaikothrips flaviantennatus*, and *Teuchothrips longus* respectively, within 10–15 days after the gall-maker infesting the young leaf, the cecidozoa get established with eggs and immature stages in the gall, and in about a fortnight, the predatory inquiline *Androthrips flavipes* and the predatory anthocorid bug *Montandoniola moraguesi* enter the gall system. Simultaneously, coccids, and *Drosicha* sp. (Megasodiae), also occupy the gall. The hymenopteran internal eulophid parasite, *Tetrastichus thripophonus* too enters the gall community by parasitising the II larvae of gall thrips. After 25–30 days of gall development, *Chiracanthisum melanostomae* (Culubionidae: Arachnida), the tritonymphs of *Euryolpium indicum* Murthy and Ananthakrishnan (Olpidae: Chelonethi), and some microlepidopteran larvae gain entry into the gall. While the gall-maker, coccids, and larvae of Microlepidoptera feed on the gall-tissue, the inquiline-predatory thrips and the anthocorid bugs feed not only on the gall-thrips, but also on *Androthrips*. *Euryolpium* predates on the gall-forming thrips, *Androthrips, Montandoniola*, and the spiders. The spiders in turn feed on pseudoscorpions. With the decline in number of all these gall inhabitants, i.e., when the gall becomes senescent, the detritous feeding psocids come into the picture.

In the more compact leaf roll/fold/rosette gall types like *Ventilago maderaspatana, Memecylon edule* and the rosette and inflorescence galls of *Acacia leucophloea* induced by *Schedothrips orientalis, Crotonothrips danta-*

hasta and *Thilakothrips babuli* respectively, the gall community structure is slightly different from the less compact and open types. Because of the more-compact nature, only small-sized arthropods inhabit the gall after the cecidozoa have established their populations. The large anthocorid bug and the spiders generally do not occur in these galls. Populations of pre-datory-inquiline, internal parasite, larvae of Microlepidoptera, pseudoscor-pions, and psocids, enter the gall sequentially. Unlike the less-compact type, spiders spin their webs near the galls and feed on thrips as they leave the senescent galls.

In the closed types, as in *Calycopteris* galls, with the building up of *Austrothrips* populations, a weak population of the inquiline (*Androthrips ramchandrai*) alone occurs. On maturation and drying up of the gall, the gall-maker population declines, and the inquiline population builds up. In completely dried galls, which by then become open gall-systems due to the exit holes of thrips, and due to weathering, the thrips are completely replaced by a 'successori' of species of *Crematogaster* and *Camponotus* (For-micidae: Hymenoptera), coccids, spiders, mites, psocids, etc. Similar pat-terns of succession have been reported by Mound (1971) as well.

In these gall community systems, the gall tissue functioning as the pri-mary producer occupies the first trophic level; the central species, viz., the gall-maker., and coccids and the larvae of microlepidoptera are primary consumers occupy the second trophic level; the inquiline predator acts as the primary carnivore in the third trophic level; while the secondary consu-mers like the anthocorid bugs, spiders, pseudoscorpions and the hymenop-teran parasites occupy the fourth trophic level. The psocids that feed on the decomposing gall tissues and the dead animal remains act as detrivores.

REFERENCES

1. Ananthakrishnan, T.N. 1954. The bionomics of *Mimusops* gall thrips *Arrhenothrips ramakrishnae* Hood. *Agra. Univ. J. Res.* (*Sci.*) 3 : 463–474.
2. Ananthakrishnan, T.N. 1964. A contribution to our knowledge of the Tubulifera (Thysanoptera) from India. *Opsuc. ent.* Suppl. 25 : 1–120.
3. Ananthakrishnan, T.N. 1967. Studies on new and little known Indian Thysanoptera. *Oriental Ins.* 1 : 113–138.
4. Ananthakrishnan, T.N. 1969. New Indian gall thrips. *Senck. Biol. Frankfurt.* 50 : 179–194.
5. Ananthakrishnan, T.N. 1969a. *Indian Thysanoptera*, CSIR, Zoological Monograph No. 1. New Delhi, CSIR, Publications and Information Directorate. 171 pp.
6. Ananthakrishnan, T.N. 1971. Further studies on Indian gall thrips—I. *Marcellia* 37 : 11–127.
7. Ananthakrishnan, T.N. 1972. Further studies on Indian gall thrips—II. *Marcellia* 37 : 3–20.
8. Ananthakrishnan, T.N. 1973. Further studies on Indian gall thrips—III. *Oriental Ins.* 7 : 539–546.
9. Ananthakrishnan, T.N. 1973a. *Thrips: Biology and Control.* New Delhi, Macmillan India. 120pp.

10. Ananthakrishnan, T.N. 1976. New gall thrips of the genus *Crotonothrips* Thysanoptera. *Oriental Ins.* 10 : 411–419.

11. Ananthakrishnan, T.N. 1976a. New species of *Liothrips* inhabiting *Schefflera* galls from India wtth a key to known species of *Liothrips* from *Schefflera* species. *Ceylon. J. Sci. (Biol. Sci.)* 7 : 23–28.

12. Ananthakrishnan, T.N. 1978. Thrips galls and gall thrips. *Zool. Surv. India Tech. Monogr.* 1 : 1–69.

13. Ananthakrishnan, T.N. 1979. Biosystematics of Thysanoptera. *Ann. Rev. Entomol.* 24 : 159–183.

14. Ananthakrishnan, T.N. 1980/81. On some aspects of thrips galls. *Bull. Soc. Bot. Fr. (Actual. Bot.)* 127 : 31–34.

15. Ananthakrishnan, T.N. 1981. Thrips-plant gall association with special reference to patterns of gall diversity in relation to varying thrips populations. *Proc. Indian Natn. Sci. Acad.* 47 : 41–46.

16. Ananthakrishnan, T.N. 1983. Adaptive strategies in ceicidogenous Thysanoptera. *Curr. Sci.* 52 : 77–79.

17. Ananthakrishnan, T.N. 1983a. Unique aspects of thrips galls. In *Biology of Galls Induced by Insects and Acarina*, ed. O. Rohfritsch and J.D. Shorthouse, New York, Praeger Press (In Press).

18. Ananthakrishnan, T.N., Muraleedharan, N. 1974. Studies on the *Gynaikothrips—Liophlaeothrips—Liothrips* complex from India. *Oriental Ins. Suppl.* No. 4 : 85 pp.

19. Ananthakrishnan, T.N., Swaminathan, S. 1977. Host-parasitic and host-predator interactions in the gall thrips *Schedothrips orientalis* Anan. (Insecta : Thysanoptera) *Entomon* 2 : 247–251.

20. Ananthakrishnan, T.N., Varadarasan, S. 1977. *Androthrips flavipes* Schmutz (Insecta : Thysanoptera) a predatory inquiline in thrips galls. *Entomon* 2 : 105–107.

21. Anthony, M. 1974. Cecidogénèse comparée de deux galles de coccides (*Gallococcus anthonyae* Beardsley et *Gallococcus secundus* Beardsley) developpées sur le *Shorea curtisii* Dyer. ex King (Dipterocarpacée). *Marcellia* 38 : 99–144.

22. Bagnall, R.S. 1928. Thysanoptera. On some Samoan & Tongan Thysanoptera (Tubulifera) with special reference to *Ficus* gall causers and their inquilines. *Brit. Mus. (Nat. Hist.)* 7 : 55–76.

23. Bagnall, R.S. 1929. On some new genera and new species of Australian Thysanoptera (Tublifera) with special reference to gall species. *Marcellia* 25 : 184–204.

24. Baudys, E., Krotochvil, J. 1941. Prispevek K. Poeznani nasich Thysanopterocecidi. *Sborn ent. odd. Zemsk. Mus. Praze* 19 : 136–147.

25. Costa Lima, A. 1935. Tisanopterocecidios do Brasil. *O. campo Julho* : 25–29.

26. Costa Lima, A. 1937. Insetos do Brasil. *O. campo Julho* : 50–57.

27. Costa Lima, A. 1938. Insectos do Brasil. Capitulo XXI Ordem Thysanoptera. 1st Tome Serie da Escola Nacional de Agronomia, Rio de Janeiro Dicatica, No. 2 : 405–452.

28. Cederholm, L. 1963. Ecological studies on Thysanoptera. *Opusc. ent. Suppl.* 22 : 215 pp.

29. Docters van Leeuwen, W. 1929. Ein neuer Typus eines Thysanopterocecidiums. *Marcellia* 26 : 3–5.

30. Docters van Leeuwen, W. 1956. The aetiology of some thrips galls found on leaves of Malaysian *Schefflera* sp. *Acta Bot. Neerland.* 5 : 80–89.

31. Froggatt, W.W. 1927. The bubble leaf gall thrips (*Kladothrips* sp.). In *Forest Insects and Timber Borers* 107 pp. Sydney.

32. Heming, B.S. 1978. Structure and function of the mouthparts in larvae of *Haplothrips verbasci* (Osborn) Thysanoptera : Tubulifera, Phlaeothripidae). *J. Morph.* 156 : 1–38.

33. Karny, H. 1911. Uber Thrips-Gallen and Gallen-Thripse. *Centralbl. Bakt. Parasinfekt. Abt.* 30 : 556–572.
34. Karny, H. 1912. Gallenbewohnende Thysanopteren aus Java. *Marcellia* 11 : 115–169.
35. Karny, H. 1913. Uber gallen bewohnende Thysanopteren. *Verh. Zool. Bot. Ges. Wien.* 63 : 6–12.
36. Karny, H. 1923. VIII Ueber die tiergeographischen Beziehungen der Malayischen Thysanopteren. *Treubia* 3 : 326–380.
37. Karny, H. 1928. Beitrage zur Malayischen Thysanopteren Fauna. Ein Neuer Javanischer *Gynaikothrips*. *Treubia* 10 : 33–44.
38. Krishnamurthy, K.V., Raman, A., Ananthakrishnan, T.N. 1975. On the morphology of the ceratoneon thrips galls of *Schefflera racemosa* Harms. *Marcellia* 38 : 179–184.
39. Lewis, T. 1955. Two interesting British records of Thysanoptera. *J. Soc. Br. Ent.* 5 : 110–113.
40. Mani, M.S. 1964. *Ecology of Plant Galls*. The Hague: Dr. W. Junk Publishers : 434 pp.
41. Morison, G.D. 1957. A review of British Glasshouse Thysanoptera. *Trans. R. Ent. Soc. Lond.* 109 : 467–520.
42. Moulton, D. 1927. New gall-forming Thysanoptera of Australia. *Proc. Linn. Soc. NSW.* 52 : 153–160.
43. Moulton, D. 1929. An interesting new thrips from Australia. *Trans. R. Soc. S. Aust.* 53 : 264–266.
44. Moulton, D. 1935. New species of thrips from South Western Australia. *J. Proc. R. Soc. West Aust.* 21 : 97–100.
45. Moulton, D. 1942. Seven new genera of Thysanoptera from Australia and New Zealand. *Bull. Sth. Calif. Acad. Sci.* 41 : 1–13.
46. Mound, L.A. 1970. Intragall variation in *Brithothrips fuscus* Moulton with notes on other Thysanoptera induced galls on *Acacia* phyllodes in Australia. *Ent. Mon. Mag.* 105 (July/August/September).
47. Mound, L.A. 1971. The complex of Thysanoptera rolled leaf galls in *Geijera*. *J. Austral. Ent. Soc.* 10 : 83–97.
48. Mound, L.A. 1971a. Gall forming thrips and allied species (Thysanoptera: Phlaeothripidae) from *Acacia* trees in Australia. *Bull. Br. Mus. (Nat. Hist.) Ent.* 25 : 389–466.
49. Mound, L.A., Heming, B.S., Palmer J.M. 1980. Phylogenetic relationships between the families of recent Thysanoptera (Insecta). *Zool. J. Linn. Soc.* 69 : 111–141.
50. Muraleedharan, N., Ananthakrishnan, T.N. 1971. Bionomics of *Montandoniola moraguesi* (Puton) (Heteroptera : Anthocoridae) a predator on gall thrips. *Bull. Ent.* 12 : 4–10.
51. Priesner, H. 1928. Uber Australische Thysanoptera. *Sitz. Acad. Wiss. Wien.* 137 : 643–659.
52. Priesner, H. 1949. Genera Thysanopterorum. *Bull. Soc. Fouad. Ent.* 33 : 31–157.
53. Priesner, H. 1953. On the genera and allied to *Liothrips* of the oriental Fauna I (Thysanoptera). *Treubia* 22 : 357–380.
54. Priesner, H. 1960. A monograph of the Thysanoptera of Egyptian deserts. *Publ. Inst. Desert* 13 : 541 pp.
55. Priesner, H. 1968. On the genus allied to *Liothrips* of the oriental fauna II (Insecta : Thysanoptera). *Treubia* 27 : 175–285.
56. Raman, A. 1981. *Morphological studies on some galls induced by thrips (Thysanoptera : Insecta) from south India*. Univ. of Madras, Ph.D. Thesis.
57. Raman, A., Ananthakrishnan, T.N. 1983. On the developmental morphology of

the rosette galls of *Acacia leucophloea* Willd. (Mimosaceae) induced by *Thilako-thrips babuli* Ramk. (Thysanoptera: Insecta). *Proc. Ind. Acad. Sci.* 92 : 343–350.

58. Raman, A., Ananthakrishnan, T.N. 1983a. Studies on some thrips (Thysanoptera: Insecta) induced galls. 1. Developmental Morphology, *Proc. Ind. Natn. Sci. Acad.* 49 B: 313–358.

59. Raman, A., Ananthakrishnan, T.N. 1983b. Studies on some thrips (Thysanoptera: Insecta) induced galls. 2. Fine-structure of the nutritive zone. *Proc. Ind. Natn. Sci. Acad.* (In Press).

60. Raman, A., Ananthakrishnan, T.N., Swaminathan, S. 1978. On the simple leaf galls of *Casearia tomentosa* Roxb. (Samydaceae) induced by *Gyanikothrips flaviantennatus* Moulton (Thysanoptera: Phlaeothripidae). *Proc. Ind. Acad. Sci.* 87 : 231–242.

61. Ramakrishna, T.V. 1928. A contribution to our knowledge of the Thysanoptera of India. *Mem. Dept. Agr. Ind. Ent. Ser.* 10 : 215–316.

62. Rey, L. 1979. Particularités ultrastructurales des cellules nourriciéres de la galle provoquée par *Neuroterus guercusbaccarum* Linn. sur *Quercus pedunculata* Ehrh. *Beitrage Biol. Pflanzen* 54 : 315–327.

63. Rohfritsch, O. 1966. Action specifiques des galles de Chermisidae sur la maturation et l'ouverture des galles. Mise en evidence par la realisation des galles mixtes. *C.R, Acad. Sci. Paris* 262 : 370–372.

64. Rübsaamen, E.H. 1902. Mitteilung uber die von Herrn J. Born-Muller in orient gesammelten Zoocecidien. *Zool. Jahab. Abt. Syst.* 16 : 243–283.

65. Takahashi, R. 1934. Association of different species of thrips in their galls (In Japanese). *Botany Zool. Tokyo* 2 : 1827–1836.

66. Titschack, E. 1969. Der *Tarothrips*, ein neuer schadinsekt in Deutschaland. *Sonderdr. Azz. Schadlingsk. Pftschutz* 42 : 1–6.

67. Uichanco, H.B. 1919. A biological and systematic study of the phillipine plant galls. *Phil. J. Sci.* 14 : 527–554.

68. Varadarasan, S. 1979. *Bioecological investigations on some gall-inhabiting Thysanoptera (Insecta : Arthropoda)*. Univ. of Madras, Ph.D. Thesis.

69. Varadarasan, S., Ananthakrishnan, T.N. 1981. Population of dynamics and prey/predator/parasite relationships of gall forming thrips. *Proc. Ind. Natn. Sci. Acad.* 47 : 321–340.

70. Varadarasan, S., Ananthakrishnan, T.N. 1982. Biological studies on some gall thrips. *Proc. Ind. Natn. Sci. Acad.* 48 : 35–43.

71. Wahlgren, E. 1945. Gall Bildende Thysanoptera. *Opusc. ent.* 11–126.

72. Washburn, J.O., Cornell, H.V. 1981. Parasitoids, patches and phenology: their possible role in the local extinction of a cynipid gall wasp population. *Ecology* 62 : 1597–1607.

73. Westphal, E. 1977. Morphogénèse, ultrastructure et etiologie de quelques galles d'eriophyides (Acariens). *Marcellia* 39 : 193–375.

6. Gall Tephritidae (Diptera)

Amnon Freidberg

Introduction

The Tephritidae (fruit flies), with some 4000 described species, is a fairly large family of Diptera. Almost all the species of which the biology is known have phytophagous larvae. Close to 200 species have been recorded that induce the formation of galls, thus ranking the Tephritidae as the second most important cecidogenous family of Diptera after the Cecidomyiidae. Based on taxonomic considerations eventually 300–400 species will probably be found to be cecidogenous. If this estimation is correct, then there already is some biological information on half of the cecidogenous species. This is a much higher rate than that existing for the rest of the family (e.g. Foote, 1967 : 1).

Strangely enough, there is no comprehensive work dealing with or summarizing the knowledge on tephritid galls, not even for a single zoogeographical region. On the other hand, a fairly large number of works have been published which deal with various aspects of cecidogenous tephritids and their galls. Some of these works are very old, rare, or otherwise not easily accessible and a few were not available at the time of the preparation of this review. This review, therefore, does not pretend to cover all the details published to date on the subject, rather, it attempts a general approach that has not been used previously for tephritids. The main purpose of this review is to summarize: (a) the relationships between the cecidogenous tephritids and their host plants; (b) the natural history of gall tephritids; and (c) the knowledge of gall structure and formation.

The cecidogenous Tephritidae

To better understand the relationships between the gall-forming fruit flies and their host plants, one should first understand the phylogenetic relation-

Dedicated to H.K. Munro who has studied and described more tephritid galls than any student of the family, and to whom we are indebted for practically all our knowledge on the Afrotropical tephritid galls.

129

ships within both groups. Taxonomists generally agree that the higher classification of the Tephritidae is in an unsatisfactory state. This situation is the combined outcome of the large size of the group and of the fact that most existing revisions treat only single zoogeographical regions. Another inherent difficulty is that major taxonomic characters intergrade between most higher categories, including genera and subfamilies. Hering's (1947) Key is the only serious attempt at a classification of the higher categories (subfamilies and tribes) on a world-wide basis. For the sake of simplicity it is adopted here. Hering (1947) divided the Tephritidae into eight subfamilies. The Dacinae, Schistopterinae and Terelliinae are not known to contain cecidogenous species (but see Appendix). The Aciurinae and Trypetinae each contain only very few gall makers. The overwhelming majority of Tephritidae that cause the formation of galls belong to three subfamilies, the Myopitinae, Oedaspinae and Tephritinae.

The Myopitinae (see Steyskal, 1979 for a review of the genera) is the least diverse of the three. Almost all the known cecidogenous species in this subfamily belong to only two, predominantly Palaearctic genera, *Myopites* and *Urophora*. None of the Afrotropical and American *Urophora* is known with certainty to induce the formation of galls. They are most probably not congeneric with the Palaearctic species and should be transferred to other, possibly new genera. The Oriental and Australian regions and the Pacific Islands are not known to have Myopitinae.

The Oedaspinae contains more genera, though probably no more species, than the Myopitinae. The genera of Oedaspinae are more uniformly distributed over the world, but individual genera are more or less restricted to single zoogeographical regions. Despite a lack of data on the biology of many species, existing records (with only minor exceptions) indicate that cecidogenesis is a common trait of the members of this subfamily. The limits of the Oedaspinae are not clear, however. Some genera, such as *Afreutreta* and *Parafreutreta* are placed in this subfamily (Cogan & Munro, 1980 : 155) with reservations (Munro, 1939c), while other, similar genera, such as *Eutreta* and *Strobelia* are assigned to the Tephritinae (Foote, 1980).

The Tephritinae is probably the largest subfamily in the Tephritidae. Its members, as far as is known, mostly infest flower heads of various Compositae without making galls. Many genera contain cecidogenous species, but exactly how many, will remain unclear at least as long as the assignment of some border genera between the Tephritinae and Oedaspinae remains unsettled.

The scarcity of Tephritid galls in the Oriental region is probably due to the notable paucity of species of the latter three subfamilies in that region.

The existence of gall-producing members of the Aciurinae and Trypetinae represents an unusual phenomenon for these subfamilies. The Aciurinae, with about 150 species, mainly inhabit the Afrotropical and Oriental regions, and to a much lesser extent the Palaearctic and the Australian regions. Genuine Aciurinae probably are not represented in the Americas. All known host plants of the Aciurinae are within the families Acanthaceae, Labiatae and Verbenaceae, in the flowers of which the larvae develop. Only two species have been reported to produce galls on their host plants.

The Trypetinae is a very large and diverse group both with regard to adult morphology and host plant relationships. Members of this subfamily are present in all the major zoogeographical regions (excluding Antarctica), although with further study the group may be split into several subfamilies. The great majority of Trypetinae are probably frugivorous and only a few species of this subfamily have been reported as gall makers.

Host plants and biogeography

Table 1 is an alphabetical list of the 48 tephritid genera, which are known to include gall formers. Information in the table includes the size of the genus (number of species in parentheses), the number of known cecidogenous species for each of the zoogeographical regions, the known host plant genera and basic references. The taxonomic data are based on current usage, and therefore subject to subsequent changes. Introduced species are indicated by parentheses. Host plant genera belong to the family Compositae, unless otherwise indicated. Galls are usually formed on the stems or branches; galls produced on other parts (roots, flower heads etc.) are specifically mentioned. At least one reference for each zoogeographical region is given. The table does not include questionable records of cecidogeny; these are listed in the Appendix.

The most outstanding facts to emerge from the table are that almost all Tephritid galls are formed on plants of the family Compositae, and that almost all of them are stem galls. Plants belonging to only seven other families (Acanthaceae, Aquifoliaceae, Goodeniaceae, Melastomataceae, Mimosaceae, Onagraceae and Verbenaceae) have been reported to bear tephritid galls and less than 10 species of flies are involved. Within the Compositae a few host genera are especially favoured by cecidogenous tephritids, for example *Helichrysum* in the Old World, *Baccharis* and *Chrysothamnus* in the Americas and *Artemisia* and *Senecio* over the greater part of the world. Wasbauer (1972), for example, lists 11 species of four genera forming galls on various species of *Chrysothamnus* in North America.

The Palaearctic region, with 52 reported cecidogenous species plus one introduced, is the richest region in tephritid galls, and is matched only by the Nearctic (50 + 3 introduced). The Afrotropical region, with 40 species, is the richest in cecidogenous genera, but 11 out of the 18 genera recorded are monobasic (containing only a single species). The Neotropical region has 28 known cecidogenous species, but this relatively low number probably reflects insufficient knowledge rather than a real situation. The Oriental region is generally poor in Tephritidae that infest Compositae plants and

Table 1. Species distribution and hosts of genera of gall-forming Tephritidae*

	Pal	Afr	Nea	Neot	Or	Aus	Oc	Total	Hosts	Reference
Acanthiophilus Becker (13)	2							2	*Brachylaena*	Munro, 1926
Acinia Robineau-Desvoidy (16)		1						1	*Tessaria*	Kieffer & Jörgensen, 1910
Aciurina Curran (12)			9					9	*Artemisia, Baccharis, Chrysothamnus Gutierrezia, Haplopappus, Senecio, Solidago*	Wasbauer, 1972
Acronneus Munro (1)		1						1	*Senecio*	Munro, 1929a, 1939c
Actinoptera Rondani (30)	3				1			5	*Antennaria, Filago, Gnaphalium, Helichrysum*	Hendel, 1927; Munro, 1934, 1935, 1953a, 1957
Afreutreta Bezzi (4)		3						3	*Brachylaena, Senecio*	Munro, 1925, 1926, 1940
Anomoia Walker (20)								1	*Ilex* (Aquifoliaceae)	Blanchard, 1929
Callachna Aldrich (1)			1					1	*Ambrosia*	Wasbauer, 1972
Campiglossa Rondani (19)	1				1			2	*Artemisia, Helichrysum, Solidago, Vernonia*	Hendel, 1927; Munro, 1925, 1935
Cecidochares Bezzi (11)				3				3	*Ageratum, Eupatorium*	Aczél, 1949
Chrysotrypanea Malloch (2)						2		2	*Helichrysum*	Bush, 1966a
Cosmetothrix Munro (1)		1						1	*Vernonia*	Munro, 1935
Dithryca Rondani (2)	2							2	*Achillea, Santolina* (stem or root gall)	Hendel, 1927
Embaspis Munro (2)		2						2	*Helichrysum, Stenocline*	Munro, 1952a, 1953a
Euarestella Hendel (5)	2							2	*Inula, Pulicaria*	Freidberg, 1980
Eurosta Loew (11)			6					6	*Solidago* (root or stem galls)	Wasbauer, 1972
Euthauma Munro (1)		1						1	*Schistostephium*	Munro, 1949

*For explanation see section on host plants and biogeography—Abbreviations: Pal—Palaearctic, Afr—Afrotropical, Nea—Nearctic, Neot—Neotropical, Or—Oriental, Aus—Australian, Oc—Oceania

Genus				Host plants / notes		References
Eureta Loew (31)	9	4	(1) 13	In Nea. all Compositae: *Ambrosia, Artemisia, Aster, Callistrephus, Chrysothamnus, Bidens, Helianthus, Ratibida, Senecio, Solidago, Vernonia* (stem or root galls) In Neot. *Coleus, Conyza* (Compositae) *Lantana, Verbena, Stachytarpheta* (stem galls; Verbenaceae). One Neot. species was introduced into Hawaii		Stoltzfus, 1977
Isoconia Munro (22)	1			*Blepharis* (Acanthaceae)	1	Munro, 1947
Marriotella Munro (1)	1			*Helichrysum*	1	Munro, 1939b
Mesoclanis Munro (8)	1			*Chrysanthemoides, Osteospermum, Sphaeranthus*	1	Munro, 1950, 1955
Metatephritis Foote (1)	9	1		*Artemisia*	1	Fronk et al, 1964
Myopites Blot (15)				*Inula, Jasonia, Pulicaria, Sphaeranthus* (flower head galls)	9	Freidberg, 1979
Notomma Bezzi (6)	5	1		*Dichrostachys* (Mimosaceae)	1	Munro, 1952b
Oedaspis Loew (16)	5	1		*Artemisia, Schizogyne, Senecio*	6	Hendel, 1927; Munro, 1939a
Oxyna Robineau-Desvoidy (21)	2	1		*Achillea, Artemisia, Chrysanthemum, Pyrethrum* (root or stem galls)	3	Hendel, 1927; Wasbauer, 1972
Parafreutreta Munro (14)	10			*Crassocephalum, Dodonea, Senecio* (*P. foliata* Munro makes leaf petiole gall)	10	Munro, 1926, 1939c, 1940, 1952a, 1953b
Parastenopa Hendel (6)	2			*Marcetia* (Melastomataceae) (*P. elegans* Blanchard mines in galls formed by a psyllid on leaves of *Ilex* (Aquifoliaceae))	2	Bezzi & Tavares, 1916; Blanchard, 1929

	Pal	Afr	Nea	Neot	Or	Aus	Oc	Total	Hosts	Reference
Paratephritis Shiraki (9)	2							2	*Cacalia, Ligularia*	Kandybina et al. 1967
Peronyma Loew (1)			1					1	*Heterotheca*	Wasbauer, 1972
Phaeogramma Grimshaw (2)					1			1	*Bidens*	Hardy, 1980
Platensina Enderlein (22)				1				1	*Jussiaea* (Onagraceae)	Hardy, 1973
Polymorphomyia Snow (4)			2					2	*Eupatorium, Tessaria* (*P. footei* Korytkowski makes leaf galls on *Tessaria*)	Korytkowski, 1971
Procecidochares Hendel (16)	(1)		10	2	(1)	(1)	(2)	12	*Ageratum, Ambrosia, Aster, Brikellia, Chrysothamnus, Conyza, Corethrogyne, Erigeron, Eupatorium, Grindelia, Gutierrezia, Helianthus, Heterotheca, Machaeranthera, Solidago, Viguiera.* (Two species were introduced from the Neotropic to other regions).	Wasbauer, 1972; Hardy, 1980; Foote, 1967
Pseudeutreta Hendel (10)				2				2	*Baccharis*	Kieffer & Jörgensen, 1910
Pseudoedaspis Hendel (3)				1				1	*Senecio*	Kieffer & Jörgensen, 1910
Ptiloedaspis Bezzi (1)			1					1	*Artemisia*	Hendel, 1927
Rhachiptera Bigot (4)				1				1	*Baccharis*	Stuardo Ortiz, 1929
Spathulina Rondani (10)	1	3						4	*Ceruana, Elytropappus, Helichrysum, Helipterum, Matricaria, Metalasia, Phagnalon*	Munro, 1935, 1938
Stamnophora Munro (1)		1						1	*Vernonia* (flower head gall)	Munro, 1955
Stenopa Loew (2)			1					1	*Senecio*	Novak & Foote, 1975
Strobelia Rondani (7)				2				2	*Baccharis, Grindelia, Heterothalamus*	Aczél, 1949

						Host plants / notes	References
Tephritis Latreille (~130)	9	4		1	14	Flower head or stem galls on *Achillea, Arnica, Baccharis, Bellidiastrum, Cirsium, Crepis, Diotis, Doronicum, Erechthites, Hieracium, Hypochaeris, Lappa, Leontodon, Matricaria, Scorzonera, Solidago, Sonchus, Taraxacum, Tragopogon.* However most *Tephritis* spp. develop in the flower heads of many genera without producing galls	Séguy, 1934; Hendel, 1927; Wasbauer, 1972; Gourlay, 1955
Tomoplagia Coquillett (45)		1	4		5	*Aster, Perezia, Vernonia.* There are conflicting records about whether *T. obliqua* (Say) produces galls or not.	Aczél, 1949; Hering, 1938; Wasbauer, 1972
Trupanea Schrank (~200)	2	4	3	3	2	14 — *Argyroxiphium, Berkheya, Conyza, Dubatia, Elytropappus, Gnaphalium, Heterotheca, Pulicaria, Senecio*	Freidberg & Kugler (in press); Wasbauer, 1972; Hardy, 1980; Hendel, 1914; Munro, 1940, 1953b, 1964
Tylaspis Munro (5)			3		3	*Helichrysum, Othonna, Pluchea*	Munro, 1935, 1940; Hendel, 1927
Urophora Robineau-Desvoidy (105)	15	(4)			15	About 60 spp. are Palaearctic. *Urophora* spp. from other regions (Afrotropical, Americas) most probably belong to other genera. None of the latter is known with certainty to induce galls. Four Palaearctic species were introduced into North America. Almost all Palaearctic species with	

	Pal	Afr	Nea	Neot	Or	Aus	Oc	Total	Hosts	Reference
									known biology form flower head galls on Compositae; 1 species forms stem galls. Hosts: *Carduus, Carlina, Carthamus, Centaurea, Centrophylum, Cirsium, Cousinia, Echinops, Inula, Microlonchus, Onopordon, Ptilostemon, Saussurea, Serratula.*	Wangberg, 1978
Valentibulla Foote & Blanc (6)	3							3	*Chrysothamnus*	
Undetermined				1		4	1	6	Goodeniaceae; Compositae: *Olearia, Cassinia*	
Total	52	40	50	28	3	6	5	184		

the number of cecidogenous species $(3+1$ introduced) is probably more or less a true reflection of the actual situation. The situation in the Australasian and Pacific islands (Oceania) resembles that of the Oriental region, though recent findings suggest that gall forming among Hawaiian tephritids is more widespread than previously known (Prof. D.E. Hardy, personal communication). Likewise the picture obtained for Australia (6 species $+1$ introduced) is probably entirely a reflection of the state of research. Not only is the Tephritid fauna of Australia little known (except the subfamily Dacinae), but of the six recorded cecidogenous species, only one is named.

None of the larger genera (with 10 species or more) is studied well enough to state that all its contained species are cecidogenous, yet those genera that contain a high proportion of cecidogenous species may be considered as "typical" cecidogenous genera. I would expect that all or most of their member species will eventually be found to be cecidogenous. *Aciurina* (9 species out of 12 already proven cecidogenous); *Eurosta* (6 of 11), *Eutreta* (11 of 31), *Myopites* (9 of 15), *Oedaspis* (6 of 16), *Parafreutreta* (10 of 14), *Procecidochares* (12 of 16) and *Urophora* (15 out of 60; Palaearctic species only) are typical large cecidogenous genera. The ratio in *Urophora* appears small but almost every *Urophora* species (in the Palaearctic region) of which the biology has been clarified, is found to be cecidogenous. There are only one or two *Urophora* species that do not form galls. Based on the same considerations, there are also smaller genera that are typically cecidogenous, such as *Afreutreta* (2 of 4 species), *Tylaspis* (3 of 5) and *Valentibula* (3 of 6). *Actinoptera* and *Spathulina* are considered to be atypical cecidogenous genera since the number of cecidogenous and non-cecidogenous species in these genera is about equal. Most typical cecidogenous genera are restricted to a single zoogeographical region. Atypical cecidogenous genera are more widespread. Additional examples of the latter are *Tephritis* with about 130 species, and *Trupanea* with about 200 species distributed over most of the world. Each of these genera is known to contain several cecidogenous species, but most of their species with a known biology develop in flower heads of Compositae without forming galls.

A survey of the distribution of the gall-forming tephritids reveals that the great majority of the known cecidogenous species occur between 25° and 60° northern latitudes and 15° and 40° southern latitudes. In the northern hemisphere there is a marked dwindling of such species in both the northernmost and southernmost marginal 10° of this area, so that the main band where cecidogenous species abound is more or less 35° to 50°. A similar, though less marked trend may be observed in the southern hemisphere, where the main band of abundance stretches between 25° and 40° southern latitudes. There is an apparent paucity of cecidogenous species in the tropics, which is more pronounced in the Old World than in the New World. Almost all the large typical cecidogenous genera of Tephritidae are confined to the above delimited bands. The following are some examples from various zoogeographical regions. In the Palaearctic region *Urophora* (as accepted in this review) and *Myopites* (with one extension, *M. dellotoi* Munro, in Ethiopia) are clearly restricted to the northern band. *Eurosta, Aciurina* and *Procecidochares* typify this situation for the Nearctic region, the latter two genera with minor extensions into the Neotropics. In the Neotropic region *Cecidochares, Parastenopa, Polymorphomyia, Pseudeutreta, Rhachiptera*, and *Strobelia* are mostly endemic to the main band. In Africa *Afreutreta, Parafreutreta*, and *Tylapsis* are almost entirely restricted to South Africa. The only available example for Australia is *Chrysotrypanea*. The Oriental region, lacking any indigenous genus, is the best example of the paucity of cecidogenous tephritids in tropical areas.

The only typical cecidogenous tephritid genus that is a clear exception to the above mentioned rule is the American *Eutreta*. This genus at present is known to contain 31 species, 30 of which are discussed by Stoltzfus (1977). Fifteen species are Nearctic, while the rest extend more or less southward into the Neotropics, with seven species distributed around the equator and south of it.

It seems that species of typical cecidogenous genera occurring in the tropics favour high altitudes. For example, most of the Neotropical species distributed around the equator have been recorded from the Andes in Colombia, Equador and Peru, rather than from the vast lowlands of Brazil. Recently, Prof. D.E. Hardy (in litt.) informs me that the Hawaiian gall formers (mostly hitherto unrecorded as such) breed at elevations of 1000–3000 metres. Additional examples may be the high plateau of East Africa and the Chasi Hills of north east India. In all these areas the flies enjoy climatic conditions comparable to those prevailing in temperate regions.

Although the distributional pattern just sketched is substantiated by many examples, it should be kept in mind that more research is needed especially in the tropics; new information received from these areas may somewhat change the present picture.

Location and structure of Tephritid galls

Tephritid galls may be found on roots, stems and twigs, leaves and inflorescences, but the great majority of galls have been recorded from stems and twigs. The galls of most species are restricted to one part of the plant only. However, a few species apparently form galls on more than one part. *Stamnophora vernoniicola* (Bezzi) (Munro, 1955 : 416) and *Procecidochares australis* Aldrich (Wasbauer, 1972 : 31) were reported to form galls on stems as well as on flower heads of their respective hosts (*Vernonia* spp. and *Conyza canadensis*). *Procecidochares anthracina* (Doane) was reported by Wasbauer (1972 : 38) to form galls on both the stem base and roots of its host, *Erigeron* sp. Stoltzfus (1977 : 379) assumes that overwintering galls of *Eutreta caliptera* (Say) are crown or root galls, while the galls of the second generation are produced in the top half of the plant stems.

1) STEM GALLS

These are the most widespread and common of all galls. They are characteristic of almost all the genera, with the exception of *Myopites*. Even the genus *Urophora*, whose species mainly attack the flower heads, contains at least one species that produces galls on stems.

Both acrocecidia and pleurocecidia (sensu Mani, 1964 : 66) have been described. Acrocecidia are galls on the growing vegetative tip, while pleurocecidia develop at a distance from the growing tip. Galls on the growing tip of the main shoot are usually described adequately, but descriptions of galls which occur some distance from the growing tip of the main shoot sometimes lack important details. These latter galls may actually be acrocecidia which develop on axillary buds (Fig. 1). Careful study of fresh galls, especially during their early stages of development, could help in the proper interpretation of such galls.

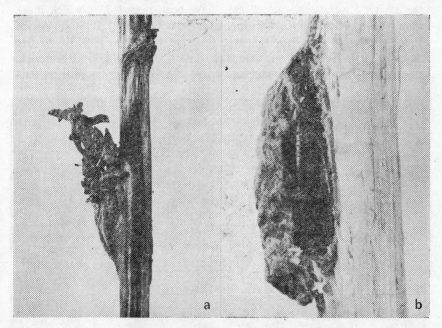

Fig. 1. Gall of *Oedaspis trotteriana* on stem of *Artemisia monosperma*.
A. Intact gall. B. Longitudinal section, showing the puparium.
(Length about 10 mm).

Many stem galls are rather conspicuous objects on the plants, even to the extent that galled plants give the impression of a bloom (Stuardo Ortiz, 1929 : 347). Others, however, are not easy to detect. Some stem galls are small, almost indistinct swellings, and a few are partly or wholly subterranean. A detailed description is available for subterranean galls formed by *Eurosta reticulata* Snow on *Solidago* sp. (Thompson, 1907 : 71). These are hollow cylinders (20 × 5 mm) formed from aborted growing shoots, with only the tip projecting above the ground. *Eurosta elsa* Daecke (Novak *et al.*, 1967 : 147) and *Eutreta novaeboracensis* (Fitch) (Stoltzfus, 1977 : 385) form galls on the rhizomes of their respective hosts (*Solidago juncea* and *S. altissima*). Some species form galls on the crown, for example, *Eutreta frontalis* Curran on *Aster laevis* (Stoltzfus, 1977 : 381), or on short shoots near the base of the plant, for example, *Stenopa vulnerata* (Loew) on *Senecio aureus* (Novak and Foote, 1975 : fig. 22). By far, most species form their galls higher on the plant, some preferring the tips of the highest twigs.

The shape of the galls varies much, from simple, slight swellings of the stem, to irregular enlargements, to rather complex structures. The shape and size sometimes is dependent on geography, location on the plant and the plant condition (Zwölfer *et al.*, 1970). Most galls are monothalamous (containing only one cavity), and most of these contain only one larva in the cavity. Several 'communal' galls have been recorded, in which more than one larva develop together in the same cavity (Fig. 2). In South Africa some species of *Parafreutreta* were reported by Munro (1953b) to live 'socially' in galls formed on various species of *Senecio*. In South America, *Pseudeutreta baccharidis* Kieffer and Jörgensen forms similar galls on *Bac-*

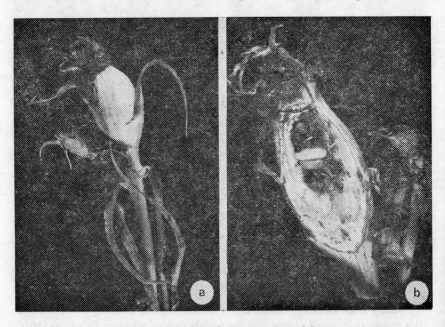

Fig. 2. Communal gall of *Tephritis hurvitzi* on stem of *Scorzonera syriaca*.
A. Intact gall. B. Longitudinal section, showing larvae and puparia.
(Length about 30 mm).

charis salicifolia (Kieffer and Jörgensen, 1910: 370, fly recorded as *Aciura baccharidis*), and *Trupanea novarae* (Schiner) forms such galls on an undetermined Compositae (Bezzi and Tavares, 1916: 165, fly recorded as *Trypanea majuscula* n. sp.). In the Palaearctic region *Paratephritis transitoria* (Rohdendorf) forms large, communal galls on *Cacalia hastata* (Kandybina *et al.*, 1967: 116) and *Tephritis hurvitzi* Freidberg forms such galls on *Scorzonera syriaca* (Fig. 2) and *Tragopogon longirostris* (Freidberg, 1980: 28). In North America *Procecidochares australis* forms communal galls on *Conyza canadensis* and *Heterotheca subaxillaris* (Benjamin, 1934: 23; first host as *Erigeron pusillus*) and another species of *Procecidochares* forms similar galls on *Chrysothamnus nauseosus* ssp. *albicaulis* (Wangberg, 1980: 412).

The largest published number of larvae living together in a communal gall is 25 (Kandybina *et al.*, 1967), however, Freidberg observed up to 40 larvae in a single gall (unpublished observation on *T. hurvitzi*). It is possible that in at least some of the communal galls, the larvae are initially confined each to its own cell, later eating away the partitions to form a common cavity. According to Kandybina *et al.* this is what happens in the case of *P. transitoria*.

Stem galls of only a few species are polythalamous (containing more than one larva, each permanently confined to its own cell or cavity) throughout their development. Examples of polythalamous stem galls are those made by *Tomoplagia vernoniae* Hering on *Vernonia tweediana* (Hering, 1938: 415), and those made by *Cecidochares connexa* (Macquart) on *Eupatorium* sp. (Bezzi and Tavares, 1916: 157). The most widespread and best studied polythalamous stem galls are those made by the European tephritid, *Urophora cardui* (Linnaeus) on *Cirsium arvense*. The shape of this gall varies from globular to fusiform and rather elongate. The average length of the galls varies between 17 and 27 mm, and the average width between eight and 15 mm (Zwölfer *et al.*, 1970). Between one and ten larvae develop in a single gall, thus, in certain cases the galls of this species are actually monothalamous. Other cases of apparent gradation between monothalamy and polythalamy occur. *Eurosta solidaginis* (Fitch) usually forms monothalamous galls, but quite often double and triple galls occur, which obviously are compound (several single galls overlapping) monothalamous galls. Yet it is quite common to find what looks like a single gall containing two or three larvae in separate cells (Uhler, 1951: 35). Munro's (1940: 80) description of the gall of *Parafreutreta fluvialis* Munro formed on *Senecio* sp. is another example: "Each swelling is found to consist of four to seven locules, each with a single puparium and eventually each with its own exit. Whether these are definitely compound galls cannot be decided yet, it may simply be that a few otherwise single galls have grown together."

Some of the largest stem galls are the previously mentioned galls of *P. transitoria* (70 mm long and 30–50 mm in diameter), the galls of *Parafreu-*

treta leonina Munro (60 × 25 mm; Munro, 1953b : 226) and the galls of
Eurosta solidaginis on *Solidago canadensis* (20–40 mm long by 20–30 mm
wide; Wasbauer, 1972 : 83). The smallest galls are hardly wider than the
stem, or only a few millimetres long. The great majority of galls do not
attain 20 mm in length, and many do not reach 10 mm.

Stem galls may be symmetrical or unilateral, smooth externally or with
tubercles or ribs, bare, pubescent or even wooly. Many bear leaves which
are more or less modified from normal leaves, sometimes only broadened
at their bases, often imbricated, or arranged in such a way that a rosette
gall is formed.

The walls of the galls are usually described as thin or thick, without
giving precise measurements. However, in most measured galls the thickness of the wall is about 1–1.5 mm (Fig. 3). When the gall is still fresh, the
wall may be soft, succulent, fleshy or brittle; upon scenescence the gall
usually becomes lignified and hardened. The gall cavity has often been
described as large or small, sometimes compared with the size of the puparium. Usually it is much larger than the puparium, which lies loose therein;
sometimes, as in the gall of *Trupanea lignoptera* Munro formed on *Berkheya
carlinopsis* ssp. *magaliesmontana*, the walls of the gall become firmly glued
around the puparium (Munro, 1929b : 399). Depending on the species, the
cavity is straight and vertical, curved or irregular. Sometimes, as in some
communal galls, it contains a loose, spongy mass of fibres, and in other
cases, it was reported to contain frass.

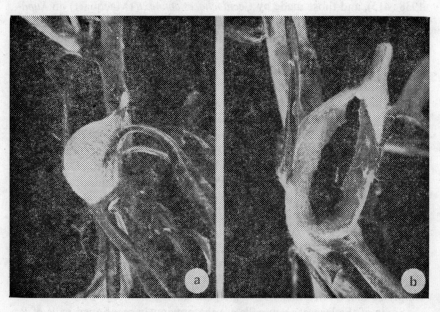

Fig. 3. Gall of *Spathulina sicula* on stem of *Phagnalon rupestre*. A. Intact gall.
B. Longitudinal section, showing the puparium. (Length about 10 mm).

In a very few stem galls a permanent opening, through which the adults escape, seems to be present. Thus, in the galls of *Spathulina arcucincta* Bezzi var. *simplex* Bezzi produced on *Metalasia muricata*, the top does not appear to be quite closed (Munro, 1938 : 426), and in the galls of *Trupanea dumosa* Munro produced on *Elytropapppus rhinocerotis* the outer end is not closed, but is guarded by outwardly directed hairs (Munro, 1940 : 86). A different exit strategy was observed in the galls of *Urophora cardui* formed on *Cirsium arvense* (Lalonde and Shorthouse, 1982 : Zwölfer, 1967b). The channels bored by the larvae of this species are filled with a spongy tissue of callus cells, which temporarily block the passage out. When the gall is wetted by rain and melting snow, the callus tissue deteriorates, separates from the surrounding lignified cells and can be easily removed, thus enabling the adults to escape. The most general exit strategy seems to be the one in which the larva, before entering diapause or pupation, scrapes a certain spot in the wall of the gall, leaving only a thin layer. The emerging adult finds no difficulty in breaking through this thin layer. The location of the exit varies in the different species from the side wall to the top.

The number of galls per branch or plant can vary from one to several tens. Gourlay (1955 : 5) states that about 40 per cent of the terminal vegetative shoots of *Erechtites atkinsoniae* in New Zealand were galled by *Tephritis fascigera* Malloch. In North America, *Procecidochares polita* (Loew) sometimes produces as many as 25 galls at the end of a stalk of *Solidago altissima* (Wasbauer, 1972 : 83). In South Africa plants of *Othonna pallens* contain fifty to one hundred galls of *Tylaspis maraisi* Munro (Munro, 1935 : 34). The highest record for a single plant is probably the 135 galls of *Spathulina sicula* Rondani ($=S.$ *tristis* Loew) found in Israel on a small plant of *Phagnalon rupestre* (Freidberg, unpublished record).

Although most tephritid stem galls are produced on composite plants, a few have been recorded from plants belonging to other families. In the genus *Eutreta* most species form galls on stems and roots of Compositae. However, *Eutreta sparsa* (Wiedemann) has been reported to form galls on *Stachytarpheta cayennensis* (Verbenaceae) (Stoltzfus, 1977 : 389), and *E. xanthochaeta* Aldrich has been reported from galls on the same plant as well as from *Lantana camara, Verbena bonariensis, V. litoralis* and *Stachytarpheta jamaicensis* (all Verbenaceae) (Stoltzfus, 1977 : 390). Munro (1947 : 126) recorded and described galls made by *Isoconia atricomata* Munro on *Blepharis transvaalensis* (Acanthaceae). Hardy (1973 : 301) recorded a gall made by *Platensina acrostacta* (Wiedemann) on *Jussiaea* sp. (Onagraceae). Bezzi and Tavares (1916 : 157) recorded a gall made by *Parastenopa marcetiae* Bezzi and Tavares on *Marcetia* sp. (Melastomataceae).

In his revision of the genus *Notomma*, Munro (1952b : 336) described galls produced by *N. galbanum* Munro on twigs of *Dichrostachys glomerata* (Mimosaceae). Additional host records are not available for this genus, but Kugler and Freidberg (1975 : 60) and Freidberg (unpublished record) have substantiated the association of other species of *Notomma* with the Mimosaceae. It is possible that more gall-forming species will be found breeding on other Mimosaceae, such as *Acacia* species.

Blanchard (1929) described *Anomoia ogloblini* (under *Hamouchaeta*) and stated (p. 4) that the larvae produced dark coloured swellings on the tender shoots of *Ilex paraguariensis* (Aquifoliaceae). This is a very unusual record despite the fact that fruits of various *Ilex* species are infested by another tephritid, *Myoleja limata* (Coquillett) (Wasbauer, 1972 : 51).

Lastly, Bush (1966a : 120) recorded stem galls on *Goodenia ovata* (Goodeniaceae) in Australia, produced by an undescribed genus and species of Oedaspinae.

Before leaving the subject of stem galls, a few unusual galls on Compositae should be mentioned. In South Africa *Cosmetothrix discoidalis* (Bezzi) forms an irregular gall on *Vernonia anisochaetoides*, 15–25 mm long, 7.5–13.5 mm wide, with a leaf or a twig on one side (Munro, 1925 : 54; fly recorded as *Afreutreta discoidalis*). The young larva burrows downwards in the pith of a young stem. From time to time the larva closes off the upper portion of the cavity with a small plug. During this period the presence of the larva is indicated by a moderate swelling of the stem. Eventually the larva settles in one place, usually opposite a leaf, and the gall then develops rapidly and becomes conspicuous. The narrow cavity in the stem above the main cavity is also shut off with a plug of fibrous material.

In South America *Strobelia baccharidis* Rondani has been studied on its hosts, *Grindelia pulchella* and *Baccharis salicifolia* (Kieffer and Jörgensen, 1910 : 372, 397), and in more detail on *G. discoidea* (Bruch and Blanchard, 1940). The larva develops at the apex of a twig, feeding on a rich plant exudate, which is subsequently excreted as a mass of foam, in which the larva remains. In the beginning the foam is soft and viscid, 12–18 mm in diameter. Later it dries, shrinks to about half of the original size, darkens and becomes spongy. The external walls of the whole structure are made of the unmodified leaves and the foam. There is a single larval cavity, separated from the external wall by a very thin internal wall. Kieffer and Jörgensen claim that the structure is a gall, but Blanchard, not being able to observe any histological modification, disagrees. A somewhat similar structure was reported to be made by *Rhachiptera limbata* Bigot on *Baccharis rosmarinifolia* (Stuardo Ortiz, 1929; fly recorded as *Percnoptera angustipennis* Philippi). However, in this case the larva is said to penetrate more deeply inside the twig.

Another noteworthy case is that of the inquiline gall-fly, *Afreutreta frauenfeldi* (Schiner) (Munro, 1926 : 26). The larvae of this species were found consuming the young and succulent galls made by a gall midge, *Afrodiplosis tarchonanthi* Felt (Diptera: Cecidomyiidae) on the tree, *Tarchonanthus camphoratus*. They later pupate in the remains of the gall, while the original inhabitant sometimes dies. This association, however, is facultative, since swellings made by the fly larvae themselves were discovered on the twigs of this plant, and also on twigs of *T. trilobus* (Munro, 1935 : 38). The only other recorded cases of inquiline gall-tephritids are mentioned in the section on leaf galls and in the Appendix (*Dacus*).

2) ROOT GALLS

A relatively small number of genera contain species that form galls on roots of plants. The holarctic *Oxyna* and the American *Eutreta* and *Eurosta* contain most, if not all, the species of this group. All root galls recorded so far seem to be located at or very near to ground level. In Europe *Oxyna flavipennis* (Loew) forms one to several, monothalamous, pea-sized, almost globular galls on the crown of *Achillea millefolium* (Hendel, 1927 : 166), whereas *O. nebulosa* (Wiedemann) forms similar, though polythalamous, galls on *Chrysanthemum* spp. (Hendel 1927 : 168). Galls of *Eurosta comma* (Wiedemann) on *Solidago juncea* in North America are potato-like both externally and in internal consistency. They are 20–25 mm long and 12 mm in diameter, rounded and rather smooth, and the larval cavity is almost spherical. Galls of *E. reticulata* on the same host are much larger (up to 45 × 25 mm) and occasionally compound (Wasbauer, 1972 : 84).

3) LEAF GALLS

Only six species have been recorded to develop in galls formed on leaves, and all but one of these on composite plants. The most extensively studied and the only illustrated are the galls produced by *Paratephritis fukaii* Shiraki on the petioles of *Ligularia tussilaginea* (Ito, 1947). From Ito's figure it appears that galled petioles are swollen along 40 or 50 mm, and that the galls are probably communal. Munro (1939c : 161) described *Parafreutreta foliata* Munro, reared from petioles of low fleshy leaves of *Senecio erubescence*. Korytkowski (1971) described *Polymorphomyia footei* Korytkowski, reared from leaf galls on *Tessaria integrifolia* in Peru. Blanchard (1929 : 465) states that the larva of *Parastenopa elegans* Blanchard (recorded as *Mesaraelia elegans*) mines in galls that are formed by a psyllid, *Metaphalara spegazziniana* Lzr., on the leaves of *Ilex paraguariensis* (Aquifoliaceae). In addition to being an inquiline, this species is peculiar in attacking this unusual host, especially in view of the fact that another species of *Parastenopa* attacks a Melanostomataceae host. There is a doubtful record of a tephritid forming monothalamous leaf galls, 7 mm long, on *Celmisia coriacea* in New Zealand (Lamb; 1960 : 124). Lastly, Bess and Haramoto (1958 : 544) reported the occasional galling of petioles and midribs of leaves of *Eupatorium adenophorum* by *Procecidochares utilis* Stone, a species that normally galls the stems of this host.

4) FLOWER HEAD GALLS

There are no tephritid galls exclusively on flowers, and those associated with flowers all incorporate other parts of the flower heads as well. All tephritid flower head galls occur on Compositae. The main genera of this ecological group are the Palaearctic *Myopites* and *Urophora*, as well as the monobasic Afrotropical *Stamnophora*, all three belonging to the subfamily Myopitinae. There are also several records of galls produced by some members of the large and widespread genus, *Tephritis* (Séguy, 1934; Berube, 1978b), but these galls are much less specialized than those produced by the Myopitinae (Shorthouse, 1980). The American genus *Procecidochares*, most members of which produce stem galls, should also be mentioned here since there are at least two records of *Procecidochares* species producing flower head galls (Benjamin, 1934 : 23; Wangberg, 1980 : 413).

Galls produced by species of *Myopites* are among the most elaborate of all tephritid galls (Fig. 4). They have been studied by various authors, including Séguy (1934) and recently by Freidberg (1979). The galls are restricted to plants of the tribe Inuleae, in particular to the genera *Inula* and *Pulicaria*. They are usually polythalamous, with up to 40 tunnels in a single flower head. The larval cavities run more or less vertically inside the swollen receptacle. The top of each cavity is covered by a hollow and modified achene, through which the larva penetrates into the receptacle in the early stages of the infestation. Later, these achenes lignify and protrude chimney-like from the upper surface of the receptacle. In mature galls the cavities inside the achenes are separated from the receptacle cavities by a few membranes of unknown origin (Freidberg, 1979).

The formation and structure of *Urophora* galls were studied by Persson (1963) and found essentially similar to the *Myopites* galls described above. However, from several other studies of various *Urophora* species (Varley, 1947; Zwölfer, 1967a, 1970, 1972) it appears that the main difference between the galls of the two genera lies in the relation between the receptacle and the achenes through which the larvae penetrate. In *Urophora* galls the receptacle may become swollen or remain flat. In both cases the receptacle tissue seems to engulf the larval chamber, which is restricted to the achene, or sometimes, penetrates deeply into the receptacle. Almost all *Urophora* galls are formed on plants of the tribe Cynareae (Fig. 5). The flower heads of infested Cynareae remain more or less closed by the surrounding bracts, and the galls of *Urophora*, unlike those of many *Myopites* species, remain concealed.

The gall produced by *Stamnophora vernoniicola* in flowers of *Vernonia abyssinica* and *V. leplolepis* is more or less globose, up to 30 mm in diameter, and with a neck at the top (Munro, 1955 : 416). It is polythalamous and may contain over 60 tunnels, each with a larva. The long axis of each tunnel points toward the middle of the gall top, which is closed during early stages of development. Later, the top seems to soften, and the flies emerge through a common opening.

The gall of *Tephritis dilacerata* (Loew) on *Sonchus arvensis* is an

Fig. 4. Flower head galls of *Myopites* spp. (size is given as diameter of receptacle): a. Gall of *Myopites apicatus* on *Pulicaria dysenterica* (the gall on the right side was cut open, showing the tunnel and a larva; about 10 mm). b. Gall of *M. cypriacus* on *P. arabica* (about 5 mm). c. Gall of *M. cypriacus* on *Inula graveolens* (about 6 mm). d. Gall of *M. stylatus* on *I. viscosa* (about 8 mm). e. Gall of *M. variofasciatus* on *I. crithmoides* (about 5 mm).

148

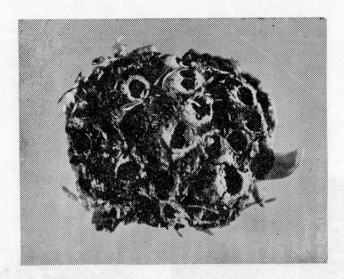

Fig. 5. Cross-section of flower head gall of *Urophora macrura* on *Cartha-mus tenuis*, showing many separate larval cells (diameter about 10 mm).

unopened, broad, button-like flower head. This form is due to the lateral expansion of the involucral bracts, which are generally vertical in normal flower heads (Shorthouse, 1980). Sèguy (1934) mentioned galls formed by several other species of *Tephritis*, but did not describe them. It is possible that these are also simple galls like those of *T. dilacerata*.

GALL HISTOLOGY AND MORPHOGENESIS

There is a notable deficiency of information on the fine structure and morphogenesis of tephritid galls. Most observations on the histology of tephritid galls are restricted to remarks made on the existence of a nutritive tissue around the larval cavity and on the hardening of the surrounding tissues (Kandybina *et al.*, 1967; Varley, 1947; Zwölfer, 1972). Cook (1903: 422) compared the histology of the stem gall formed by *Eurosta solidaginis* on *Solidago* with that of an unaffected stem. Several differences were noted. The walls of the outer parenchymatous cells of the gall were much thickened, and large intercellular spaces occurred. The sclerenchyma and tracheary tissues were reduced, while the fibrous tissue increased. The inner parenchyma cells were reduced in size and increased in amount, and throughout that tissue tubes occurred that were associated with small bundles of fibrous tissue. These tubes were not found in sections of normal stems and were considered important in the nutrition of the larva.

Shorthouse (1980) studied the morphogenesis of galls induced by *Tephritis dilacerata* in flower heads of *Sonchus arvensis*. In these galls special nutritive cells do not appear, indicating a rather simple gall, in comparison to galls formed by many other tephritids. During gall development, the area between the base of the bracts and the peduncle (referred to as involucral receptacle) enlarges laterally. The cells of the involucral receptacle become more vacuolate, and intercellular spaces become more prominent. The larvae feed on florets and receptacle tissue.

Lalonde and Shorthouse (1982) studied the histology of the stem galls formed by *Urophora cardui* on *Cirsium arvense*, in conjunction with the exit strategy of this species. The growth of this gall (and probably of many others) is due to proliferation and enlargement of parenchymatous tissues of the pith adjacent to the feeding larvae. A distinctive feature of this particular gall is the callus tissue associated with the exit strategy discussed above.

Uhler (1951) studied the morphogenesis of galls formed by *Eurosta solidaginis* on *Solidago canadensis*. He found that galls were first noticeable 21–27 days after the deposition of eggs, and 10–20 days after the eggs hatched. The galls grew rapidly, attaining their full size in about a month. Larval growth was much slower, the larvae requiring about three-and-a-half months to attain their full size. Accurate data from other species are not available.

Life history

The life history of the Tephritidae was discussed by Christenson and Foote (1960) in their review on the biology of fruit flies. However, gall-forming species were not included in this review. Numerous life history data of gall-forming species are scattered in the literature, while monographic works on single species are relatively few (Uhler, 1951; Varley, 1947; Wadsworth, 1914; Zwölfer, 1967a, 1970, 1972; Zwölfer et al., 1970). These life history data do not show considerable differences from non-cecidogenous species, hence only their salient features will be given here.

Males and females of apparently all the species recognize their host plant and meet on it exhibiting a "rendezvous-behaviour" sensu Zwölfer (1974a). Males are territorial; females of some species are autogenic. Courtship involves complex visual stimuli and responses (Zwölfer, 1974b; Berube, 1978a; Wangberg, 1978, 1980, and many others) and sometimes gustatory behaviours (see Freidberg, 1982 for a summary of the known cases). Zwölfer (1974a) showed that meeting at the specific rendezvous place was an important reproductive isolation mechanism for eight *Urophora* species, that, deprived of their host plants, mated in various interspecific combination, never or rarely observed in the field.

The gravid female, probably using visual, tactile and chemical stimuli, searches the host plant for oviposition sites, where she deposits one or several eggs per each site. Eggs are usually deposited into the tissue, where the larvae subsequently feed. Eggs of *Aciurina ferruginea* (Doane) are laid in the sticky exudate covering the axillary buds of the host, *Chrysothamnus viscidiflorus* (Tauber and Tauber, 1967 : 908); the larva subsequently forms a gall in the axillary bud. The eggs of *Procecidochares utilis* are inserted between the young, paired leaves at the tip of the shoots, and the galls are formed along the growing shoots (Bess and Haramoto, 1958 : 544). The eggs of European *Urophora* species are usually placed on the young florets, inside the space roofed by the capitulum bracts. It is interesting to note that both Wadsworth (1914) and Varley (1947) reported of an unusually long route traversed by the ovipositor of females *Urophora jaceana* (Hering) (in Wadsworth recorded as *U. solstitialis*) inside the flower heads of *Centaurea* spp. According to their descriptions and illustrations the ovipositor first is pushed down between the bracts. When it reaches the receptacle, it turns upwards, intruding the receptacle and the florets, eventually reaching the space where the eggs are deposited.

Eggs of gall-forming tephritids usually hatch within several days after oviposition. Eggs of *Aciurina ferruginea* (Tauber and Tauber, 1967) and *Procecidochares* spp. (Wangberg, 1980) overwinter. *Urophora* spp. are unusual in that the first larval instar is spent inside the egg shell (e.g. Persson, 1963; Varley, 1947). The duration of the three larval instars varies among the species, as well as is dependent upon environmental factors such as temperature. Usually the larva commences feeding immediately after hatching and continues until it is fully grown. However, *Stenopa vulnerata* overwinters as a first or second-instar larva, which renews feeding in spring (Novak and Foote, 1975). When fully grown, the larvae of many species scrape the wall of the gall and prepare an exit for the adults. Many species overwinter as fully-grown larvae or prepupae, which rest inside the galls, with their heads directed downwards. In the next season, before the appearance of adults, they turn, with their heads directed towards the exit, and pupate. The duration of the pupal period is usually less than a month, thereafter the adults emerge. Pupation, as a rule, is inside the gall.

Table 2 presents some life history data for six representative species.

Special adaptations

An interesting question is how to identify a tephritid as a gall former, without actually rearing it from its gall. It can sometimes be done by determining the species taxonomically and checking the literature, but this is a task for a specialist, and does not necessarily mean that the answer will be readily available. There are several anatomical and behavioural traits, that, if considered with caution, may be useful for characterizing tephritid gall formers. Many gall formers are relatively robust with broad frons, short, broad thorax and, in the female, with a robust, conical oviscape (ovipositor sheath) and specialized aculeus (ovipositor). The latter often has an arrow-shaped tip or fine pre-apical serration (e.g. Stoltzfus, 1977). The exact

**Table 2. Life history data of six gall-forming Tephritidae in the
northern hemisphere
(numbers represent ranges or average)**

Species and reference	Duration of the various stages					Adult activity period
	Incubation (days)	Larva (months)	Pupa (days)	Preoviposition period (days)	Adult longevity (days)	
Procecidochares utilis (Bess and Haramoto, 1958)	3–8	20 days	14–21	0	14–21	throughout the year (multivoltine)
Procecidochares sp. A* (Wangberg, 1980)	9–10 months	1–2	30	?	10–20	June–July (univoltine)
Stenopa vulnerata (Novak and Foote, 1975)	a few	8–9	24 incl. prepupa	15	36–70	July–October (univoltine)
Eurosta solidaginis (Uhler, 1951)	4–6	10 incl. prepupa	37	0–1	4–12	May–June (univoltine)
Urophora jaceana (Varley, 1974)	12 incl. 1st instar larva	10 incl. prepupa	approx. 30	2	7	July–August (univoltine)
*Spathulina sicula*** (Freidberg, 1978)	14–30	approx. 2	10–20	5–6	30	Nov.–Dec.; Feb.–May (bivoltine)

*Rough estimates based on seasonal occurrence.
**Rough estimates for various stages of 2nd generation. Aestivating stage not known.

biological significance of these modifications is not clear. Another characteristic trait is the strong tendency of adults to become greasy after being killed and pinned. This phenomenon may be correlated with a large amount of fat accumulated in the body (Munro, 1940: 87), but why should there be a larger amount of fat in gall formers than in other species?

A peculiar character that occurs in some genera (all apparently referrable to the Oedaspinae) is a small, non-functional proboscis. The females of these species are apparently autogenic, i.e., emerging from the pupa with mature eggs in the ovaries. Moreover, the adults of some species have been reported to be short-lived flies. It seems that all these characteristics are connected with the fact that these species do not feed as adults, but there is no answer to the question as to why such characteristics occur in this group and not in others. Another unexplained behavioural trait is the ability to feign death by falling to the ground when disturbed. This has been observed in at least two species, *Eutreta diana* (Osten Sacken) (Dr. G.J. Steck, personal communication) and *Oedaspis trotteriana* Bezzi (Freidberg, unpublished observations).

Observations on the reproductive behaviour of gall tephritids are scanty. Nevertheless there is a disproportionately large amount of literature recording the existence of mating trophallaxis (sensu Freidberg, 1981) in this group of tephritids. Such trophallaxis, which occurs between the mates shortly before or after copulation, has been recorded in three species of gall formers which constitute half the number of known cases of trophallaxis for the whole family. These species are: *Stenopa vulnerata* (Novak and Foote, 1975), *Eutreta novaeboracensis* (Stoltzfus and Foote, 1965; recorded as *E. sparsa* (Wiedemann)) and *Spathulina sicula* (Freidberg, 1982, recorded as *S. tristis* (Loew)).

Larvae of tephritids which develop in galls are usually shorter and wider than other tephritid larvae, especially those that develop in fleshy fruits. Perhaps larvae of the latter group, which may crawl relatively long distances both in the fruit and outside of it, require a more vermiform shape. Larvae which develop in galls often restrict their movement to a distance not much longer than their own body. Since pupation takes place inside the gall, there is no need for an elongate, agile body.

Ecology

Tephritid ecology was reviewed by Bateman (1972), but he did not deal with gall-forming species. As summarized in the section on biogeography; gall-forming Tephritidae occur in all major zoogeographical regions excluding Antarctica. Some species mainly inhabit lowlands: other prefer montane habitats, for example, *Tephritis hurvitzi*, found between 1000 and 2000 m altitude (Freidberg, 1980 : 28); still others are found from about sea level to an altitude of 3000 m, e.g. *Eutreta angusta* Banks (Stoltzfus, 1977 : 377). Tephritid galls are found in various habitats, including river banks and forests, however their greatest diversity is probably to be found in steppe and steppe-like habitats.

Most gall-forming tephritids are oligophagous, although many are hitherto known as monophagous. This degree of specificity is higher than what is indicated in Table 1, because different species of a genus sometimes utilize host species belonging to different genera. The high specificity of gall-forming tephritids is well documented and has also been checked experimentally, mainly using oviposition tests (Berube, 1978b; Peschken and Harris, 1975; Zwölfer, 1967a, 1970, 1972). This specificity tends to decrease interspecific competition, although plants sometimes bear galls formed by two or more species (Fronk *et al.*, 1964; Wangberg, 1978, 1980). In the latter case, segregation of the infected parts of the plant and different seasons of activity probably act to decrease competition.

Fecundity of most gall-forming tephritids apparently is low if compared with that of frugivorous species. Zwölfer (1967a : 8) counted up to 90 eggs in the ovaries of mature *Urophora sirunaseva* (Hering) females, while Varley (1947) calculated the average fecundity of females *U. jaceana*, a closely related species, as 52 and 70 for two years respectively. Varley also tested the effect of various factors on the fecundity of *U. jaceana*, and counted eggs laid by the females in a variety of conditions. The range found was between 0 and 205, and the average was usually well below 50. Uhler (1951 : 18) found in oviposition experiments with *Eurosta solidaginis* conducted in a greenhouse, an average of about 12 eggs per female. Dissected females contained between 205 and 276 well-developed eggs in their ovaries, which made Uhler expect higher fecundity rates in uncrowded field conditions. In similar dissections of virgin females Milne (1940 : 101) found an average of only 73 eggs per female. These few direct observations on low fecundity are supported by more, indirect observations. Thus, Novak and Foote (1975 : 45) reported of very low infestation rates of *Stenopa vulnerata* on *Senecio aureus*—only four galls found in several hundred examined stems, and data from many other species corroborate this one.

Egg survival (determined by per cent hatch) was high for *E. solidaginis* tested in various temperatures, and often reached 100 per cent (Uhler, 1951 : 26). Egg survival of *Urophora jaceana* was 91.1 and 84.7 per cent respectively in two different years (Varley, 1947). It is possible that high egg survival like that is characteristic of other gall-forming tephritids as well. The main cause for the mortality of eggs is probably lack of fertilization (Varley, 1947).

Larval and pupal mortality is probably the most important factor regulating populations of gall tephritids. Parasitism (by parasitic wasps), predation (both by invertebrates and vertebrates) and possibly fungi and climatic factors are the most important environmental causes for larval and pupal mortality. Parasitism of gall tephritids has been reported in the literature numerous times (e.g. Bess and Haramoto, 1958; Uhler, 1951; Varley, 1947), although determinations of the parasitoids were not always given. The following are the most important families of parasitic wasps recorded from tephritid galls: Ichneumonidae, Braconidae, Cynipidae, Eulophidae, Eupelmidae, Eurytomidae, Pteromalidae and Torymidae; of these the chalcidoids are by far the most common.

Parasitoids of gall tephritids may be ecto- or endoparasitic, solitary or gregarious, larval or larval-pupal (possibly also pupal) parasitoids. Parasitized larvae often pupate earlier than intact larvae. Infestation rates in field samples vary from zero to 100 per cent, often being around 50 per cent. Usually more than one parasitoid species develops in galls of a given tephritid species. Thus, Fulmek (1968 : 754) recorded 20 species of wasps parasitizing *Urophora cardui* over its whole range, almost all of which were chalcidoids. The biology of some of the tephritid parasitoids has been studied (e.g. Claridge, 1961; Varley, 1947).

Varley (1947) reported various invertebrate predators of *Urophora jaceana*, some of which were reported again by others. These predators include moths' caterpillars, gall-midge larvae, mites and host-feeding chalcidoid females. Zwölfer (1972) and Zwölfer *et al.* (1970) reported of the mortality of *Urophora* species possibly caused by staphilinid beetles, by a dipteron, *Lonchaea* sp. (Lonchaeidae) and by snails, the latter destroying young galls. Uhler (1951) reported of two galls of *Eurosta solidaginis* invaded by clerid beetles, which apparently consumed the pupae of the fly. Milne (1940) reported of a species of *Mordellistena* (Coleoptera: Mordellidae) predaceous on *E. solidaginis* in its gall, but Uhler (1951) reported of *M. unicolor* Lec. not as a predator, but as an inquiline. Other inquilines found by Uhler were undetermined Cecidomyiidae. All these associations, except those with caterpillars, seem to be rare or only locally prevalent. At this point it is interesting to note the variety of relationships, rare as they are, between Cecidomyiidae and gall tephritids. Gall midges can be inquilines in tephritid galls, they can prey on tephritid larvae, but they can also host in their own galls inquiline tephritids (Munro, 1926 : 26).

Vertebrate predators of gall tephritids in their galls are mice and birds (Varley, 1947; Milne, 1940; Miller, 1959; Stoltzfus, 1977; Bess and Haramoto, 1958). Actual predation by vertebrates has apparently never been observed, but is inferred from the type of damage caused to the gall and other circumstantial evidence. The birds presumed to damage the galls are woodpeckers and nuthatches, and the toll they take may be considerable (Milne, 1940).

The causes of adult mortality are not well known; spiders and ants were mentioned by Zwölfer *et al.* (1970), and sphegid wasps by Stoltzfus (1977 : 390).

Phenology and number of generations

Probably the single most important factor determining the phenology of gall-forming Tephritidae is the corresponding phenology of the vulnerable parts of their host plants. It seems that susceptibility of appropriate plant parts is very short in some cases where the appearance of adults in the field is restricted to only about a month in a particular region. Although this situation is common among the Tephritidae, in the gall formers, perhaps more than in any other group, it is a strong manifestation of the intimate association between the flies and their host plants. Many, perhaps most, gall makers are univoltine, however in some cases the susceptibility of the host to ovipositing females is extended over a long period, permitting the existence of a second or perhaps even a third generation. The occurrence of multivoltine gall formers is sometimes associated with a sequential change in choice of host part, e.g. from flower head galls to stem galls in *Procecidochares australis* (Huettel and Bush, 1972 : 468). In very few cases is there a shift from one host to another during the course of the season that makes possible the extension of the species activity beyond the usual duration. *Myopites cypriacus* Hering breeds in Israel in flower head galls on *Pulicaria arabica* and *Inula graveolens* (Fig. 4). In the former plant it develops during spring and summer, and in the latter during the fall, thus producing at least three generations per year (Freidberg, 1979: 23). Upon discovering similar cases special attention must be paid to the correct specific determination of the broods from the different host plants, since they may actually be distinct, but closely related, species. Colour differences between the broods indeed occur in *M. cypriacus* (Freidberg, 1979), but likewise they occur between the spring and summer generations of a related species, *M. eximia* Séguy which develops on a single host, *I. crithmoides* (Hering, 1943). Since records of tephritid galls from areas close to the equator are rather few and fragmentary, no conclusion can be drawn as to their phenology in these areas.

Adult activity in both the northern and southern hemispheres usually takes place during spring, summer and fall, depending on the condition of the host. Judgement of the seasonality should be based on field records only, since in the laboratory adults sometimes emerge in undue seasons. Although recorded for both hemispheres (Bush, 1966a: 120; Freidberg, 1982: 384; Munro, 1926: 32, 1957: 910) winter activity is quite unusual.

Damage to plants and economic importance

The Tephritidae have long been regarded primarily as pests of many important fruit and vegetable crops. Only rather recently have the beneficial members of the family gained recognition as agents in the bio-control of weeds. The cecidogenous species form the majority of such beneficial or potentially beneficial species. Since the economic importance of a species is often a reflection of the damage it causes to its host plant, it is necessary to first review the existing information on the kinds of damage caused to plants by gall-forming tephritids. Then, the actual and potential bio-control agents among the Tephritidae will be surveyed.

Information on the effects of gall formation on the plants can be conveniently divided into qualitative and quantitative data. The qualitative knowledge is based mainly on observations in the field, mostly on non-economic species. Reports of such observations vary from evaluations claiming no apparent ill-effects (Wangberg, 1978 : 481; Stoltzfus, 1977 : 392 and others), through statements about the occurrence of restricted damage (Wangberg, 1980 : 414; Berube, 1978a : 78 and others) to observations on the death of plants as a consequence of galling (Frick, 1972 : 630; Bess and

Haramoto, 1958 : 546). Bess and Haramoto observed an array of injuries caused to *Eupatorium adenophorum* by *Procecidochares utilis*, which sometimes eventually led to the death of plants. In an area in Hawaii where the fly was abundant all shoots were attacked several times during their growth. Buds and growing tips died because of intense oviposition. Decay appeared following the entry of microorganisms through the emergence holes. The overall effect was a reduced stem and foliage growth as well as seed production, and in certain circumstances the death of branches and entire plants.

Quantitative observations as well as experimentation are available for several species, notably those of bio-control merit. Peschken and Harris (1975) found that roots of single and double-galled plants of *Cirsium arvense* attacked by *Urophora cardui* weighed 65 and 78 per cent respectively less than roots of plants without galls. The stem and leaves of galled plants were found to weigh 47 and 58 per cent less than the respective parts of ungalled plants. Uhler (1951 : 38), on the other hand, did not find significant differences in the height between galled and ungalled plants of *Solidago canadensis* attacked by *Eurosta solidaginis*.

Damage to plants which are galled in their flower heads has bearing on the reproductive capacity of these plants. According to Berube (1978a : 78), buds of *Sonchus arvensis* attacked by *Tephritis dilacerata* rarely bloom, and the seed is almost never produced in such buds. Wadsworth (1914), who studied *Urophora jaceana* (as *U. solstitialis*) on *Centaurea nigra*, found a reduction of nearly 50 per cent in the number of seeds produced in galled heads, compared with the number produced in ungalled heads. Further, in a germination experiment he found about 60 per cent reduction in the percentage of germination of seeds taken from galled heads, as compared with seeds taken from ungalled heads. Zwölfer (1972) found about 80 per cent reduction in the number of healthy achenes per head in flower heads of *Cirsium vulgare* galled by *Urophora stylata* (Fabricius) compared with ungalled heads. However, there was no difference in germination between healthy achenes taken from galled and ungalled heads. In another series of observations Zwölfer found that the presence of five to six *U. stylata* cells in large heads was enough to eliminate all viable achenes in these heads. However, in the case of *U. sirunaseva* attacking *Centaurea solstitialis*, Zwölfer (1976a) found that the effect of gregarious attacks (several cells in a flower head) appeared to be relatively lower than that of a single attack.

Four species of *Urophora*, two species of *Procecidochares* and one species of *Eutreta* have already been introduced outside their normal ranges as a part of bio-control campaigns against various weeds. The most remarkable success in weed control by an introduced tephritid is that of *Procecidochares utilis*. This Mexican species has already been introduced into several regions of the world to fight its sole host, the weed, *Eupatorium adenophorum*. The fly was first introduced into Hawaii in 1945. Bess and Haramoto studied it there (1958, 1959) and gave a vivid description (1958) of the situation before and after the introduction. The following is a citation of that description: *"Eupatorium adenophorum*, known as pamakani in Hawaii, is a tropical America plant which was introduced into Hawaii about 1860. Following its introduction it thrived and spread, forming dense thickets up to 10 feet tall on valuable grazing lands. Prior to 1945 labour and funds were commonly used to reclaim infested ranch land through the mechanical removal of this weed. In 1945 a tephritid gall fly, *Procecidochares utilis* Stone was introduced from Mexico into Hawaii to combat *E. adenophorum*. The fly increased rapidly and soon many plants which had been heavily and persistently attacked were weakened and died. Due to the optimism over its eventual success in the control of the weed, further mechanical control measures were abandoned by 1948. The fly was an outstanding success." *Procecidochares utilis* was subsequently introduced into Australia and New Zealand to combat the same plant (Hoy, 1960).

Another species of *Procecidochares*, *P. alani* Steyskal (described by Steyskal (1974) from Mexico) was introduced into Hawaii for the biological control of *Ageratina riparia* (Hardy, 1980).

Four *Urophora* species have been introduced from Europe into North America to control weeds which originated in Europe. *Urophora affinis* Frauenfeld was released in Canada in 1971 in order to control diffuse knapweed, *Centaurea diffusa*, and is now established on this plant as well as on *C. maculosa*. Diffuse knapweed, accidently introduced from eastern Europe into North America, now covers about 65,000 acres in British Colombia and is spreading eastward (Shorthouse and Watson, 1976). Wherever established, it becomes dominant, and its spiny foliage reduces the availability of desirable forage species. The larvae of *U. affinis* induce the formation of a thick nutritive layer inside their galls, a kind of damage thought to divert energy from the plant's normal processes of growth and reproduction. In addition, each larva displaces about 2.5 seeds in the flower heads of diffuse knapweed (Shorthouse and Watson, 1976). *Urophora quadrifasciata* Meigen is another species presently studied with regard to *C. diffusa*.

Urophora cardui was studied by Zwölfer (1967b), Zwölfer *et al.* (1970) and by Peschken and Harris (1975) who considered this species a promising candidate for the biological control of Canada thistle, *Cirsium arvense. U. cardui* was recently introduced into various regions of Canada, but has not become established in Western Canada (Peschken *et al.*, 1982). Lalonde and Shorthouse (1982) suggested that inadequate moisture in this area impedes breakdown of the plugs that block the gall exits, thus interfering with the normal emergence of the adults.

Additional *Urophora* species introduced into North America are *U. jaceana* to control *Centaurea nigra* and *U. sirunaseva* to control *C. solstitialis* (Wasbauer, 1972 : 140). Another species of this genus, *U. stylata* Fabricius is still under study, with the prospect of using it to control *Cirsium arvense* and *C. vulgare* (Zwölfer, 1972).

Two other cecidogenous tephritids, *Tephritis dilacerata* and *T. formosa* (Loew) both from Europe, are presently being studied in Canada with the intention of releasing them there for the control of the sow-thistle, *Sonchus arvensis*. (Berube, 1978a, b; Shorthouse, 1980). This plant has been introduced from Europe or Asia and is a troublesome weed in cultivated fields.

The only species belonging to this beneficial group which does not infest Compositae hosts is *Eutreta xanthochaeta*. This species was introduced from Central America into Hawaii to control *Lantana camara* (Verbenaceae), an escaped shrub originally introduced into Hawaii as an ornamental (Swezey, 1923). Although *E. xanthochaeta* is not considered important as a check on the growth of *Lantana* (Perkins and Swezey, 1924), its introduction together with seven other insects helped to bring the shrub under control (Stoltzfus, 1977).

A completely different utilization of tephritid galls is that mentioned by Kandybina *et al.* (1967). According to them extracts of the galls produced by *Paratephritis transitoria* on *Cacalia hastata* in the U.S.S.R. are widely used for tannery.

The only case of a gall-forming tephritid causing damage to crops is probably that of *Urophora macrura* (Loew). This species was listed by Avidov and Kotter (1966; as *Urophora* sp.) as one of eight species of Tephritidae that were reared from safflower (*Carthamus tinctorius*). However, its damage was negligible.

Evolution

The ability to cause formation of galls apparently arose several times in the Tephritidae. The two main avenues of discussing the evolution of gall tephritids are through the taxonomy and phylogeny of the gall formers and through the types of associations between them and their host plants. Evidence from palaeontology is non-existent for this family (Bush, 1966b), and genetic information is very scarce (see Bush, 1966a for a review).

Almost all gall-forming tephritids belong to one of three currently recognized subfamilies, Myopitinae, Oedaspinae and Tephritinae. The Tephritinae is one of the largest subfamilies of the Tephritidae, and is well represented over all major zoogeographical regions. Most of its member species infest flower heads of Compositae without inducing the formation of galls. The Oedaspinae is a much smaller subfamily and not well defined. Although world-wide in distribution, it is unevenly distributed, being very diverse in North and South America and possibly also in the Afrotropical region. Most of its member species are apparently gall formers. Taxonomically, various cases of intergradation between the Tephritinae and Oedaspinae occur, and Bush (1966a) postulated a common origin for both subfamilies. This assumption was mainly based on the occurrence of female heterogamety in gall-forming species of both subfamilies. Sex determination by male heterogamety is apparently more widespread in the Tephritidae. It should, however, be kept in mind that female heterogamety does not occur in all Tephritinae, and that in the Oedaspinae, it is not necessarily associated with the formation of galls (Bush and Huettel, 1970).

The Myopitinae, a small and relatively well-defined subfamily, is predominantly an Holarctic group, occurring also in the Neotropical and Afrotropical regions, and apparently lacking from the Oriental, Australian and Oceanic regions. It may be a sister group of the Terelliinae, another subfamily of Compositae feeders, but one which does not contain gall formers. Both subfamilies may be relatively young, since they are strongly associated with the Cynareae, a highly evolved tribe in the Compositae. The Palaearctic members of the Myopitinae, as a rule, are gall formers, but in the other regions gall forming is apparently rare. The relationships between the Myopitinae (plus Terelliinae) and the Tephritinae are not clear, but it is possible that they diverged from the main stem of the Tephritinae before the Oedaspinae appeared. The ability to form galls appeared in the Myopitinae only after they split from the Terelliinae. Information on the sex determination mechanism of the two subfamilies will probably shed more light on the phylogenetic relationships between them and the others that infest Compositae.

The possible origin and evolution of tephritid galls, from the stand-point of the structure and location of the galls on the plants, was discussed by Steck (1981 : 16). Steck suggested that the gall makers are derived from seed-receptacle feeders. The first step which occurred was the galling of the flower head, the next step being a transition to stem galls. Steck argued that root and stem gall formers in Compositae do not actually attack the mature root or stem tissue. Instead, they oviposit into axillary or terminal buds, and the shoot may, or may not, grow beyond that point. To support his theory Steck brings two examples of more or less the entire gamut of transitions shown by existing taxa. The first example is the genus *Urophora*, which contains species that develop in flower heads without galling them, species that gall the flower heads, and at least one species that forms galls on stems. The second example, and there are several similar ones, is *Procecidochares australis* which forms galls in either flower heads or stems.

While generally accepting the above scheme, the main reservation concerns the details of the first step. Flower head galls are rarely associated with the Tephritinae and Oedaspinae. Flower head galls formed by species of *Tephritis* (Tephritinae) are simple and probably primitive, unlike the structurally complex and probably highly evolved stem galls induced by members of these two subfamilies. Flower head galls of the Myopitinae are widespread and complex. However, in the latter subfamily there has hardly been any further evolution along the lines suggested above. That stage of evolution, namely stem galls, is best shown by present day Tephritinae and Oedaspinae. It is, therefore, suggested that evolution of gall for-

160

mation in the Tephritidae has been taking place mainly along two independent lines. The first line, represented in the Myopitinae, involves a transition from ungalled to galled flower heads. A further transition is shown only by one species, *Urophora cardui* a stem-gall maker, which probably is a most recent development of this line. The second line, represented by the gall-forming Tephritinae and Oedaspinae, first involves a transition from breeding in ungalled flower heads to breeding in stem galls. The actual transition could have taken place from oviposition into young buds of flower heads to oviposition into young vegetative buds, both observed in *Paracantha culta* (Wiedemann), the larvae of which are flower head breeders (Steck, 1981 : 18). *Spathulina euarestina* Bezzi var. *piscatoria* Munro offers another example supporting this assumption. This species usually forms terminal stem galls on its host, but one puparium was found in an apparently ungalled flower head (Munro, 1938 : 427). *Spathulina*, *Actinoptera* and other atypical cecidogenous genera, all containing some species that induce galls on stems and some that develop in ungalled flower heads, offer similar examples on a generic level.

The rarely occurring root and leaf galls probably constitute the products of further evolutionary steps arising from stem galls. Contemporary species that facultatively gall either roots or stems provide evidence for the possible evolution of root galls. *Procecidochares utilis*, a species that normally induces the formation of stem galls, and only occasionally of leaf galls, demonstrates the possible evolutionary route of leaf galls.

Gall formation in the Aciurinae and Trypetinae apparently arose separately from the two main lines suggested above. As these are rare and not well studied phenomena, nothing else can be said about their evolution.

Figure 6 shows a graphic representation of the possible two important evolutionary routes in the development of tephritid galls. Route No. 1 is represented by the Myopitinae, and route No. 2-by the Tephritinae and Oedaspinae.

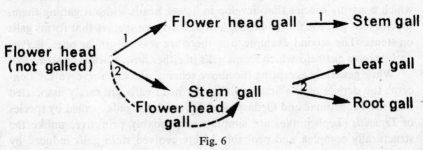

Fig. 6

Acknowledgements

I would like to extend my sincere thanks to Dr. R. H. Foote and Dr. L. Knutson (USDA, Washington D.C.) whose assistance made available to me much of the literature cited in this review. I am greatly indebted to Dr. G.J. Steck (Texas A & M University), who reviewed an early draft of the manuscript, and to Prof. J. Kugler, Prof. D. Gerling and Prof. D. Wool (Tel Aviv University) who reviewed a later draft. Mrs. T. Feler and Mrs. F. Kaplan (Tel Aviv University) are thanked for their technical help. My neighbour, Mrs. Eda Edenburg is acknowledged for translating Spanish texts, and my wife, Pnina, for translating from Russian. The photographic work was done by Mr. A. Shoob and Miss L. Maman.

REFERENCES

1. Aczel, M. 1949. Catalogo de la familia 'Trypetidae' (Dipt. Acalypt.) de la region Neotropical. *Acta Zool. Lilloana* 3 : 177–328.
2. Avidov, Z. and Kotter, E. 1966. The pests of safflower *Carthamus tinctorius* L. in Israel. *Scr. Hierosolymitana Heb. Univ. Jerus.* 18 : 9–26.
3. Bateman, M.A. 1972. The ecology of fruit flies. *Annu. Rev. Entomol.* 17 : 493–518
4. Benjamin, F.H. 1934. Descriptions of some native trypetid flies with notes on their habits. *US Dep. Agric. Tech. Bull.* 401 : 95 pp.
5. Berube, D.E. 1978a. Larval descriptions and biology of *Tephritis dilacerata* (Dip. : Tephritidae) a candidate for the biocontrol of *Sonchus arvensis* in Canada. *Entomophaga.* 23 (1) : 69–82.
6. Berube, D.E. 1978b. The basis for host plant specificity in *Tephritis dilacerata* and *T. formosa* (Dipt. : Tephritidae). *Entomophaga.* 23 (4) : 331–337,
7. Bess, H.A. and Haramoto, F.K. 1958. Biological control of Pamakani, *Eupatorium adenophorum*, in Hawaii by a tephritid gall fly, *Procecidochares utilis*. I. The life history of the fly and its effectiveness in the control of the weed. *Proc. Int. Congr. Entomol.* 4 (1956) : 543–548.
8. Bess, H.A. and Haramoto, F.K. 1959. Biological control of Pamakani, *Eupatorium adenophorum*, in Hawaii by tephritid gall fly, *Procecidochares utilis*. 2. Population studies of the weed, the fly, and the parasites of the fly. *Ecology* 40 (2) : 244–249.
9. Bezzi, M. and Tavares, J.S. 1916. Alguns muscideos cecidogenicos do Brazil. *Broteria. Ser. Zool.* 14 (3) : 155–170.
10. Blanchard, E.E. 1929. Descriptions of argentine Diptera. *Physis* (B. Aires) 9 : 458–465.
11. Bruch, C. and Blanchard, E.E. 1940. Observaciones Biologicas sobre *Strobelia baccharidis* Rondani (Dipt.: Trypetidae) y description de dos Himenopteros (Chalcid). *Rev. Mus. La Plata* (N.S.), *Secc. Zool.* 2 : 85–98.
12. Bush, G.L. 1966a. Female heterogamety in the family Tephritidae (Acalyptratae, Diptera). *Am. Nat.* 100 (911) : 119–126.
13. Bush, G.L. 1966b. The taxonomy, cytology and evolution of the genus *Rhagoletis* in North America (Diptera, Tephritidae). *Bull. Mus. Comp. Zool.* 134 (11) : 431–562.
14. Bush, G.L. and Huettel, M.D. 1970. Cytogenetics and descriptions of a new North American species of the Neotropical genus *Cecidocharella* (Diptera : Tephritidae). *Ann. Entomol. Soc. Am.* 63 (1) : 88–92.
15. Christenson, L.D. and Foote, R.H. 1960. Biology of fruit flies. *Annu. Rev. Entomol.* 5 : 171–192.
16. Claridge, M.F. 1961. Biological observations on some eurytomid (Hymenoptera: Chalcidoidea) parasites associated with Compositae, and some taxonomic implications. *Proc. R. Entomol. Soc. Lond. Ser.* (A) *Gen. Entomol.* 36 : 153–158.
17. Cogan, B.H. and Munro, H.K. 1980. 40. Family Tephritidae. In *Catalogue of*

162

the Diptera of the Afrotropical Region, Ed. R.W. Crosskey, pp. 518–554. London : British Museum (Natural History), 1437 pp.

18. Cook, M.T. 1903. Galls and insects producing them. Parts 3, 4, and 5. *Ohio Nat*. 3 : 419–436.

19. Efflatoun, H.C. 1927. On the morphology of some Egyptian trypaneid larvae (Diptera), with descriptions of some hitherto unknown forms. *Bull. Soc. R. Entomol. Egypt* 11 : 17–50.

20. Foote, R.H. 1967. A catalogue of the Diptera of the Americas South of the United States. Departmento de Zoologia, Secretaria da Agricultura, Sao Paulo, 57 : 1–91.

21. Foote, R.H. 1980. Fruit fly genera South of the United States (Diptera : Tephritidae). *US Dep. Agric. Tech. Bull*. 1600 : 1–79.

22. Foote, R.H. and Blanc, F.L. 1959. A new genus of North American fruit flies (Diptera : Tephritidae). *Pan-Pac. Entomol*. 35 (3) : 149–156.

23. Freidberg, A. 1974. Descriptions of new Tephritidae (Diptera) from Israel. I. *J. Entomol. Soc. South Afr*. 37 : 49–62.

24. Freidberg, A. 1978. *Reproductive behaviour of fruit flies*. Tel Aviv University. Unpublished Ph. D. dissertation (in Hebrew).

25. Freidberg, A. 1979. On the taxonomy and biology of the genus *Myopites* (Diptera : Tephritidae). *Isr. J. Entomol*. 13 : 13–26.

26. Freidberg, A. 1980. Descriptions of new Tephritidae (Diptera) from Israel. II. *J. Wash. Acad. Sci*. 70 (1) : 25–28.

27. Freidberg, A. 1981. Mating behaviour of *Schistopterum moebiusi* Becker (Diptera : Tephritidae). *Isr. J. Entomol*. 15 : 89-95.

28. Freidberg, A. 1982. Courtship and post-mating behaviour of the Fleabane Gall Fly *Spathulina tristis* (Diptera : Tephritidae). *Entomol. Gen*. 7 (2) : 273–285.

29. Freidberg, A. and Kugler, J. *Fauna Palaestina, Insecta-Diptera : Tephritidae*. Jerusalem : The Israel Academy of Sciences and Humanities. In press.

30. Frick, K.E. 1972. Third list of insect that feed upon Tansy Ragwort, *Senecio jacobaea*, in Western United States. *Ann. Entomol. Soc. Amer*. 65 (3) : 629–631.

31. Fronk, W.D., Beetle, A.A. and Fullerton, D.G. 1964. Dipterous galls on the *Artemisia tridentata* complex and insects associated with them. *Ann. Entomol. Soc. Amer*. 57 : 575–577.

32. Fulmak, L. 1968. Parasitinsekten der Insektengallen Europas. *Beitr. Entomol*. 18 (7/8) : 719–952.

33. Gourlay, E.S. 1955. Notes on exotic and New Zealand Insects. *N. Z. Entomol*. 1(5) : 3–5.

34. Hardy, D.E. 1973. The fruit flies (Tephritidae-Diptera) of Thailand and bordering Countries. *Pacific Insects Monogr*. 31: 353 pp.

35. Hardy, D.E. 1980. Family Tephritidae—The fruit flies. In: *Insects of Hawaii*, Ed. Hardy, D.E. and M.D. Delfinado, Vol. 13. Diptera: Cyclorrhapha III, pp. 28–102

36. Hendel, F. 1914. Die Bohrfliegen Sudamerikas. Abhandlungen und Berichte des Konigl. *Zool. Anthrop—Ethno. Mus. Dresden* 14(3) : 1–84.

37. Hendel, F. 1927. 49. Trypetidae. In: *Die Fliegen der palaearktischen Region*, Ed. E. Lindner, pp. 221. Stuttgart: Schweizerbart'sche Verlagsbuchhandlung.

38. Hering, E.M. 1938. Neue Palaearktische und exotische Bohrfliegen. 21. Beitrag zur Kenntnis der Trypetidae (Dipt.). *Deutsche Entomol. Z*. 1938(2) : 397–417.

39. Hering, E.M. 1943. Generationsverschiedenheiten bei *Myopites eximia* Seguy (Dipt.) (44. Beitrag zur Kenntnis der Trypetidae). *Deutsche Entomol. Z*. 1943(3–4): 127–128.

40. Hering, E.M. 1947. Neue Gattungen und Arten der Fruchtfliegen. *Siruna Seva* 6:1–16.

41. Hoy, J. M. 1960. Establishment of *Procecidochares utilis* Stone (Diptera: Trype-

tidae) on *Eupatorium adenophorum* Spreng, in New Zealand. *N. Z. J. Sci.* 3 (2): 200–208.

42. Huettel, M.D. and Bush, G.L. 1972, The genetics of host selection and its bearing on sympatric speciation in *Procecidochares* (Diptera; Tephritidae). *Entomol. Exp. Appl.* 15: 465–480.

43. Ito, S. 1947. *Paratephritis fukaii* et sua Galla (Trypetidae, Diptera). *Collecting and Breeding* 9(5): 97–98, 101 (In Japanese).

44. Kandybina, M.N., Kovalev, O.V. and Richter, V.A. 1967. *Paratephritis transitoria* Rohd. (Diptera, Tephritidae) forming galls on the tassel flower (*Cacalia*) in the Maritime Province of the U.S.S.R. *Rev. Entomol. URSS* 46(1):113–116(In Russian).

45. Keiffer, J.J. and Jorgensen, P. 1910. Gallen und Gallentiere aus Argentinien. *Centrabl. Bakteriol. Parasitenk. und Infektionskr.* (2) 27: 362–444.

46. Korytkowski, C.A. 1971. A new Cecidogenous species of the genus *Polymorphomyia* Snow (Diptera: Tephritidae). *Proc. Entomol Soc. Wash.* 73(4):446–449.

47. Kugler, J. and Freidberg, A. 1975. A list of the fruit flies (Diptera: Tephritidae) of Israel and nearby areas, their host plants and distribution. *Isr. J. Entomol.* 10:51–72.

48. Lalonde, R.G. and Shorthouse, J.D. 1982. Exit strategy of *Urophora cardui* (Diptera: Tephritidae) from its gall on Canada thistle. *Can. Entomol.* 114:873–878.

49. Lamb, K.P. 1960. A check list of New Zealand plant galls (Zoocecidia). *Trans. R. Soc. N. Z.* 88:121–139.

50. Mani, M.S. 1964. *Ecology of Plant Galls*. The Hague: Dr. W. Junk. 434 pp.

51. Mani, M.S. 1973. *Plant galls of India*. New Delhi: MacMillan. 354 pp.

52. Martelli, G. 1911. Descrizione e prime notizie di un nuovo zoocecide *Ceratitis savastani* (Mosca del cappero). *Mem. Cl. Sci. R. Accad. Zelanti Ser.* 3A. 7 : 1–8.

53. Miller, W.E. 1959. Natural history notes on the goldenrod ball gall fly, *Eurosta solidaginis* (Fitch), and on its parasites, *Eurytoma obtusiventris* Gahan and *E. gigantea* Walsh. *J. Tenn. Acad. Sci.* 34(4):246–251.

54. Milne, L.J. 1940. Autecology of the golden-rod gall fly. *Ecology* 21(1):101–105.

55. Munro, H.K. 1925. Biological notes on South African Trypaneidae (fruit-flies). I. *South Afr. Dep. Agric. Entomol. Mem.* 3:39–67.

56. Munro, H.K. 1926. Biological Notes on the South African Trypaneidae (Trypetidae: Fruit-flies). II. *South Afr. Dep. Agric. Entomol. Mem.* 5:17–40.

57. Munro, H.K. 1929a. Biological notes on the South African Trypetidae (Fruit-flies, Diptera). III. *South Afr. Dep. Agric. Entomol. Mem.* 6:9–17.

58. Munro, H.K. 1929b. New Trypetidae from South Africa (Dipt.). *Bull. Entomol. Res.* 20 (4):391–401.

59. Munro, H.K. 1934. A study of the genus *Actinoptera* Rond. (Diptera, Trypetidae) in the Ethiopian Region. *Trans. R. Entomol. Soc. Lond.* 82 (1):99–105.

60. Munro, H.K. 1935. Biological and systematic notes and records of South African Trypetidae (Fruit-flies, Diptera) with descriptions of new species. *South Afr. Dep. Agric. Entomol. Mem.* 9:18–59.

61. Munro, H.K. 1938. A revision of the African species of the genus *Spathulina* Rond. (Diptera, Trypetidae). *Trans. R. Entomol. Soc. Lond.* 87 (18): 417–430.

62. Munro, H.K. 1939a. Studies in African Trypetidae, with description of new species. *J. Entomol. Soc. South Afr.* 1:26–46.

63. Munro, H.K. 1939b. Some new species of South African Trypetidae (Diptera), including one from Madagascar. *J. Entomol. Soc. South Afr.* 2:139–153.

64. Munro, H.K. 1939c. On certain South African gall-forming Trypetidae (Diptera) with descriptions of new species. *J. Entomol. Soc. South Afr.* 2:154–164.

65. Munro, H.K. 1940. Further South African gall-forming Trypetidae (Diptera) with descriptions of new species. *J. Entomol. Soc. South Afr.* 3:76–87.

66. Munro, H.K. 1947. African Trypetidae (Diptera). *Mem. Entomol. Soc. South. Afr.* 1:1–284.

164

67. Munro, H.K. 1949. A new gall-forming trypetid. *J. Entomol. Soc. South. Afr.* 12: 130–133.

68. Munro, H.K. 1950. Trypetid flies (Diptera) associated with the Calendulae, plants of the family Compositae in South Africa. I. A bio-taxonomic study of the genus *Mesoclanis. J. Entomol. Soc. South. Afr.* 13:37–52.

69. Munro H.K. 1952a. Les Trypetides, Dipteres cecidogenes de la serie *Eutreta Oedaspis.* A propos de deux nouvelles especes Malagaches. *Mem. Inst. Sci. Madagascar, Ser.* E 1(1):217–225.

70. Munro, H.K. 1952b. A remarkable new gall-forming trypetid (Diptera) from Southern Africa, and its allies. *South Afr. Dep. Agric. Entomol. Mem.* 2(10): 329–341.

71. Munro, H.K. 1953a. Notes sur les Trypetides de Madagascar et description de nouvelles especes cecidogenes (Diptera). *Mem. Inst. Sci. Madagascar, Ser.* E 4:543–552.

72. Munro, H.K. 1953b. Records of some Trypetidae (Diptera) collected on the Bernard Carp Expedition to Barotseland, 1952, with a new species from Kenya. *J. Entomol. Soc. South. Afr.* 16(2):217–226.

73. Munro, H.K. 1955. The influence of two Italian entomologists on the study of African Diptera and comments on the geographical distribution of some African Trypetidae. *Boll. Lab. Zool. Gen. Agrar. 'Filippo Silvestri'* 33:410–426.

74. Munro, H.K. 1957. Trypetidae. British Museum (Natural History). *Ruwenzori Expedition* 1934–35 Vol. 2(9):853–1054.

75. Munro, H.K. 1964. The genus *Trupanea* in Africa. An analytical study in biotaxonomy. *South Afr. Dept. Agric. Tech. Serv. Entomol. Mem.* 8:1–101.

76. Novak, J.A. and Foote, B.A. 1975. Biology and immature stages of fruit flies: The genus *Stenopa* (Diptera: Tephritidae). *J. Kans. Entomol. Soc.* 48(1):42–52.

77. Novak, J.A., Stoltzfus, W.B., Allen, E.J. and Foote, B.A. 1967. New host records for North American fruit-flies. *Proc. Entomol. Soc. Wash.* 69(2):146–148.

78. Perkins, R.C.L. and Swezey, O.H. 1924. The introduction into Hawaii of insects that attack *Lantana. Bull. Hawaii Sugar Plant. Assoc. Exp. Stn.* 1–83.

79. Persson, P.I. 1963. Studies on the biology and larval morphology of some Trypetidae (Dipt.). *Opusc. Entomol.* 28:33–69.

80. Peschken, D.P. and Harris, P. 1975. Host specificity and biology of *Urophora cardui* (Diptera: Tephritidae). A biocontrol agent for Canada thistle (*Cirsium arvense*). *Can. Entomol.* 107:1101–1110.

81. Peschken, D.P., Finnamore, D.B. and Watson, A.K. 1982. Biocontrol of the weed Canada thistle (*Circium arvense*): Release and development of the gall fly *Urophora cardui* (Diptera: Tephritidae) in Canada. *Can. Entomol.* 114:349–357.

82. Pruthi, H.S. and Bhatia, H.L. 1940. A New Pest (*Acanthiophilus helianthi* Rossi, Trypetidae of safflower in India. *Indian J. Agric. Sci.* 10:110–115.

83. Ross, H. 1911. *Die Pflanzengallen (Cecidien) Mittel und Nordeuropas.* Gustav Fischer: Jena. 350 pp.

84. Schwitzgebel, R.B. and Wilbur, D.A. 1943. Diptera associated with ironweed, *Vernonia interior* Small in Kansas. *J. Kans. Entomol. Soc.* 16(1):3–14.

85. Séguy, E. 1934. Faune de France. 28. *Dipteres (Brachyceres),* Paris: Paul Lechevalier. 832 pp.

86. Shinji, O. 1939. On the Trypetidae of North-Eastern, Japan, with description of new species. *Insect World* 43:288–291, 320–324, 352–355 (In Japanese).

87. Shorthouse, J.D. 1980. Modification of the flower heads of *Sonchus arvensis* (family Compositae) by the gall former *Tephritis dilacerata* (order Diptera, family Tephritidae). *Can. J. Bot.* 58(14):1534–1540.

88. Shorthouse, J.D. and Watson, A.K. 1976. Plant galls and biological control of weeds. *Insect World Digest* 3(6):8–11.

89. Steck, G.J. 1981. *North American Terelliinae (Diptera: Tephritidae): Biochemical systematics and evolution of larval feeding niches and adult life histories*. Texas University at Austin. Unpublished Ph.D. dissertation.

90. Steyskal, G.C. 1974. A new species of *Procecidochares* (Diptera: Tephritidae) causing galls on stems of Hamakua Pamakani (*Ageratina riparia*: Asteraceae) in Hawaii. *US Dep. Agric. Cooperative Inst. Rep.* 24(32):639–641.

91. Steyskal, G.C. 1979. Taxonomic studies on fruit flies of the genus *Urophora* (Diptera: Tephritidae). *Entomol. Soc. Wash.* 61 pp.

92. Stoltzfus, W.B. 1977. The taxonomy and biology of *Eutreta* (Diptera: Tephritidae). *Iowa State J. Res.* 51(4):369–438.

93. Stoltzfus, W.B. and Foote, B.A. 1965. The use of froth masses in courtship of *Eutreta*. *Proc. Entomol. Soc. Wash.* 67(4):263–264.

94. Stuardo Ortiz, C. 1929. Observationes sobre las agallas blancas de *Baccharis rosmarinifolia* Hook. y el diptero que las produce. *Rev. Chil. Hist. Nat.* 33:345–350.

95. Swezey, O.H. 1923. Records of introduction of beneficial insects into the Hawaiian Islands. *Proc. Hawaii Entomol. Soc.* 5:299–304.

96. Tauber, M.J. and Tauber, C.A. 1967. Reproductive behaviour and biology of the gall-forming *Aciurina ferruginea* (Doane) (Diptera: Tephritidae). *Can. J. Zool.* 45:907–913.

97. Tavares, J. da S. 1902. As zoocecidias Portuguezas. *Brotèria-Rev. Sci. Nat.* 1:3–142.

98. Thompson, M.T. 1907. Three galls made by cyclorrhaphous flies. *Psyche* 14:71–74.

99. Uhler, L.D. 1951. Biology and ecology of the goldenrod gall fly, *Eurosta solidaginis* (Fitch). *Cornell Univ. Agric. Exp. Stn. Mem.* 300:1–51.

100. Varley, G.C. 1947. The natural control of population balance in the knapweed gall-fly (*Urophora jaceana*). *J. Anim. Ecol.* 16(2):139–187.

101. Wadsworth, J.T. 1914. Some observations on the life-history and bionomics of the knapweed gall-fly *Urophora solstitialis* Linn. *Ann. Appl. Biol.* 1(2):142–169.

102. Wangberg. J.K. 1978. Biology of gall-formers of the genus *Valentibulla* (Diptera: Tephritidae) on rabbitbush in Idaho. *J. Kans. Entomol. Soc.* 51(3):472–483.

103. Wangberg, J.K. 1980. Comparative biology of gall-formers in the genus *Procecidochares* (Diptera: Tephritidae) on rabbitbush in Idaho. *J. Kans. Entomol. Soc.* 53(2):401–420.

104. Wasbauer, M.S. 1972. An annotated host catalog of the fruit flies of America north of Mexico (Diptera: Tephritidae) Bureau of Entomology, Department of Agriculture, Sacramento, California, Occasional Papers No. 19. 172 pp.

105. Zwölfer, H. 1967a. *Urophora siruna-seva* (Hg) (Dipt. Trypetidae), a potential insect for the biological control of *Centaurea solstitialis* L. in California. *Commonw. Inst. Biol. Control. Eur. Stn. Weed Project for the Univ. Calif. Rep.* No. 5, 35 pp.

106. Zwölfer, H. 1967b. Observations on *Urophora cardui* L. (Trypetidae). *Commonw. Inst. Biol. Control. Eur. Stn. Weed Projects for Can. Prog. Rep.* No. 19, 11 pp.

107. Zwölfer, H. 1970. Investigations on the host-specificity of *Urophora affinis* Frfld. (Dipt., Trypetidae). *Commonw. Inst. Biol. Control. Eur. Stn. Weed Projects for Can. Prog. Rep.* No. 25, 28 pp.

108. Zwölfer, H. 1972. Investigations on *Urophora stylata* Fabr., a possible agent for the biological control of *Cirsium vulgare* in Canada. *Commonw. Inst. Biol. Control. Eur. Stn. Weed Projects for Can. Prog. Rep.* No. 29, 20 pp.

109. Zwölfer, H. 1974a. Das Treffpunkt—Prinzip als Kommunikationsstrategie und Isolationsmechanismus bei Bohrfliegen (Diptera: Trypetidae). *Entomol. Germ.* 1(1): 11–20.

110. Zwölfer, H. 1974b. Innerartliche Kommunikationssysteme bei Bohrfliegen. *Biologie in unserer Zeit*. 4(5):147–153.
111. Zwölfer, H., Englert, W. and Pattullo, W. 1970. Investigations on biology, population ecology and the distribution of *Urophora cardui* L. *Commonw. Inst. Biol. Control. Eur. Stn. Weed Projects for Can. Prog. Rep*. No. 27, 17 pp.

APPENDIX

This section deals with ten doubtful records of tephritid galls. Its main purpose is to encourage researchers to try and verify these doubtful records of cecidogeny. The list is arranged alphabetically by generic names. For each species the reference, the doubtful record and my opinion are given.

Acanthiophilus helianthi (Rossi)—Hendel (1927 : 204): Flower head galls. The extensive studies of Avidov and Kotter (1966) and Pruthi and Bhatia (1940) show that this is not a gall former.

Capparimyia savastani (Martelli)—Martelli (1911 : 3): Gall former in the flower buds of *Capparis spinosa*. During my own studies on this species I reared it from, and bred it on, buds of *Capparis*, but never found galls. The main diet of the larvae is the anthers, which they may consume entirely.

Chaetostomella vibrissata (Coquillett)—Shinji (1939 : 289): Stem gall on *Arctium lappa*. This plant is probably the real host of the fly, but gall making is practically unknown in the Terelliinae, to which this species belongs.

Dacus ciliatus Loew—(Mani, 1973 : 160): Inquiline on stem galls produced by *Bimba tombii* Grover (Diptera : Cecidomyiidae) on *Coccinia indica* (Curcurbitaceae). As far as is known all Dacinae including *D. ciliatus* are frugivorous.

Dacus cucurbitae Coquillett—Mani (1973 : 160): Inquiline on stem galls produced by *Bimba tombii* on *Coccinia indica*. Also recorded as primary gall maker on stems of *Cucumis sativus*. As far as is known all Dacinae including *D. cucurbitae* are frugivorous.

Euarestella iphionae (Efflatoun)—Efflatoun (1927 : 39): Galls on branches of *Iphiona mucronata*. During the course of my studies I failed to find galls on this plant; instead I found puparia in the flower heads, which I strongly suspect to belong to this species. In addition, the only closely related congener, *E. kugleri* Freidberg is a flower head infestor (Freidberg, 1974 : 56), while other congeners, which do from galls, are not really closely related (Freidberg, 1980 : 25).

Euphranta connexa (Loew)—Ross (1911 : 301; fly recorded as *Ortalis connexa*): In swollen fruits of *Cynanchum vincetoxicum*. This is indeed the known host, but the record is doubtful because there are no other records of galls produced by tephritids on fruits.

Hypenidium graecum Loew—Tavares (1902 : 37): Flower-head gall on *Lactuca viminea*. My own observations indicate that a gall may not be formed.

Tomoplagia obliqua (Say)—Aczél (1949 : 244): Gall maker on *Vernonia* spp. I tend to accept the more detailed studies of Benjamin (1934 : 33) and Schwitzgebel and Wilbur (1943), who reared this species from the seeds of *Vernonia* spp., where they did not produce galls. Three other species of *Tomoplagia* have been reared from galls, while other species of this genus have been reared from fleshy fruits, a most unusual situation for a tephritid genus.

Valentibula californica (Coquillett)—Novak *et al.* (1967 : 148): Recorded as a flower head infestor. In view of the records of this species as a stem gall former (Foote and Blanc, 1959 : 151; Wangberg, 1978 : 472), and because I consider that this genus has affinities with the Oedaspinae, I tend to think that the former record is a mistake.

7. Biology of Gall Midges

M. Skuhravá, V. Skuhravý, and J.W. Brewer

Introduction

Gall midges are small, fragile flies, usually unnoticed except by the specialist, but the large number of species, the wide diversity of host plants they attack, and their role in various ecosystems, make them much more important than their appearance might suggest.

The name, gall midge, is derived from the ability of the larvae to produce galls, or abnormal plant growths, on various organs of plants. The Latin name of the family, Cecidomyiidae, comes from the word "gall," in Latin "cecidium".

The body of adult gall midges generally varies from 0.5–3 mm in length, but occasionally species may be as long as 8 mm, or less than 0.5 (Fig. 1). They usually have long antennae. The wing veins are reduced in number with only three or four long veins normally present. Tibial spurs are absent. The larvae vary somewhat in their habits but most species are either phytophagous, producing galls on various plants, mycophagous (fungivorous), feeding on fungi, or zoophagous, feeding on invertebrates, especially insects.

As adults, gall midges live in various environments, in forests, fields, meadows, near streams and on stands along the riverside. Adults can be captured on windows of buildings, or they can be seen flying to lights at night. Gall midges have been collected at sea level, on mountains at elevations of 3,000 m a.s.l., in sandy deserts and in the region above the arctic circle.

Taxonomically, gall midges belong to the class Insecta, the order Diptera, the suborder Nematocera, to the super-family Mycetophiloidea and to the family Cecidomyiidae. At the present time the family Cecidomyiidae consists of three subfamilies: Lestremiinae, Porricondylinae and Cecidomyiinae.

Figure 1. Adult *Monarthropalpus flavus* (Schrank) after emergence from pustule
gall on the leaf of *Buxus sempervirens* L.

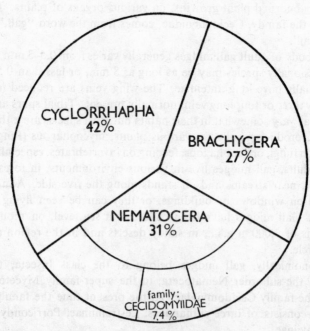

Figure 2. Frequency of the family Cecidomyiidae in Nearctic Region based on
data in Stone et al., 1965: A Catalogue of the Diptera of America
North of Mexico.

The family Cecidomyiidae is one of the largest families of Diptera, not only from the point of view of regional fauna, but also from the standpoint of continents (e.g. Europe and North America), zoogeographical regions and the entire world. Hendel (1937) ranked Cecidomyiidae with 3,000 species as one of the three largest families of Diptera in the world, viz. Asilidae (4,000 species) and Syrphidae (3,000). Also, Stone, et al. (1965) considering only Diptera of the Nearctic Region ranked Cecidomyiidae, with more than 1,200 species, among the three largest families, viz. Tipulidae (1,460 species) and Tachinidae (1,280 species) (Fig. 2).

History

The first gall midge species was described by the famous Swedish naturalist Linné under the name *Tipula juniperina* (now *Oligotrophus juniperinus*) in the tenth edition of "Systema Naturae" in 1758. During the first half of the 19th century many species of gall midges were described by various authors but generally the descriptions were inadequate by modern standards, and without biological data, for example those of Meigen, Schrank, Stephans, and Haliday. The first biological data were given in descriptions of *Cecidomyia poa* Bosc., and *Cecidomyia destructor* Say. During that time also appeared the papers of Vallot, Bouché, Schwägrichen, Dufour, Hartig and Kollar, dealing with biology (at that time called "history of life") of various gall midge species.

In 1840 Rondani published his paper on the classification of Diptera in which he divided gall midges into two groups, viz. Cecidomyiinae and Lestremiinae, and also established a number of new genera. Afterwards there appeared the extensive works of Bremi, H. Loew and Winnertz with descriptions of many new species and data on their biology and host plants. Good descriptions of life cycles and behaviour of gall midges were given in papers of F. Löw, Mik, and Wachtl.

Two European entomologists, Kieffer and Rübsaamen, conducted research on cecidomyiids for about 30 years, from about 1890 until the 1920's. Both of them published descriptions of many new species and genera. Kieffer's (1913) "Family Cecidomyiidae" in Genera Insectorum and Rübsaamen-Hedicke's (1926–1939) "Die Zoocecidien, durch Tiere erzeugte Pflanzengallen Deutschlands und ihre Bewohner" (Hedicke published this work after the death of Rübssamen, on the basis of Rübsaamen's material) established the groundwork for the study of the family Cecidomyiidae until the present time.

In America the study of gall midges began with the work of Osten Sacken, followed by that of Walsh, Beutenmuller, Felt and Pritchard.

In India the study of cecidogenous insects began in the 1920's with the papers of Ramachandra Rao. Mani, working since 1926, has had a great impact on gall research in India and throughout the world. He founded a cecidological school in India where many scientific workers study various problems of cecidology. In about 1930, the general direction of cecidological research turned from taxonomy and zoocecidology to biological studies of pests and their control. The most important and comprehensive work dealing with problems of economically important gall midges was written by Barnes (1946–1956) in seven volumes. Barnes, who followed the research works of Theobald and Edwards, founded a modern school of biological research of gall midges in England, where both crop pests and beneficial insects were studied.

Also important in the study of cecidogenous organisms was the work of Docters van Leeuwen whose "Gallenboek" is now in the second edition (1982) and that of Buhr (1964–1965) "Bestimmungstabellen der Gallen (Zoo- und Phytocecidien an Pflanzen Mittel- und Nordeuropas". These works have influenced cecidological research in various countries of Europe for a long time.

At the present time there are several centres of research of gall midges with emphasis in various areas of study including biotaxonomy and biosystematics, ecology and biology of gall midge pests, and those species which may be used in the biological control of pest insects. Some species of gall midges serve as model organisms in various areas of basic biological research such as ecology, physiology and genetics. It is clear that research on gall midges has made, and continues to make, great contributions to the study of many areas of biological science.

Morphology and variability of diagnostic characters

Adult gall midges have holoptic eyes with the eye bridge relatively broad, or rarely, the eyes may be divided into two or three separate eyes. Ocelli are present in most Lestremiinae, but absent in Porricondylinae and Cecidomyiinae. Antennae usually have 12–14 flagellomeres but may range from 6–40. The flagellomeres are covered with microtrichia and whorls of long setae or circumfila of various forms. Mouthparts consist of one pair of palps with one to four segments.

There is one pair of small, broad wings with reduced venation, covered with microtrichia, scales and macrotrichia. The costa vein (C) usually has a break just beyond the insertion of the R_5 vein. The R_5 vein is unbranched, the M vein is free, or it may be branched (in Lestremiinae), or absent (in Porricondylinae and Cecidomyiinae). The Cu vein is free or branched. Cross veins are present in Lestremiinae, Porricondylinae and some Cecidomyiinae. Some extreme forms of gall midges are brachypterous but much variation occurs. Individuals of the same species may be found with wings ranging from slightly smaller than normal to very small wings, to forms that are completely wingless (apterous). In some species there is sexual dimorphism in regard to wing form. The wings of males are normal but the females are brachypterous or even apterous.

The legs are usually long and slender with the first tarsomere (metatarsus) much shorter than the second, except in Lestremiinae. Tarsal claws may be toothed or untoothed. The empodium is usually well-developed and the pulvilli are normally very short.

The male terminalia (also called hypopygium), consisting of a pair of symmetrical clasping gonocoxites and gonostyles with a median aedeagus and various accessory structures, show differences at both the generic and specific levels and thus are very important in the taxonomy of the group.

The ovipositor is usually long and protrusible, ending with lamellae (cerci) which may be two- or three-segmented, or fused to form a single median lobe or a chitinized aciculate structure. One or two sclerotized spermathecae may be present in females of some primitive groups (Fig. 3).

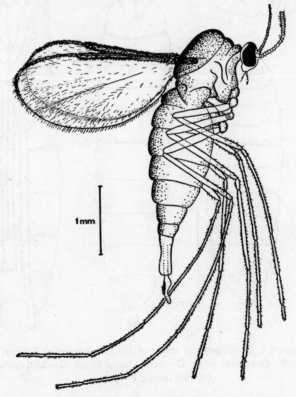

1 mm

Figure 3. Adult female of *Lasioptera arundinis* Schiner.

There may be three, four or even five larval instars. The first instar larvae have no spatula sternalis but do have a specialized respiratory system. The mature larva usually has an elongate-cylindrical body, it is two to five mm long and orange, red, yellow or white in colour. The head capsule and mouthparts are reduced. The prothorax usually has a characteristic median sclerotized organ on the ventral side, the spatula sternalis. In Lestremiinae, the anus is found terminally on the last abdominal segment but in the other subfamilies it usually occurs on the ventral side of the last segment.

The pupa is usually of the mumien-pupa form, on which may be found all the morphological structures of the future adult. These structures, especially the antennae, may be used for taxonomic purposes, i.e., for the identification of species.

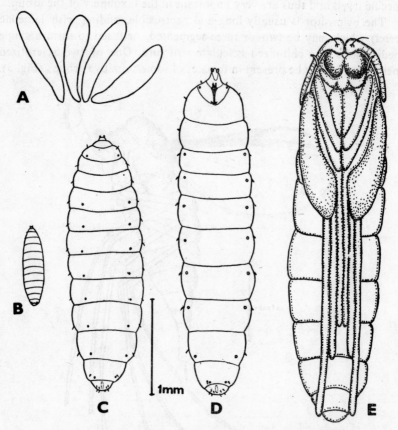

Figure 4. Developmental stages of *Lasioptera arundinis* Schiner:
A—eggs; B—first larval stage; C—second larval stage; D—third larval stage (mature larva); F—pupa.

The eggs of gall midges are usually very small, elongate in shape and of various colours including white, red, yellow and orange. There have been some attempts to use egg shape and morphology in the taxonomy of various species but these have not been very successful.

The major diagnostic characters of gall midges used in the determination of various taxonomic level may be found on larval, pupal or adult stages.

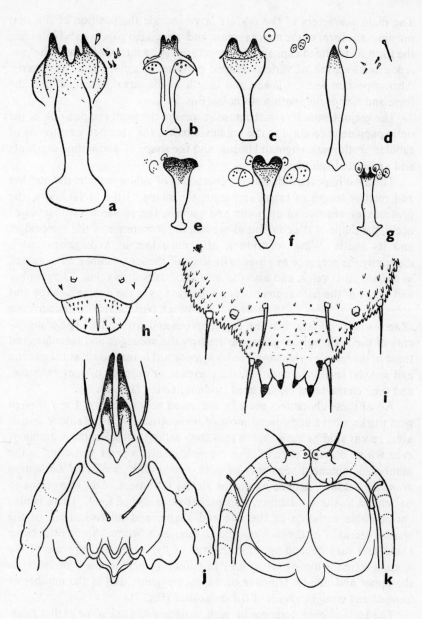

Figure 5. Morphological characters of larvae and pupae of various species of gall midges, spatula sternalis: a—*Asphondylia miki* Wachtl; b—*Lasioptera arundinis* Schiner; c—*Dasineura kellneri* (Henschel); d—*Kiefferia pericarpiicola* (Bremi); e—*Haplodiplosis marginata* (von Roser); f—*Clinodiplosis cilicrus* (Kieffer); g—*Giraudiella inclusa* (Frauenfeld); anal segments of larvae: h—*Lasioptera arundinis* (Schiner); i—*Clinodiplosis cilicrus* (Kieffer); head structures of pupae: j—*Giraudiella inclusa* (Frauenfeld); k—*Lasioptera arundinis* Schiner.

The main characters of the mature larva include the position of the anal opening, the form of the anal segment and associated papillae and spinulae, the form of the head capsule, the structure of the integument and the presence and location of various shaped papillae, spinulae, setae and warts. Also important are the shape and length of the larval antennae, and the form and size of the spatula sternalis (Fig. 5).

The major characters of the pupal stage are the form and position of the optic papillae in relation to the antennae bases, the presence or absence of tubular prothoracic stigmatal horns, and the shape of abdominal segments and associated spinulae.

The most important diagnostic characters of adults include: the number and relative length of tarsal segments, the shape of the tarsal claws, the presence or absence of empodia and pulvilli, the presence or absence of ocelli, the shape of the compound eye and the occurrence of the eye-bridge, and its width. Wing venation is also important as a diagnostic tool, especially the presence or absence of some of the longitudinal veins as well as certain cross veins, and also the bifurcation of some veins. The number and shape of the palp segments are important as is the number, shape and accessory structures of the antennal segments, particularly the occurrence of sensorial processes, bristles, setae, spines and circumfila. The components of the male hypopygium, particularly the aedaegus and lamellae, and those of the female ovipositor, are very important in taxonomy at the generic and specific levels. In addition, the presence or absence of spermathecae, and their characteristics, are used in identification (Fig. 6).

All of these characters must be evaluated very carefully. Even though past workers have considered most of these structures as relatively invariable, recent studies have shown that they exhibit considerable variability, even within a single species. For example, there is great variation in the number of antennal segments of both males and females of *Lasioptera arundinis* Schiner, depending on the size of the insect. And size seems to be related to the availability, and perhaps quality, of food. There is also considerable variation in the size and shape, and sclerotization, of the spatula sternalis of *L. arundinis* and *L. hungarica* Möhn. Apparently these characters vary depending upon the age of the larvae and the use of the spatula sternalis during larval life. In addition, there is much variation in the upper and middle lamellae of the hypopygium, and in the number of hooked and straight spines of the ovipositor (Fig. 7).

The lack of consideration of such variation has led some earlier taxonomists to describe several species where only one exists. More recently, studies on the extent of variation of these characters have begun to resolve such problems. For example, Skuhravá (1973) studied the variability of a number of taxonomic characters of the genus *Clinodiplosis* Kieffer described from various host plants and demonstrated that 34 species of this genus were actually only one with a broad host range and a high level of variation

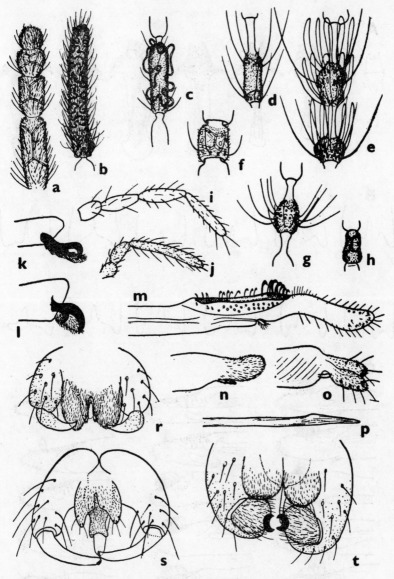

Figure 6. Morphological characters of adults gall midges: flagellomers of
antennae: a—*Asphondylia miki* Wachtl, male; b—female; c—*Kiefferia pericarpiicola* (Bremi), male; d—*Clinodiplosis cilicrus* (Kieffer), female; e—male;
f—*Lasioptera arundinis* Schiner, male; g—*Dasineura kellneri* (Henschel),
male; h—female; palpi. i—*Clinodiplosis cilicrus* (Kieffer); j—*Asphondylia
miki* Wachtl; tarsal claws: k—*Asphondylia miki* Wachtl; l—*Lasioptera
arundinis* Schiner; ovipositors: m—*Lasioptera arundinis* Schiner, n—*Dasineura kellneri* (Henschel), o—*Clinodiplosis cilicrus* (Kieffer), p—*Asphondylia miki* (Wachtl); male terminalia: r—*Dasineura kellneri* (Henschel);
s—*Clinodiplosis cilicrus* (Kieffer); t—*Asphondylia miki* Wachtl.

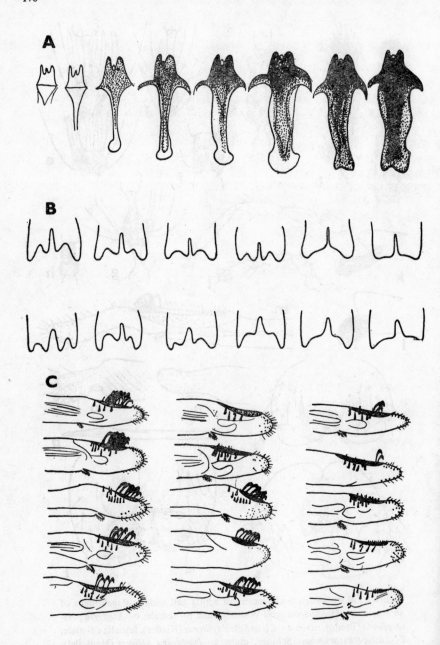

Figure 7. Variability of some morphological characters of gall midges: A—variability in shape, size and degree of sclerotization of spatula sternalis of the larvae of *Lasioptera arundinis* Schiner; B—variability on shape of middle lamella of male terminalis of *Clinodiplosis cilicrus* (Kieffer); C—variability in shape and number of straight and hooked spinae on end of ovipositor of *Lasioptera hungarica* Möhn.

in taxonomic characters. It is probable that additional studies of this type will reduce the number of species in other groups as well.

Life cycles

Adults may be collected in the field from the beginning of March until the end of September but the majority appear during June and July. Individual adults have also been collected during the winter months but this is very rare.

The life cycle of most gall midges begins early in the spring with the emergence of adults from the pupal case, followed by flight and mating. A short time after mating the females locate their host plant, host animal, or other oviposition site and lay their eggs. The life span of an individual adult is very short, a few hours, or at the most a five days. The males perish shortly after mating, the females die soon after egg laying, with the main role of their life complete, the production of the next generation. In contrast to the very short life span of adults, the larvae live for quite a long time, several months, or rarely, several years.

Gall midges may have one, two or several generations per year, depending upon the species, with the number being a result of genetic factors. The species that are polyvoltine, having more than one generation per year, may vary in the number of generations that occur, from one year to another depending upon abiotic factors.

Gall midges usually overwinter in the larval stage, either in the soil or at the site of development, the latter usually being galls on various plants.

Usually, the life cycle of a gall midge lasts only one year. In some species, however, it can last two years (e.g. *Taxomyia taxi* Inchbald), or very rarely, even three years (e.g. *Plemeliella abietina* Seitner). Also, larvae of some species, which usually have one or two generations a year and normally hibernate in the soil can remain in the larval stage for extended periods. For example, some larvae of the wheat gall midges *Contarinia tritici* (Kirby) and *Sitodiplosis mosellana* (Géhin), which normally have one generation per year, have been reported to remain in the soil for up to twelve years.

Larvae of gall midges that live in the soil, either in the upper layers (the litter) or from beneath various types of plants, may be divided into two types. The first type requires the soil as part of their habitat. The second type only uses the soil as an overwintering site, and it is not a vital part of their life cycle. Thus the relationship of gall midge larvae to the soil may be expressed by the terms "soil inhabiting" which means that the larvae actually go through a part of the developmental process in the soil, and "soil overwintering", where the larvae use the soil only for a hibernation site. The basic types of life cycles that occur in the gall midge group are shown in Fig. 8.

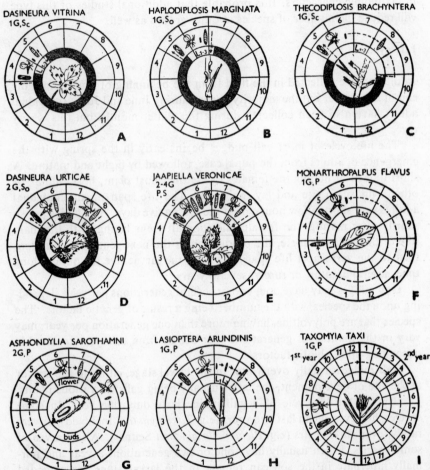

Figure 8. Types of life cycles of phytophagous gall midges with one (A, B, C, F, H), two (D, G), or several generations a year (E) and with two years per generation (I); black part of inner cycle means overwintering of larval stage in the soil, white part —development in galls on host plants above soil; G—number of generations a year; S_o—pupation or overwintering in the soil without cocoon; S_c—pupation or overwintering in the soil in cocoon; P—pupation takes place in the gall on the plant; P, S—pupation takes place in the gall on the plant in summer, overwintering in the soil; L_{1-3}—larval stages.

Biological groups

On the basis of larval feeding habits, gall midges may be divided into three major biological groups. These are:

1) Mycophagous (=fungivorous): those that feed upon fungi, including xylophagous species;

2) Phytophagous: those that feed upon plants, and
3) Zoophagous: those that feed upon other animals, especially insects (Fig. 9).

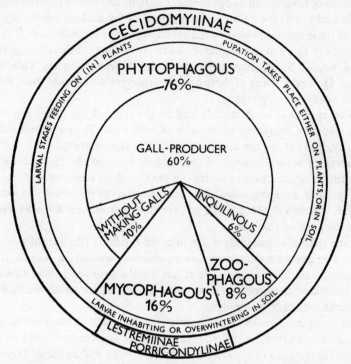

Figure 9. Frequency of the gall midges (Family Cecidomyiidae) of Palaearctic Region in three biological groups.

Independent of this biological grouping is the so-called free-living species, a category used often, but incorrectly, by various authors as a biological group. These species include a great number of gall midges known only from the adult stage whose biology is completely unknown and so they cannot yet be placed in a true biological group. The majority of these species are known only from a single collection of one specimen caught at the type locality and not reported again since that time. From a total of about 2,200 species of Cecidomyiids which have been described from the Palaearctic Region, the biology of only 75 per cent is well known. The habits of the remainder are completely unknown. In this group must be also included the so-called xylophilous gall midges, the adults of which seem to like wood and woody material, and which were caught flying near such material. The biology of such species is also completely unknown.

1) MYCOPHAGOUS GALL MIDGES
The mycophagous gall midges include those whose larvae are associated

in various ways with fungi (*Mycophyta*). This group includes about 240 species, or 16 per cent of those in the Palaearctic Region, and they are associated with various fungal groups.

Larvae of several gall midge species live in or on various types of mildew, moulds and rusts (the parasitic fungi), feeding by sucking on the mycelia. Most of these species belong to the genus *Mycodiplosis* Rübsaamen. About 20 species of this genus were described from various species of fungi including *Erysiphe*; *Sphaerotheca*, *Puccinia*, and *Uromyces* in the Palaearctic Region. However, many of these may be synonymous because these midges are not always monophagous.

Other species of cecidomyiids feed on the saprophytic fungi. For example, larvae of *Campylomyza mucoris* (Kieffer) and *Peromyia minutissima* Mamaev, and others, are associated with this fungal group.

Larvae of several species of gall midges feed on the mycelium of the larger fungi, which develop under the bark of dead trees, or stumps, or in other types of decaying wood. For example, larvae of *Winnertzia maxima* Mamaev develop in the mycelia of *Auricula mesenterica* Retz. ex. Hooki, under the bark of *Quercus*.

Some species of gall midges live in or on the body (thallus) of the larger fungi. For example, larvae of *Camptodiplosis auriculariae* Barnes live on *Hirneola* (=*Auricularia*) *auricula* L. on *Sambucus nigra* L. and *Ulmus* sp. Also cultivated mushrooms (*Agaricus* (=*Psalliota*) *hortensis*) are damaged by several species of gall midges.

Larvae of two species of midges even produce galls on the body of larger fungi. *Brachyneurina peniophorae* Harris causes small conspicuous irregular galls on the fungus *Peniophora cinerea* (Fr.) Cooke and larvae of *Mycocecis ovalis* Edwards produces small circular, blister-like swellings on the surface of the fungus *Hypochnus fuscus*.

The group of mycophagous gall midges also includes the phytosaprophagous midges, whose larvae live in decaying plant tissues, such as fallen leaves, decaying clover heads or mummified fruits. For example, larvae of *Clinodiplosis cilicrus* (Kieffer) and various species of the genus *Mycodiplosis* live in old, decaying blossoms of various plants of the family *Asteraceae* (=*Compositae*), and larvae of *Winnertzia conorum* Kieffer develop in fallen cones of various coniferous trees.

Sometimes it is difficult to determine whether the midge larva feeds on the fungus itself, or on the decaying plant tissues infected by the fungus.

In the group of mycophagous gall midges should also be included the so-called xylophagous gall midges, whose larvae inhabit ligneous tissues in various stages of decay since such decay is partly produced by a number of fungal species.

2) PHYTOPHAGOUS GALL MIDGES

The phytophagous gall midges include those whose larvae feed on plant tissues. These larvae live in, or on, various organs of many host plant species where they feed by sucking sap from the tissues. They may feed without causing galls but most induce development of some type of abnormal plant growth, or gall. The phytophagous gall midges also include the inquilinous species.

The majority of the cecidomyiids found in the Palaearctic Region are phytophagous. This includes about 1,200 species or some 76 per cent of those described. Most of these, about 60 per cent, produce galls, but some 10 per cent live on, or in, plants without doing so. A few, about 6 per cent, live in galls of other cecidomyiids, or other insects, as inquilines.

All gall-producing cecidomyiids belong to the subfamily *Cecidomyiinae*. There are no representatives of the gall-making group in the other two subfamilies, the *Porricondylinae* or the *Lestremiinae*.

a) *Non-gall Forming Phytophagous Gall Midges*

Larvae of these gall midges develop on, or in, various organs of the host plant but do not evoke gall development. They live in the same types of situations as do the gall-forming species, i.e. under the leaf sheaths of grasses, under the bark of living, injured or dying trees, in resin masses, in inflorescences, in or on seeds, and cones. This group includes only a small portion of the total number of gall midges known from the Palaearctic Region, about 150 species or 10 per cent.

Larvae of several species of the genus *Mayetiola* Kieffer live on stems of various grasses. It is interesting that this genus includes species whose larvae inhabit grasses without causing galls (*Mayetiola schoberi* Barnes), species which evoke small swellings at the stem base (*M. festucae* Ertel) and species which produce very well-formed galls (*M. poae* (Bosc)). Also, larvae of *Antichiridium caricis* (Rübs.), *Brachydiplosis caricum* Rübs., and *Coquillettomyia caricis* (Möhn) develop under the leaf sheaths of various species of *Carex* without causing galls. Larvae of some species of the genus *Dasineura* Rondani (formerly included in the genus *Rabdophaga*) develop in the tissues of various species of *Salix* but no abnormal growths are produced. These insects may live under the bark of trunks or roots of willow, or sometimes near the bud (*D. justini* Barnes and *D. medullaris* (Kieffer)). Other species in the same genus induce well-formed galls.

Larvae of some species of the genus *Resseliella* Seitner develop under the bark of injured trees but do not produce galls. For example, *R. oculiperda* (Rübs.) develops at sites of root grafts on roses, apples and other fruit trees, and thus may interfere with graft development but does not induce abnormal growth. Other species that develop under the bark without causing apparent galls include *R. ribis* (Marik.) on *Ribes nigrum* L. and *R. crataegi* (Barnes) on *Crataegus oxycantha* L.

Several species in the genera *Cecidomyia* Meigen and *Resseliella* Seitner develop in coniferous resin. For example. larvae of *C. magna* (Möhn) and

C. mesasiatica (Mamaev) develop in the resin on trunks of spruce, *Picea excelsa* Link.

Some gall midges cause economically important damage to seeds and flowers even though they do not cause galls to develop. Probably the most serious of these pests are those that attack cereals, especially *Contarinia tritici* (Kirby) and *Sitodiplosis mosellana* (Géhin) which develop in the seed. The seeds of various other grasses are similarly damaged by other species of the genera *Contarinia* Rondani and *Dasineura* Rondani. Flower heads, and thus the seed, of various species of the family *Asteraceae* are damaged by larvae of several species of gall midge including: *Jaapiella compositarum* (Kieffer), *Contarinia hypochoeridis* (Rübs.), *Macrolabis achilleae* Rübs., *M. cirsii* (Rübs.), *Contarinia tanaceti* (Rübs.), *C. chrysanthemi* (Kieffer) and *C. dipsacearum* Rübs.

b) *Gall Makers*

Larve of many species in the subfamily *Cecidomyiinae* possess the ability to evoke a specific growth response in host plant tissue leading to the formation of an abnormal structure, or gall. The gall midges, process of gall formation and types of galls on various host plant structures are treated in a separate section (see page 187).

c) *Inquilinous Gall Midges*

The inquilinous gall midges include those whose larvae are phytophagous but neither evoke galls or live free in plant tissue. Instead, they live in galls produced by other insects, mostly gall midges but also those produced by wasps (*Cynipidae*), flies (other than *Cecidomyiidae*), or psyllids (*Psyllidae*). These gall midges have sometimes been called symbionts or commensals. These terms, however, do not accurately describe the relationship between such species and their hosts.

Females of inquilinous species lay their eggs on, or in, galls produced by other species. Larvae of the inquilinous species feed by sucking sap from the tissue of the gall, like those of the gall producer. Inquilinous larvae do not injure the larvae of the gall maker but they may weaken them indirectly by reducing the amount of food available in the gall. As far as one knows, inquilinous larvae do not benefit the host larvae in any way.

Although inquilinous species do not cause galls, some may influence the size, or shape of the gall of their hosts. For example, galls of *Kiefferia pericarpiicola* (Bremi) are different in both size and shape when the inquiline *Trotteria umbelliferarum* Kieffer is also present.

Inquilinous gall midges are not very common. Only about 90 species (6 per cent of the total) are known from the Palaearctic Region. Inquilinous species occur in various genera, usually with most of the genus being typical gall formers and only a few inquilines. However, this may be a result of the general lack of knowledge of biology of many gall midge species, as noted earlier. It is possible that the number of inquilinous species may increase as one learns more about the biology of gall midges. At the present time there is tremendous variation in the level of occurrence of inquilinous species in genera in the Palaearctic Region, and no doubt, other regions as well (Table 1).

Table 1. Number of species, and number and per cent of inquilinous species of gall midges in the Palaearctic Region.

Genus	Number of species	Number of inquilinous species	Per cent of total
Dasineura	250*	15	6
Stefaniola	67	4	6
Arnoldiola	8	3	8
Lasioptera	45	6	13
Jaapiella	31	4	13
Ametrodiplosis	12	3	25
Macrolabis	31	9	30
Tricholaba	4	3	75
Schueziella	1	1	100
Amerhapha	1	1	100
Prolauthia	1	1	100
Chelobremia	1	1	100

Several species of gall midges may live in one gall. In such cases it is difficult to determine which larvae are the gall formers and which are inquilinous. For example, in galls from the swollen unopened flowers of *Sambucus nigra* L. there are often larvae of three species of gall midges: orange larvae of *Placochela nigripes* (F.Lw.), yellow larvae of *Contarinia lonicerearum* (F.Lw.) and white larvae of *Arnoldiola sambuci* (Kieffer). The biological relationship of these three species is not presently understood.

Inquilinous gall midges are sometimes known only from one collection, which may not only cause taxonomic difficulties, as already noted, but may also make the ecological position of the insect uncertain. Sometimes these insects are collected in emergence cages along with adults of the gall producers. Usually, in such cases, there are many adults of one species, generally the gall maker, and a single adult of the inquilinous species. In such cases it is not difficult to determine the gall maker. However, at times the problem is more difficult because numerous specimens of both species emerge. For example, in the swollen leaflet galls of *Rosa canina* L. there are numerous gregarious larvae of *Wachtliella rosarum* (Hardy) which causes the gall, but also numerous larvae of *Macrolabis luceti* Kieffer, which is inquilinous.

The majority (about 90 per cent) of inquilinous gall midges live in galls of cecidomyiids. The remainder live in galls of cynipids, other flies, and psyllids.

3) ZOOPHAGOUS GALL MIDGES

Zoophagous gall midges include those species whose larvae actively attack their victim (i.e., which act as predators) or those that develop inside the body of other animals (i.e., internal parasitoids), in either case resulting in death of the host animal.

Zoophagous larvae generally live upon larvae of other gall midges, either free or in their galls. They feed by sucking body fluids from their prey. These larvae may also attack aphids, free-living or gall-forming mites, coccids, psyllids, thrips and various other invertebrates. Obviously it may be difficult to determine whether adult midges emerging from galls of other insects are zoophagous or inquilinous species.

In the Palaearctic Region there are about 130 species of gall midges in various genera that feed zoophagously (i.e. about 8 per cent of the total). All of these belong to the subfamily Cecidomyiinae. The most important and best known of the zoophagous species are those that prey upon aphids. Aphidophagous larvae develop very quickly. For example, *Aphidoletes aphidimyza* (Rondani) completes development in about seven days, during which time one larva will consume 60–80 aphids. Thus these gall midges may be important in biological control of aphids. However, only a few species of such midges are known in the Palaearctic Region and generally distribution seems to be limited. For example only one of four species of the genus *Aphidoletes* Kieffer, *A. aphidimyza* (Rondani), is widespread in both the Palaearctic and Nearctic Regions.

About 100 species of zoophagous gall midges have been reported to prey on other cecidomyiid species in the Palaearctic Region. Most of these, reared from galls of other gall midges, belong to the genus *Lestodiplosis* Kieffer. However, many species will probably be found to be synonymous since recent work in other groups has demonstrated that gall midges are not as monophagous as was once thought. It is possible that the present 90 species in this genus will be reduced to three or four, as in the case of the genus *Aphidoletes* Kieffer.

Little is known about the acariphagous gall midges, whose larvae attack mites, either free-living or gall-causing forms. However, these species do attack and kill mites much like the predatory forms discussed above. Observations indicate that one larva of *Therodiplosis persicae* Kieffer could destroy about 30 *Tetranychus* mites per day. Thus this gall midge group might be important as biological control agents of these pests. About 30 species of acariphagous gall midges are known in the Palaearctic Region, mostly in the genus *Arthrocnodax* Rübs. However, as in other groups, the actual number of true species will probably be much lower after a modern revision is completed.

Larvae of many gall midge species are coccidophagous, i.e. they prey on various species of the family *Coccidae*. Harris (1967) summarized the knowledge of coccid-eating gall midges in the world. These midges are widespread mainly in the Afrotropic and Oriental Regions. Only about 10 species are known to occur in the peripheral areas of the Palaearctic. The distribution of this biological group is probably related to that of the *Coccidae*, which are more common in those two regions.

Only two species of cecidomyiids are known to feed on psyllids. Adults of *Lestodiplosis liviae* (Rübs.) were reared from galls of *Livia juncorum* Latr. The larvae probably feed as predators on the psyllids which cause these galls. Larvae of the second species, *Endopsylla agilis* Meijere, are endoparasites which develop inside the abdomen of *Psylla foersteri* Fl. on *Alnus*.

Two gall midge species are probably thrips feeders. Adults of *Adelgimyza tripidiperda* Del Guercio and *Diplosis fleothripetiperda* Del Guercio were found on an olive tree, *Olea europea* L. infested by *Liriothrips oleae* Costa. The biology of these species remains unknown and even the generic position of the midge is uncertain.

Zoophagous gall midges also attack invertebrates other than insects. For example, larvae of *Lestodiplosis vasta* (Möhn) prey on the milliped *Polyxenus lagurus* (L). (*Diplopoda*).

Only a few gall midges are known to be endoparasitic on other insects, and only four such species are known from the Palaearctic Region. Larvae of *Endaphis perfidus* Kieffer develop inside the abdomen of the aphid *Drepanosiphum platanoides* Schrank which feeds on *Acer pseudoplatanus* L. Larvae of *Occuloxenium compitale* Mamaev were described as endoparasites of *Aphis pomi* L., and *Endopsylla agilis* Meijere develops inside the abdomen of *Psylla foersteri* Fl., as noted above. Larvae of *Endopsylla endogena* (Kieffer) develop inside the body of *Stephanitis pyri* Fabr. (*Tingidae*). Unfortunately, almost nothing is known about the biology of these species.

Galls and gall-producing gall midges

1) PROCESS OF GALL FORMATION

The structure of a gall depends mainly upon the species of the gall-inducing organism, the species of the plant attacked, and the organs on which the galls are produced. Thus the galls caused by a specific gall former are generally quite distinct in regard to size, shape and morphology and are normally quite easy to distinguish from those caused by other species. Two species of gall formers may even attack the same structure of the same host plant but cause distinctly different galls. For example, the gall midges *Janetiella coloradensis* Felt and *Pinyonia edulicola* Gagné both attack needles of *Pinus edulis* Engel. Yet the two galls differ greatly morphologically and are easily distinguished. The production of such different appearing galls on the same plant structure is probably related in part to the developmental stage of the tissue at the time of attack, and of course, to the fact that two different species of insects are involved.

The importance of the stage of development of host tissue has been clearly demonstrated on *Thecodiplosis brachyntera* (Schwägr.). With this species it is possible to artificially vary the stage of development of the host plant *Pinus sylvestris* L. at the time of attack. The results showed that early infestations, at the beginning of May, resulted in the development of nor-

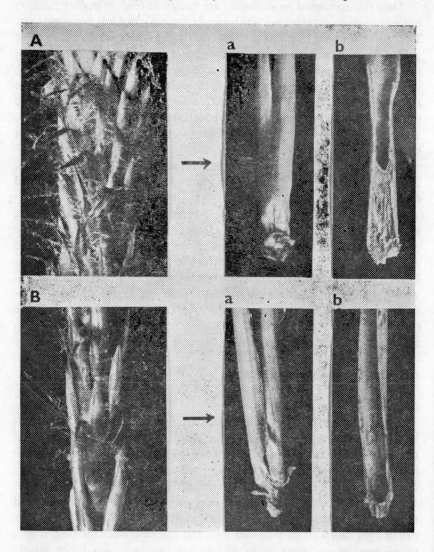

Figure 10. Two types of gall formation of *Thecodiplosis brachyntera* (Schwägr.) on needles of *Pinus sylvestris* L. influenced by the developmental stage of plant tissue at the time of attack: A—the shoot at the time of early infestation (on 15.5.); a— the basal part of a pair of needles grown together for about 6 mm; b—larval chamber is situated several mm from the base of needles (section); B—the shoot infested 6 days later (21.5.); a—the basal part of the needles is coalesced for 3 mm; b—the larval chamber is situated close to the base of the needles.

mal galls, i.e., the needles were coalesced for about 7 mm and greatly shortened, being less than half the normal length. Infestations that occurred two weeks later in plant development resulted in much different galls. There was little coalescence and the needle length was nearly normal (Figs. 10 and 11).

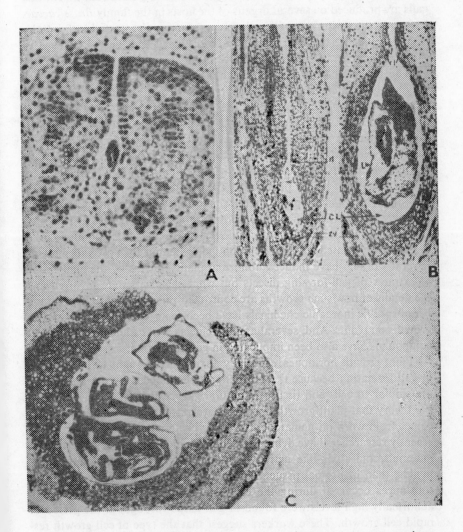

Figure 11. Anatomy of gall of *Thecodiplosis brachyntera* (Schwägr.): A—cross-section of the base of needle of *Pinus mugo* Turra ssp. *mughus* (Scop.) Domin.; B—longitudinal section of the needle attacked by larva of *Thecodiplosis brachyntera*; L—larva, tn—nutritive tissue, Cl—larval chamber, ZV—zone of vascular system, RL—needle dividing zone; left—3 weeks, right—8 weeks; C—cross-section of needle with central larval chamber developed including three young larvae of *Thecodiplosis brachyntera*.

The organ of the host plant that is attacked is also important in determining gall morphology and the same insect species may cause very different galls when it attacks different organs of the same plant. For example, the gall midge *Contarinia nasturtii* (Kieffer) has several generations a year and galls are produced on several organs of the hosts in the family *Brassicaceae*. Larvae of the spring generation develop on young shoots and leaf petioles where they cause these structures to be distorted and twisted. Larvae of the summer generation attack the flowers which are galled and never open.

The process of gall formation, or the level of occurrence, sometimes may be also influenced by other factors. For example, *Lasioptera arundinis* Schiner cannot produce galls on the main shoot of its host *Phragmites communis* Trin. Rather, females lay their eggs only on lateral shoots, which do not occur on undamaged plants. However, large numbers of lateral shoots develop after destruction of the main shoot by other invertebrates, and *L. arundinis* uses such shoots for its development (Fig. 12).

Insect galls of higher plants are generally thought to be caused by the introduction of some type of chemical substances by the causative insect. However, authorities differ in their opinion as to whether each species of gall maker releases a different cecidogen (gall-inducing compound) or if there is one related group of compounds common to most gall makers. Several workers have induced abnormal plant growths by the application of extracts of gall-forming insects but what chemicals were involved and the details of how such growths are controlled remain unresolved. A number of specific inorganic chemicals have been reported to produce gall-like growths on plants. And, several amino acids as well as adenine-containing compounds have also been implicated in gall formation.

Plant growth hormones, especially auxins, are thought to be involved in gall formation because the normal growths induced by these materials are similar to cells and tissues in natural galls. Much of the evidence for the theory that plant growth hormones are responsible for gall formation comes from work on nematode, fungal or bacterial galls. However, more recently researchers have demonstrated much higher levels of gibberlin-like materials, and auxin-like activity in tissue from galls caused by cecidomyiids, than in normal tissue. The highest levels of these plant growth substances occurred during the early stages of gall formation although abnormally high levels were found in gall tissue throughout the period of rapid cell growth. These workers suggest that the type of cell growth responsible for the galls studied, i.e. hyperplasia and hypertrophy, support the idea that plant growth-controlling substances are important in gall formation. Thus it seems likely that plant growth chemicals are responsible for the production of insect galls. However, it is not known whether the gall inducing organism produces these materials, or somehow induces the plant to produce them itself. In any case it is clear that through some mechanism gall-forming insects are able to direct plant growth very precisely.

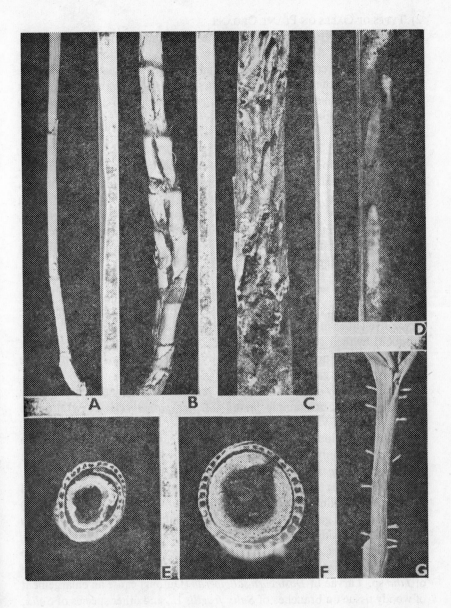

Figure 12. Development of gall midge *Lasioptera arundinis* Schiner on reed, *Phragmites communis* Trin.: A—uninfested side shoot; B—infested shortened side shoot; C—larval chambers in black mass of fungus of the genus *Sporothrix*; D—larvae situated in fungus mass; E—cross-section of uninfested side shoot; F—cross-section of infested side shoot with central situated fungus mass; G—exuviae of pupae remain on host plant after emergence of adults.

2) Types of Galls on Plant Organs

Galls may be produced on all parts of the host plant, viz. on roots, stems, buds (both flower and leaf), leaves, flowers and inflorescences, fruits and seeds. However, an individual gall midge species is usually quite specific as to the part of the plant it attacks. In some cases, though, a species that has two generations per year may produce galls on different organs of the same host plant species.

a) Root Galls

Gall midges do cause galls on roots of plants but they are not common. The galls are usually created on swollen underground buds. For example, larvae of *Ametrodiplosis auripes* (F. Löw) produce galls on roots of *Galium mollugo* L., *Dasineura galeobdolontis* (Winnertz) on *Galeobdolon luteum* L., and *Geocrypta braueri* (Handlirsch) on *Hypericum perforatum* L. Such galls may be more common than usually thought but the difficulty of observing them may make them appear rare.

b) Stem Galls

Gall midge larvae produce various types of galls on plant stems. The larvae live either in the tissues of the stem directly, or in a hollow cavity produced by their gall-making activities. A stem gall can consist of one or many cavities (or none) and in one cavity may live one, two or many larvae of one or several gall midge species (Fig. 13).

Stem galls may occur on one or two year herbacious plants, or on woody shrubs or trees. For example, larvae of *Geocrypta galii* (H. Loew) produce galls on stems of various species of *Galium, Acodiplosis inulae* (H. Loew) on stems of various species of *Inula, Dasineura sisymbrii* (Schrank) on species of the family *Brassicaceae,* and *Lasioptera eryngii* (Vallot) on stems of *Eryngium campestre* L. There are, of course, many other examples.

Galls on stems may assume a variety of shapes but are generally round or elongate swellings. For example, larvae of *Lasioptera rubi* (Schrank) produce rounded galls on shoots of *Rubus* sp. and those of *Dasineura salicis* (Schrank) cause similar rounded, or fusiform, swellings on twigs of various species of *Salix. Contarinia tiliarum* (Kieffer) evoke swellings in the shape of rounded tumours on young shoots and stalks, leaves and inflorescences of *Tilia* sp. Larvae of *Dasineura saliciperda* (Dufour) cause an enlargement of woody tissue on branches of *Salix fragilis* L., and other species of *Salix,* which result in the bark falling away from the infested area.

More unusual galls are caused by larvae of *Mayetiola poae* (Bosc.) which produce stem swellings on *Poa nemoralis* L. followed by the development of adventitious roots in the form of white frills. Also, larvae of the genus *Planetella* Westwood produce various corn-shaped galls on the stems of *Carex* spp. Galls of *Giraudiella inclusa* (Fraunfeld) are interesting because they develop inside the stem of *Phragmites communis* Trin. The lateral

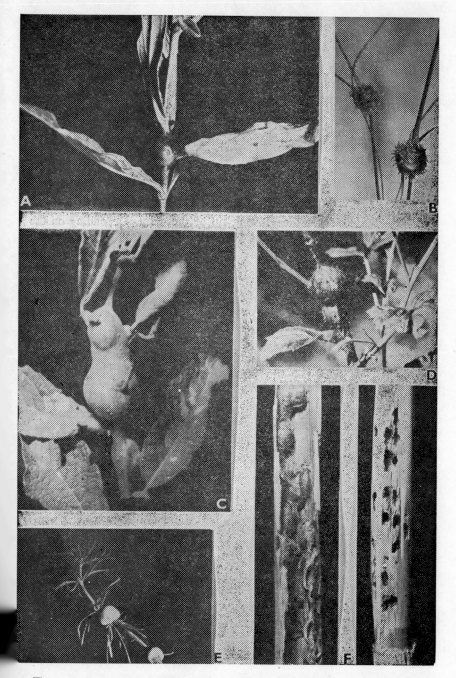

Figure 13. Stem galls of gall midges: A—*Acodiplosis inulae* (H. Loew) on *Inula britannica* L.; B—*Mayetiola poae* (Bosc) on *Poa nemoralis* L.; C—*Dasineura salicis* (Schrank) on *Salix aurita* L.; D—*Lasioptera rubi* (Schrank) on *Rubus* sp.; E—*Geocrypta galii* (H. Loew) on *Galium* sp.; F—*Giraudiella inclusa* (Frauenfeld) on *Phragmites communis* Trin.; on the left: galls in the inner wall of stem; on the right: galls pecked by birds.

shoots of *P. communis*, which develop after destruction of the vegetative top by other insects, are the location of galls caused by larvae of *Lasioptera arundinis* Schiner, which appear as a thickening of these lateral shoots. *Haplodiplosis marginata* (von Roser) produces saddle-shaped enlargements of the stem of various species of *Gramineae*, mostly on cereals. Also, larvae of *Orseolia oryzae* (Wood-Mason) and *O. oryzivora* Harris and Gagné produce galls that damage stems of *Oryza sativa* L.

c) Bud Galls

Gall midges attack both the flower and leaf buds. These may be either laterally situated or at the growing tip of the plant. Depending on the species, the larvae may destroy the bud totally, often producing a cavity in which they live, or they may live among the young leaves without destruction of the growing point. Attack of the apical bud at the growing point, however, usually leads to a reduction of shoot development and coalescence of growing leaves to produce the so-called "rosette gall". It is sometimes difficult to determine if such a gall is a bud gall, a stem gall or a leaf gall (Fig. 14).

Buds of various host plants may be totally destroyed by a number of species in the genus *Asphondylia* H. Loew. For example, larvae of *A. sarothamni* H. Loew produce bud galls on *Sarothamnus scoparius* (L.) Wimm. which destroy the bud. Similar damage is caused by *A. cytisi* Frauenfeld on various species of the genus *Cytissus*, *A. ononidis* F. Löw on *Ononis spinosa* L., and *A. pruniperda* Rondani on various species of the genus *Prunus*. Often the type of damage, or even the structure of the plant that is attacked, changes from the generation to the next. For example, the larvae of some species of *Asphondylia* attack buds in the spring but the summer generation of the same species develops in flower or fruit galls that they produce.

Larvae of *Dasineura kellneri* (Henschel) develop in slightly swollen buds of *Larix decidua* Mill., covered with a resin mass. One larva develops in each bud, which is totally destroyed. Larvae of *D. medicaginis* (Bremi) produce a swelling of the axillary buds on *Medicago sativa* L., but here several larvae develop in each bud. *Rhopalomyia tanaceticola* (Karsch) produces great fleshy galls on axillary buds of *Tanacetum vulgare* L. and similar galls are produced by larvae of *R. millefolii* (H. Loew) on *Achillea millefolium* L. Both species also cause similar galls on other organs of their host plants, i.e., on the surface of leaves and even on individual flowers in the flower head.

d) Leaf Bud Galls and Leaf Rosette Galls

The shape of galls developing from buds depends greatly on the developmental stage of the bud at the time of attack, thus there is considerable variation in such galls (Fig. 15). If the bud is attacked in the early stages of development it may be completely destroyed. However, if the attack comes

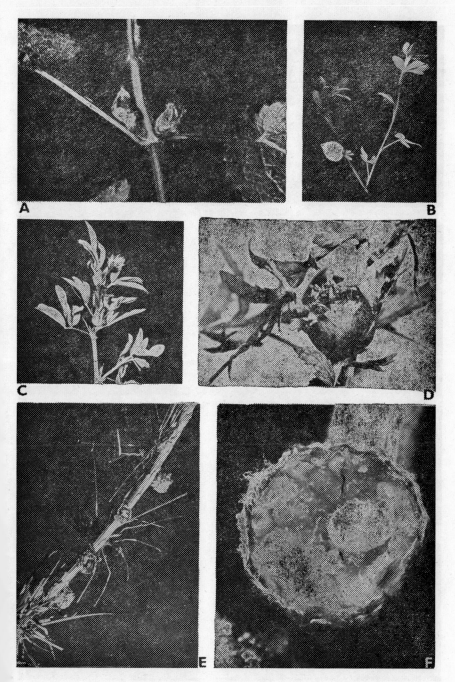

Figure 14. Bud galls of gall midges: A—*Wachtliella stachydis* (Bremi) on *Stachys silvatica* L.; B—*Asphondylia cytisi* Frauenfeld on *Cytisus supinus* L.; C—*Dasineura medicaginis* (Bremi) on *Medicago sativa* L.; D—*Rhopalomyia tanaceticola* (Karsch) on *Tanacetum vulgare* L.; E—*Dasineura kellneri* (Henschel) on *Larix decidua* Mill.; F—opened bud gall of *D. kellneri* with a single cocoon containing a pupa in the spring.

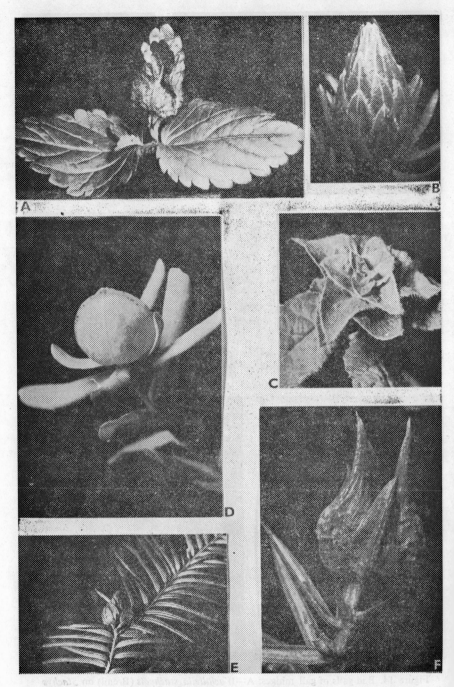

Figure 15. Leaf rosette galls of gall midges: A—*Jaapiella veronicae* (Vallot) on
Veronica chamaedrys L.; B—*Wachttiella ericina* (F. Löw) on *Erica carnea* L.;
C—*Dasineura rosaria* (H. Loew) on *Salix caprea* L.; D—*Zeuxidiplosis giardi*
(Kieffer) on *Hypericum perforatum* L.; E—*Taxomyia taxi* (Inchbald) on *Taxus
baccata* L.; F—*Oligotrophus juniperinus* (L.) on *Juniperus communis* L.

Figure 16. Leaf galls of gall midges on various host plants: A—*Sackenomyia reaumuri* (Bremi) on *Viburnum lantana* L.; B—*Mikiola fagi* (Hartig) on *Fagus silvatica* L.; C—*Dasineura corniculata* (Kieffer) on *Lamium maculatum* L.; D—*Dasineura tiliam-volvens* (Rübs,) on *Tilia platyphyllos* Scop.; E—*Paradiplosis abietis* (Hubault) on *Abies alba* Mill.; F—*Craneiobia corni* (Giraud) on *Cornus sangu̇nea* L.

after the development is more advanced the growing point may not be killed but may instead develop into a more "typical" gall. When such late infestations occur on leaf buds, the partially formed leaves coalesce to produce the so-called "rosette" gall (Fig. 15).

There is also variation in the number of larvae inside such galls. For example, only one larva of *Dasineura rosaria* (H. Loew) develops in the rosette gall it produces on *Salix alba* L. Also, one larva of *Taxomyia taxi* (Inchbald) causes a rosette gall on *Taxus baccata* L. In contrast, several larvae of *Dasineura crataegi* (Winnertz) are found in the rosette galls they produce on *Crataegus oxyacantha* L.

Superficially similar leaf bud galls are produced by two midge species on lateral leaf buds of *Hypericum perforatum* L. Several larvae of *Dasineura hyperici* (Bremi) produce the first gall type, and live gregariously among the hypertrophied leaves. Individual larvae of the second species, *Zeuxidiplosis giardi* (Kieffer) produce a similar-looking gall but close examination reveals that in this gall the bud meristem is totally destroyed and there is a central cavity inside which the larva lives.

e) Leaf Galls

Leaf galls are the most common type of galls caused by cecidomyiids. There are various types of these galls ranging from small depressions on the leaf surface, to pustule galls, to small rounded swellings, to leaf rolls, folded leaf margins, swollen vein folds, irregular folded and frilled leaf surfaces, and even new forms of leaves that are almost unrecognizable. It is possible to arrange these various types of galls in order of increasing complexity. The most primitive galls (or at least the most simple) are the small depressions on the lower surface of leaves, and the so-called pustule galls. Sometimes it is very difficult to determine whether these types of abnormalities are true galls. However, close examination usually reveals that there are changes in cells in the tissues around the insect typical of more normal galls. These simple galls are followed in complexity by those consisting of small rounded swellings which develop on one, or both, sides of the leaf. This series continues in increasing complexity to the galls which consist of large structures of various forms, on either leaf surface. The most complex galls are those which consist of completely new leaf forms.

Of course the morphology of the various leaf galls is closely connected with the growth of plant tissues in leaf buds during their embryological development, and with their position in buds during this period. For example, various types of folded leaf galls and marginal galls (those on the leaf margin), are in reality the result of hypertrophied plant tissues, rather than growth.

An example of the simplest gall type is the small depression on the lower leaf surface of *Acer pseudoplatanus* L. caused by a single larva of *Drisina glutinosa* Giard. This gall is visible on the upper leaf surface in the form of a small elevation. Larvae of *Dasineura pustulans* (Rübs.) cause similar damage on leaves of *Filipendula ulmaria* Maxim.

Pustule galls develop on leaves of various trees early in the spring. The larvae that cause these galls develop very quickly, taking only one to three weeks to complete growth. At this time, however, the gall is not visible. After the larva has dropped to the soil where it overwinters, the gall becomes dry and is visible as a brown spot on the leaf. The majority of these species have been described only on the basis of the gall and the larva. Typical pustule galls are produced by larvae of *Dasineura fraxinea* (Kieffer) on leaves of *Fraxinus excelsior* L., *D. ruebsaameni* (Kieffer) on *Carpinus betulus* L., *D. vitrina* (Kieffer) on *Acer pseudoplatanus* L., *D. tympani* (Kieffer) on *Acer campestre* L., *Physemocecis hartigi* (Liebel) on *Tilia cordata* Mill., *P. ulmi* (Kieffer) on various species of the genus *Ulmus*, and by larvae of *Mikomya coryli* (Kieffer) on leaves of *Corylus avellana* L. Larvae of *Erosomyia mangiferae* Felt produce blister galls on very young leaves of mango, *Mangifera indica* L.

Pustule type galls are also produced by larvae of the genus *Cystiphora* Kieffer on various species of the family *Asteraceae*. Larvae of *C. taraxaci* (Kieffer) produce such galls on *Taraxacum officinale* Web., those of *C. sanguinea* (Bremi) on various species of the genus *Hieracium* and larvae of *C. sonchi* (Bremi) on *Sonchus arvensis* L.

A little more complicated from the morphological point of view are the galls of *Sackenomyia reaumuri* (Bremi) (=*Phlyctidobia solmsi* (Kieffer)) which are found on leaves of *Viburnum lantana* L. The gall is a lenticular pustule of parenchymatous tissue on both leaf surfaces, slightly raised on the upper surface and covered with minute hairs. Similar galls are produced by *Anisostephus betulinum* (Kieffer) on leaves of *Betula pendula* Roth and *Loewiola centaureae* (F. Lw.) on *Centaurea jacea* L. and closely related species.

More complex galls consisting of small rounded swellings on both leaf surfaces are produced by larvae of *Dasineura urticae* (Perris) on *Urtica dioica* L. Two or more larvae live in a unilocular cavity of the gall, which opens on the upper surface of the leaf. This gall can occur also on the stem or on the flower stalks of this plant species. Larvae of *Iteomyia capreae* (Winnertz) produce small rounded swellings on leaves of willow, mainly *Salix caprea* L., and similar galls are caused by *Dasineura ulmaria* (Bremi) on leaflets of *Filipendula ulmaria* Maxim. Larvae of *Rhopalomyia foliorum* (H. Lw.) produce small inconspicuous outgrowths on leaves of *Artemisia vulgaris* L.

Large complex and conspicuous galls on the upper surface of leaves are produced by larvae of several gall midge species. The best known galls of this type in Europe are those produced by *Mikiola fagi* (Hartig) and *Hartigiola annulipes* (Hartig) on the upper leaf surface of *Fagus silvatica* L., the galls caused by *Craneiobia corni* (Gir.) on *Cornus sanguinea* L. and those of *Didymomyia tiliacea* (Bremi) on various species of the genus *Tilia*. This gall

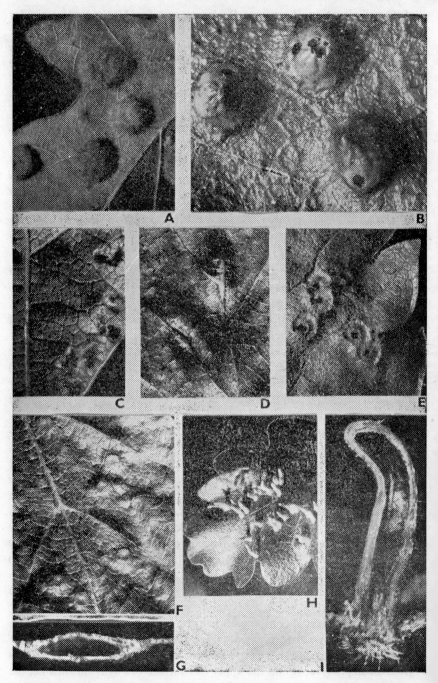

Figure 17. Leaf galls of gall midges produced on one host plant, *Quercus cerris* L.: A—*Dryomyia circinans* (Giraud); B—*Janetia cerris* (Kollar); C—*Dasineura tubularis* (Kieffer); D—*Janetia nervicola* (Kieffer); E—*Janetia homocera* (F. Löw); F—*Janetia szepligetii* Kieffer; G—cross-section of *J. szepligetii*; H—*Contarinia subulifex* Kieffer; I—cross-section of *C. subulifex*.

201

type also includes the galls produced by larvae of the genus *Harmandia* Kieffer on leaves of *Populus tremula* L. and those on leaves of *Quercus cerris* L. caused by nine other species of gall midges (Fig. 17).

Small galls on the middle or lateral veins of the leaves of *Ulmus* species are produced by larvae of *Janetiella lemei* (Kieffer). These galls appear as rounded hard swellings on the upper surface and cylindrical projections on the lower one, with the aperture at the apex (or it may also be reversed).

Leaf rolling type galls are produced by larvae of several species of gall midges including: *Dasineura tiliamvolvens* (Rübs.) on *Tilia*, *D. affinis* (Kieffer) on *Viola canina* L., *Wachtliella persicariae* (L.) on *Polygonum amphibium* L., *Dasineura filicina* (Kieffer) on *Pteridium aquilinum* (L.) Kuhn, *D. mali* (Kieffer) on *Malus silvestris* Mill., *D. populeti* (Rübs.) on *Populus tremula* L., *D. pyri* (Bouché) on *Pirus communis* L. and *D. marginemtorquens* (Bremi) on *Salix viminalis* L. Somewhat simpler galls formed by rolling the margin of the leaf without visible swelling are caused by larvae of *Dasineura kiefferiana* (Rübs.) on *Chamaenerion augustifolium* (L.) Scop.

Galls formed by irregular folded and frilled leaf surfaces are produced by larvae of *Dasineura acercrispans* (Kieffer) on *Acer pseudoplantanus* L., by *D. thomasiana* (Kieffer) on *Tilia*, by *D. plicatrix* (H. Lw.) on *Rubus caesius* L. and by *Macrolabis heraclei* (Kalt.) (= *M. corrugans* (F. Lw.)) on *Heracleum sphondylium* L. and *Chaerophyllum aromaticum* L.

Galls in which leaflets are regularly folded along the median vein are produced by larvae of *Dasineura sanguisorbae* (Rübs.) on *Sanguisorba officinalis* L., *D. viciae* on *Vicia sativa* L., *D. trifolii* (F. Lw.) on *Trifolium repens* L. and *Wachtliella rosarum* (Hardy) on *Rosa canina* L.

Larvae of *Dasineura fraxini* (Bremi) cause galls consisting of a swollen median vein without folding, on leaflets of Fraxinus *excelsior* L. and larvae of *Zygiobia carpini* (F. Löw) produce similar galls on leaves of *Carpinus betulus* L.

Larvae of *Putoniella pruni* (Kalt.) produce galls consisting of a median or lateral vein with a bag-shaped swelling on leaves of *Prunus spinosa* L. and other species of that genus. Galls consisting of a swollen median vein with the leaf folded are produced by larvae of *Dasineura acrophila* (Wtz.) on *Fraxinus excessor* L. and *D. alni* (F. Lw.) on *Alnus glutinosa* (L.) Gaertn. Galls consisting of leaflets folded along the median vein without swelling are produced by larvae of *Jaapiella medicaginis* (Rübs.) on *Medicago sativa*. Similar galls with regularly folded areas between two lateral veins are produced by larvae of *Contarinia carpini* Kieffer on *Carpinus betulus* L. and *Phegomyia fagicola* (Kieffer) on *Fagus silvatica* L.

The needle galls that are produced by various genera of gall midges, on numerous species of coniferous trees, are also leaf galls since needles are actually modified leaves. Examples of such galls include those caused by *Thecodiplosis brachyntera* (Schwägr.) on three species of *Pinus*. The gall normally consists of a pair of partially fused needles with slightly swollen bases. The needle gall caused by *Contarinia coloradensis* Felt, on *Pinus ponderosa* Laws., is more noticeable in that the needles are very short and greatly enlarged at the base, in some cases almost unrecognizable as needles.

f) Flower Galls

Flowers and inflorescences of various plants are attacked by a number of gall midge species. The result of such attacks include swollen flower buds, and/or flower heads. Attacks of such reproductive organs usually occur early in the development of the plant, normally in the bud stage. The infested organ does not develop into a normal flower, and thus does not produce seed. Some species that attack flowers may be serious agricultural pests (Fig. 18). For example, larvae of *Contarinia medicaginis* Kieffer produce flower galls on *Medicago sativa* L. and in Europe are serious pests. *Contarinia nasturtii* (Kieffer) and *Gephyraulus raphanistri* (Kieffer) produce flower galls on species of *Brassicaceae* and are pests of cultivated crops, as is *Dasineura leguminicola* (Lintner) on *Trifolium pratense* L.

Other examples of gall midges which cause galls of floral structures in Europe include: *Asphondylia verbasci* (Vallot) on flowers of *Verbascum lychnitis* L., *A. echii* (H. Lw.) on *Echium vulgare* L., *Placochela ligustri* (Rübs.) on *Ligustrum vulgare* L., and *P. nigripes* (F. Lw.) on various species of *Sambucus*. Larvae of *Contarinia craccae* Kieffer produce galls on flower buds of various species of *Vicia*, *C. loti* (de Geer) on *Lotus corniculatus* L. and other species.

Larvae of several gall midges producing galls on flower buds of mango, *Mangifera indica* L. are serious pests in India: *Dasineura amaramanjarae* Grover, *Procystiphora indica* Grover et Prasad, *Procystiphora mangiferae* (Felt) and *Erosomyia indica* Grover.

g) Fruit Galls

The number of gall midge species causing galls on fruits and seeds is much less than in the last group. These gall midges usually attack the host plant in the last stages of flowering and therefore the larvae develop in these structures. Sometimes distinct galls develop as a result of the attacks but more often the infested fruit, or seeds, are without visible damage (Fig. 18).

Larvae of several species of the genus *Asphondylia* H. Loew cause fruit galls which are the larvae of the second generation of these midges, as stated above. For example, larvae of *Asphondylia cytisi* Frauenfeld evoke galls on buds of *Cytisus supinus* L. In India, larvae of *Asphondylia pongamiae* Felt cause an ovary-gall resembling a fruit on *Pongamia glabra* Vent. Larvae of *Kiefferia pericarpiicola* (Bremi) (= *K. pimpinellae* (F. Lw.)) produce galls on fruits of various species and genera of the family *Apiaceae* (= *Umbelliferae*). Larvae of *Contarinia pyrivora* (Riley) cause damage to fruits of *Pirus communis* L. and are considered a pest of this tree. Larvae of various species of the genus *Ozirhincus* produce small fruit galls in several species of the family *Asteraceae*.

Figure 18. Flower and fruit galls of gall midges: A—*Kiefferia pericarpiicola* (Bremi) on *Pimpinella saxifraga* L.; B—*Asphondylia verbasci* (Vallot) on *Verbascum lychnitis* L.; C—*Dasineura phyteumatis* (F. Löw) on *Phyteuma orbiculare* L.; D—*Asphondylia coronillae* (Vallot) on *Coronilla varia* L.; E—*Contarinia pyrivora* (Riley) on *Pirus communis* L.; F—*Asphondylia massalongoi* Rübs. on *Ajuga chamaepytis* Schreb.

Larvae of *Dasineura brassicae* (Winnertz) develop in slightly swollen seed-pods of *Brassica napus* L. In one seed pod live 15 to 25 whitish larvae (maximum of 140 larvae). In some areas of Europe, it is a serious pest of rape and other related cultivated plants. Larvae of three species of the genus *Semudobia* Kieffer develop in seeds of various species of the genus *Betula*. Larvae of *Plemeliella abietina* Seitner develop in the seeds of *Picea excelsa* Link. and those of *Resseliella piceae* Seitner in seeds of *Abies alba* Mill. Both of these species can be pests in forestry by reducing seed production.

3) GALL MIDGES AND AMBROSIA FUNGI

Various genera of gall midges of the phytophagous biological group, either gall producers, or those that live on plants without producing galls, live in association with a special group of fungi. These fungi (*Deuteromycetes*) and the midges, live together symbiotically in the so-called "Ambrosia galls". Adults transfer the hyphae of the fungi when new galls, or plant associations, are initiated. Developing larvae feed on the mycelium of the fungus which covers the inside wall of the larval cavity. The symbiotic relationship benefits the gall midges in that the larvae use a part of the fungi for food (rather than feeding directly on the host plant). The fungi benefit since they live protected inside the plant, away from unfavourable environmental conditions. Also, the midges act to ensure wide distribution and a favourable host plant for the fungi.

Most of the gall midges that live symbiotically with ambrosia-fungi belong to the genera *Lasioptera* Meigen and *Asphondylia* H. Loew. The galls occur in various organs of the host; in bud galls, fruit galls and flower galls. Mostly the species of fungi are undetermined. For example, galls of *Lasioptera arundinis* Schiner which occur on side shoots of *Phragmites communis*, are filled with mycelia of an undetermined species of the genus *Sporothrix*. The walls of the galls of *Asphondylia pruniperda* Rondani are coated inside with mycelia of an undetermined species of Ambrosia-fungus of the genus *Shaeropis*.

The larvae of *Lasioptera hungarica* Möhn live inside the stem of the common reed *Phragmites communis* Trin. in a black mass of fungal mycelium without causing any symptom of their attack. However, birds are able to locate infested reeds and peck them open to obtain the larvae inside. Also, larvae of *Lasioptera buhri* Möhn were found inside stems of various species of the genera *Crepis*, *Tragopogon*, *Lapsana*, *Mycelis*, *Hieracium*, *Picris* (all from the family *Asteraceae*), *Melilotus* (*Fabaceae*), and *Campanula* (*Campanulaceae*), associated with such fungi.

Patterns of population changes in gall midges

Many gall midge species remain at relatively constant population levels, within a given locality, for long periods of time. These occur mainly in primary (old) ecosystems. For example, populations of *Giraudiella inclusa* (Frauenfeld) and *Lasioptera arundinis* Schiner on the reed *Phragmites communis* Trin. have remained at similar levels for the past 15 years. However, in some species, periodic fluctuations occur. Within these species there are two basically different patterns. The first is where the normal population level is low but occasional outbreaks occur. This is the situation with *Contarinia baeri* (Prell) on *Pinus sylvestris*. The second situation is the reverse of the first, that is the normal population level is high but there are periods of low populations. This pattern occurs in *Thecodiplosis brachyntera* (Schwägr.) on the genus *Pinus*. Thus, there are three basic patterns of population variation among the gall midge species for which there are population data. The first, where population levels remain relatively constant for long periods of time is probably most common. The second pattern, where populations fluctuate but numbers are low most of the time also occurs often. The third pattern, where populations fluctuate, but are commonly at high levels occur less often in nature and is usually associated with economically important species.

It is also found that a single species that occurs over a wide geographic area may exhibit different population patterns in different areas. A good example of a species showing such population variation is the gall midge *Thecodiplosis brachyntera* (Schwägr.).

Thecodiplosis brachyntera is found throughout the entire range of its main host, Scots pine, *Pinus sylvestris* L. In Europe, three types of areas can be distinguished based upon the level of occurrence of this species. These include: (1) an area of occasional occurrence, (2) an area of regular occurrence, and (3) an area of periodic outbreaks (Fig. 27). The area of occasional occurrence (1) includes Sweden, where during the last 100 years it was found at only three locations during a one-to-two-year period. It also includes England and Scotland where it was reported one year from four localities, the Soviet Union with two reports, and Eastern France and Romania with one report each during that 100-year period.

The area of regular distribution (2) of *T. brachyntera* includes Denmark and the Netherlands, where it is found mainly on *Pinus sylvestris* and *Pinus mugo* planted in landscape situations, and the territories of both German States, Austria, Czechoslovakia, Poland and Hungary.

The area of periodic outbreaks (3) of *T. brachyntera* covers a relatively limited region in central Europe and includes the German Democratic Republic, Poland, the western part of Czechoslovakia and the southern part of the German Democratic Republic. It is the area between 48 and 58 degrees north latitude and between 10 and 23 degrees east longitude. The most probable reason for the great success of *T. brachyntera* is thought to be the ideal timing of larval development with that of the needles of the host plant.

Economically important gall midges

As noted earlier in the chapter, more than 25 per cent of gall midge species are known only from the description of the adult, the larva or the gall. The biology of these species remains unknown. Biological data are available on about 75 per cent of the known gall midge species but the most comprehensive data have been obtained on the economically important gall midges, i.e., those injurious to agricultural or forestry crops. About 5–10 per cent of gall midge species occur in high numbers on plants that are important to man, including cereals, fodder, vegetable and root crops. Gall midges also attack trees and shrubs cultivated by man, the food he has harvested, his animals and his wood products.

The damage caused by gall midges is most extensive in the agricultural situations where man has created the best conditions for population growth of pests by planting monocultures over large acreages. Here, the injury caused by some gall midge species can cause yield losses of up to 50 per cent. Therefore, the biology of such species has been studied intensively and the data obtained used to reduce the damage caused by such pests. In forest situations, gall midges may also occur in great numbers but the damage is not so important because the insects usually attack buds, leaves or seeds rather than the main trunk of the tree. In only a few cases, gall midges are able to destroy forest stands. Gall midges that attack trees and shrubs are important mainly in nurseries because they reduce the growth of young seedlings.

In general, gall midges cause such growth reduction in two ways. First, they reduce the production of normal plant structures by utilizing the plant resources for production of galls. For example, the dry weight of foliage with galls may be two to five times higher than that without galls. Also, the presence of galls, especially in high numbers, probably reduces the photosynthetic ability of the leaf, so that the effect of the infestation increases with the number of galls. Secondly, the occurrence of galls causes the relocation, and perhaps loss, of chemicals needed for normal plant growth and so normal growth is reduced.

A list of the most economically important gall midge pests, along with the plants they attack and the damage they cause, is presented in Table 2.

Probably the most economically damaging species of gall midges are those that attack the cereal crops. These include *Mayetiola destructor* (Say) which attacks wheat and barley in Europe and North America, *Haplodiplosis marginata* (von Roser) which infests those crops in Europe and *Contarinia sorghicola* (Coquillett) which attacks sorghum, mainly in the tropics and sub-tropics. *Mayetiola destructor* is thought to have originated in Europe but has been transported to North America and other parts of the world. *M. destructor* was not known to occur in the United States until about the end of the 18th century but since its appearance at that time it

Table 2. Economically Important Gall Midges

Group	Host plant species	Gall midge species	Structure damaged
CEREALS	Triticum vulgare L.	Contarinia tritici (Kirby)	flowers
	Hordeum sativum L.	Sitodiplosis mosellana (Géhin)	grains
	Secale cereale L.	Mayetiola destructor (Say)	upper stem
	Avena sativa L.	Haplodiplosis marginata (von Roser)	stems
	Sorghum vulgare Pers.	Contarinia sorghicola (Coquillett)	seeds
	Panicum miliaceum L.	Contarinia panici (Plotnikov)	flowers
	Oryza sativa L.	Orseolia oryzae (Wood-Mason)	stems
		Orseolia oryzivora Harris et Gagné	stems
FODDER CROPS	Trifolium sp.	Dasyneura leguminicola (Lintner)	flowers
	Medicago sativa L.	Contarinia medicaginis Kieffer	flowers
ROOT CROPS	Brassicaceae	Contarinia nasturtii (Kieffer)	stems and other parts
	Brassica spp.	Dasyneura brassicae (Winnertz)	seed pods
	Pisum sativum L.	Contarinia pisi (Winnertz)	fruits
FRUIT TREES AND SHRUBS	Rubus sp.	Resseliella theobaldi (Barnes)	twigs
		Dasyneura tetensi (Rübsaamen)	leaves and terminal shoots
	Pyrus communis L.	Contarinia pyrivora (Riley)	fruits
	Mangifera indica L.	Erosomyia indica Grover	multiphase feeder
		Procystiphora indica Grover et Prasad	flower buds
		Dasineura amaramanjarae Grover	flower buds
		Procystiphora mangiferae (Felt)	flower buds
TREES	Pinus spp.	Thecodiplosis brachyntera (Schwägr.)	needles
		Janetiella coloradensis Felt	needles
		Pinyonia edulicola Gagné	needles

has become one of the most important pests of wheat in that country. In contrast, although once a major pest of cereals in Europe, *M. destructor* has become less important during the past 50 years.

Prior to 1956 the saddle gall midge, *Haplodiplosis marginata*, was an innocuous gall midge known mainly to cecidologists. Since that time, however, the species has become one of the major pests of wheat and barley in Europe. It appears that the major reasons for the change in pest status have been an intensification of plant production, the repeated planting of cereals without rotation, and the selection of new, highly productive, varieties that resulted in the loss of natural resistance against this pest. A similar situation has apparently caused *Contarinia sorghicola* to assume pest status on sorghum in India.

Although most gall midges that attack trees act mainly to reduce growth, as noted earlier, some cause more direct damage by killing plant parts. For example, *Thecodiplosis brachyntera* (Schwägr.), which attacks *Pinus mugo*, and *Janetiella coloradensis* Felt, which attacks *Pinus edulis,* cause needle shortening and premature needle death. Serious infestation can cause tree defoliation and sometimes death.

Gall midges and their associate plant hosts also make excellent model systems for the study of ecological principles. For example, studies on the influence of abiotic factors on insect development have been conducted on *Contarinia medicaginis* Kieffer. Basic studies on the control of diapause are possible on *Contarinia tritici* (Kirby) which has been shown to remain in this state for eleven years under some conditions. Studies have also been conducted on host range of insects using wild and cultivated grasses and the gall midges that attack them. Related studies on the polyphagous habits of some insects have been conducted using *Contarinia nasturtii* (Kieffer) as a model system. Studies on various plants and the gall midges that infest them have revealed much about the intimate association of the host and its plant chemicals and the gall midge and the environment. It is obvious that much can be learned about plant growth, its control and the factors that affect such growth, through studies utilizing gall midges and their plant hosts.

Biogeography

Although the family *Cecidomyiidae* offers a rich source of material for zoo-geographical and biogeographical studies, as mentioned above, only a few studies have been conducted on this group.

By using data collected from catalogues of Diptera from various zoogeographical regions, data of authors dealing with the distribution of gall midges of various countries, and data of the present authors an attempt was made, to analyze the distribution of gall midges in the Palaearctic Region and to determine its relationship to other biogeographical or zoogeographical regions.*

Based on these data, it is estimated that at the present time about 4,300 species of gall midges, in 560 genera, have been described in the world. However, these species are not evenly distributed among the biogeographical region (Fig. 19).

From the standpoint of number of species, Palaearctic Region is one of the richest of all the biogeographical regions. Most gall midge species were described from, and occur in, that region and its area is greater than the others. There also, the study of gall midges has its longest history and tradition.

In the Palaearctic Region the fauna of gall midges is best known in Europe, including the European part of the Soviet Union. Less well known is the fauna of Soviet Middle Asia, the Soviet Far East, and Japan. Little is known of the gall midges of Siberia and of northern Europe. Almost nothing is known about the gall midges of China, Mongolia and Korea. There are also few data about the gall midges in Turkey, Iran, Iraq and Afghanistan.

The available information on the number of gall midges species in various countries of the Palaearctic Region are summarized from Skuhravá (1985) in Table 3.

Table 3. The Number of Gall Midge Species in Several Countries of the Palaearctic Region

Norway	65	Poland	360	Bulgaria	104	Afghanistan	6
Sweden	261	Czechoslovakia	504	Greece	19	China	11
Finland	135	Switzerland	81	Cyprus	9	Mongolia	8
Denmark	160	Austria	256	Soviet Union	*	Korea	6
Great Britain	603	Hungary	240	Turkey	34	Japan	176
Netherlands	295	Romania	294	Syria	7	Marocco	16
Belgium	89	Portugal	118	Israel	25	Algeria	74
France	523	Spain	80	Jordan	3	Tunis	27
German Fed. Rep.	586	Italy	334	Lebanon	5	Libya	13
German Dem. Rep.	406	Yugoslavia	287	Iraq	9	Egypt	25
		Albania	11	Iran	9		

*European Part 524; Asian Part 330.

Obviously, the fauna of the different zoogeographical regions and countries have not been studied equally in the past, or at the present time. As with any area of science, knowledge depends in part on the location of

*Note: Udvardy (1975) presented a new classification of the biogeographical provinces of the world in which he introduced terms biogeographical realms and provinces instead of biogeographical regions and sub-regions and introduced new names, Afrotropic and Indomalayan, for the earlier terms, Ethiopian and Oriental.

research centres but also, and perhaps more importantly, on whether or not specialists in the area under consideration are present. For example, the large number of gall midge species listed for Great Britain is probably due mainly to the interest of H.F. Barnes, and his fellow cecidologists, in the gall midge fauna of that country.

Mostly the fauna of gall midges of a zoogeographical region is bound only on the territory of that region where every species, or genus, occupies one part of the territory—its distribution area. Only a few species, or genera, of gall midges are widespread, with distribution in two, or more zoogeographical regions.

From the generic point of view, there are some genera which are distributed practically all over the world; for example, *Dasineura*, *Contarinia*, *Asphondylia*, *Aphidoletes*, *Lestodiplosis*. Other genera occur only on one of the zoogeographical regions; for example, *Neolasioptera* and *Asteromyia* in Nearctic, *Alycaulus* in Neotropics, *Baldratia* in Palaearctic Regions.

From the species standpoint, it seems otherwise. For example, from about 2,000 species of gall midges that occur in the Palaearctic Region, 95 per cent are not found elsewhere and only five per cent i.e., about 100 species, occur in the Palaearctic and some other regions (Fig. 19).

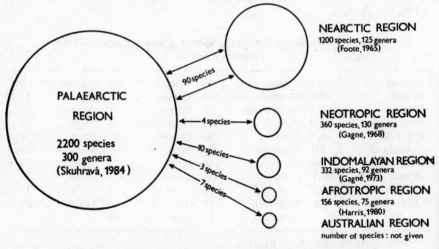

Figure 19. Number and relationships of common gall midge species among biogeographical realms (regions) of the world.

The relationships between the fauna of gall midges of six zoogeographical regions of the world are shown by number of common species, e.g., species which occur in two or more regions. The number of common species indicates a similarity of regions in faunal composition; thus, the higher the number of common species means a higher degree of similarity of the regions.

The highest number of common species (90) occurs between Palaearctic and Nearctic Regions. These two regions are often joined into the so-called Holarctic Region and show the highest degree of similarity. Only a few species of gall midges are common among Palaearctic and other regions; between the Palaearctic and Indomalayan Regions occur 10 common species, between Palaearctic and Australian Regions there are seven species, between Palaearctic and Neotropic only four species and between Palaearctic and Afrotropic Regions only three species. The fauna of gall midges of Indomalayan, Australian, Afrotropic and Neotropic Regions seems to be very distinctive.

Only few species of gall midges occur in several biogeographical regions of the world. Usually the phylogenetically older species have a larger distribution area than younger species. For example, *Lestremia cinerea* Macquart is widespread throughout three biogeographical regions of the world (Fig. 20), in contrast to *Haplodiplosis equestris* (von Roser) which is distributed only on a small area of the western part of the Euro-Siberian Subregion.

Some pest species may have widespread areas of distribution because they were secondary imported to other regions with their host plant. Some of them now seem to have nearly a cosmopolitan type of distribution. For example, the Hessian Fly, *Mayetiola destructor* (Say) was imported from Europe to Australian and Nearctic Regions. This species is now distributed throughout the USA and there is a serious pest of cereals. Another pest, *Contarinia sorghicola* (Coquillett) occupies a widespread area in the tropics and subtropics of the world.

Many species of the subfamilies *Lestremiinae* and *Porricondylinae* whose larvae are mycophagous, are also widespread both in the Palaearctic and Nearctic Regions. It is interesting to note that approximately 40 species of the gall midges of the subfamily *Lestremiinae* (20 per cent), and only 10 species of the subfamily *Porricondylinae* (4 per cent) have a holarctic type of distribution, but only 45 species (3 per cent) of the subfamily *Cecidomyiinae* have this type of distribution.

Only a few species of gall midges are distributed evenly throughout the territory of the Palaearctic Region. Most gall midges occur only on one or at most two or three subregions. Approximately 70 per cent of all gall midges known from the Palaearctic Region occur in the Euro-Siberian Subregion, 19 per cent of species in the Mediterranean Subregion, 14 per cent in the East-Palaearctic Subregion and only eight per cent in the Middle-Asian Subregion (Fig. 21).

The highest number of common species (132) occurs between the Euro-Siberian and Mediterranean Subregions. There are also many common species (40) between the Euro-Siberian and East-Palaearctic Subregions. There are a comparatively high number of common species (20) among the three subregions of the Palaearctic, viz. Euro-Siberian, Mediterranean and

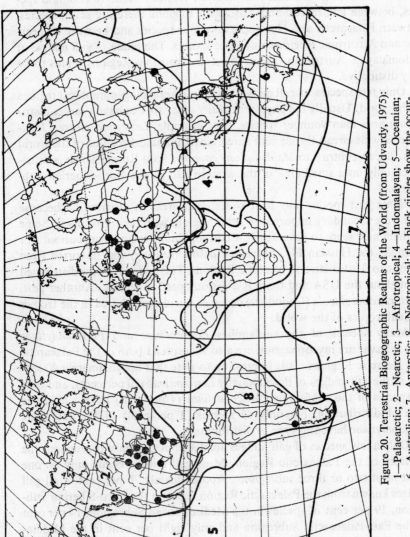

Figure 20. Terrestrial Biogeographic Realms of the World (from Udvardy, 1975):
1—Palaearctic; 2—Nearctic; 3—Afrotropical; 4—Indomalayan; 5—Oceanian;
6—Australian; 7—Antarctic; 8—Neotropical; the black circles show the occurrence of *Lestremia cinerea* Macquart in Palaearctic, Nearctic and Neotropic Realms.

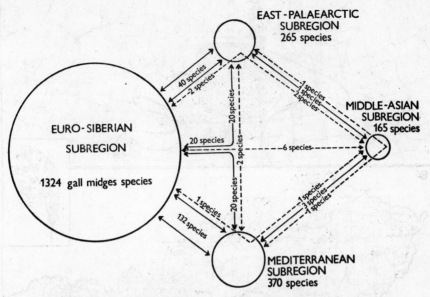

Figure 21. Number and relationships of common gall midge species among four subregions of the Palaearctic Region.

East-Palaearctic. However, there are only few species of gall midges which are common among Middle-Asian Subregion and the other three subregions.

Thus, the highest similarity seems to be among the fauna of the Euro-Siberian, Mediterranean and East-Palaearctic Subregions. The fauna of gall midges of the Middle-Asian Subregion seems to be very distinct from the fauna of other three subregions. For example, only in this subregion does one find species of the genera *Gobidiplosis, Desertovelum, Psectrosema, Kochiomyia and Halodiplosis* which seem to be endemic to this subregion.

The distribution areas of species of the subfamily *Cecidomyiinae*, whose larvae are mostly bound to their specific host plants, or animals, is only as great as the distribution of those hosts. In these cases, the species usually extends to the boundary of distribution area of their hosts. In contrast, the distribution area of some species may be smaller than the distribution area of their hosts. It probably depends upon the ecological tolerance of the gall midge species. In this case, the gall midges occupy a smaller area than the host usually laying in the centre of the host's distribution area and do not extend to the boundary. For example, nine species of gall midges whose larvae infest species of the genus *Quercus*, viz. *Q. cerris* L. and *Q. libani* L., are distributed throughout the Mediterranean Subregion and reach, with their host plant, to the northern boundary lying in the southern part of Czechoslovakia (Fig. 22).

In contrast, the gall midge *Asphondylia verbasci* (Vallot) whose larvae develop on various species of the genus *Verbascum*, occurs (in high abund-

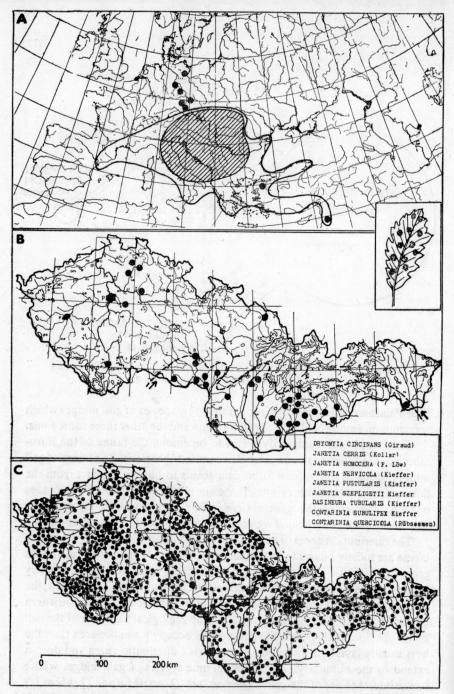

DRYOMYIA CIRCINANS (Giraud)
JANETIA CERRIS (Kollar)
JANETIA HOMOCERA (F. Löw)
JANETIA NERVICOLA (Kieffer)
JANETIA PUSTULARIS (Kieffer)
JANETIA SZEPLIGETII Kieffer
DASINEURA TUBULARIS (Kieffer)
CONTARINIA SUBULIFEX Kieffer
CONTARINIA QUERCICOLA (Rübsaamen)

0   100   200 km

Figure 22. Distribution area of nine gall midge species bound on *Quercus cerris* L. in the Palaearctic Region: A—Distribution area of *Quercus cerris* L. and closely allied species (black line); distribution of gall midges (//////); outlying occurrence of gall midges on cultivated trees and shrubs Botanical gardens (●). B—northern boundary of distribution area of both oaks (black line) and gall midges < black circles), on locations lying in the southern part of Czechoslovakia; C—for comparison, a detailed distribution of gall midges of Czechoslovakia from more than 1000 locations.

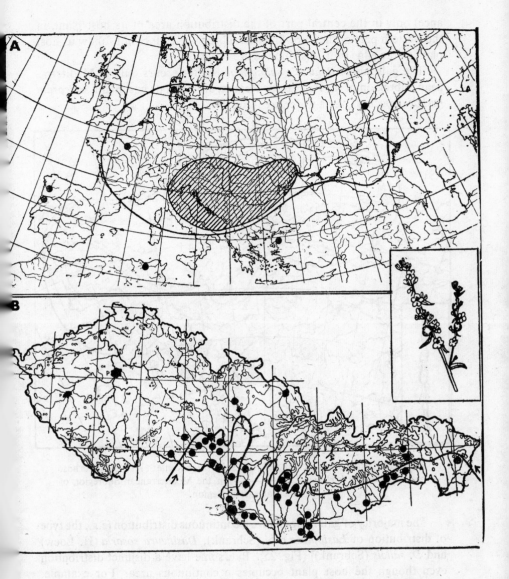

Figure 23. A—The distribution area of the gall midge *Asphondylia verbasci* (Vallot) and its host plant, *Verbascum lychnitis* L. in the Palaearctic Region; out lying occurrence on the other species of the genus *Verbascum*; B—northern boundary of distribution area of *A. verbasci* (black circles) on locations lying in the southern part of Czechoslovakia along the annual isotherme of 10 °C (black line).

ance) only in the central part of the distribution area of its host plant, in the western part of the Euro-Siberian subregion. It does not follow its host plant to the edge of its distribution area (Fig. 23).

The gall midge *Baldratia salicorniae* Kieffer occurs only in Mediterranean Subregion of the Palaearctic Region, although its host plant *Salicornia herbacea* occurs over a wider area (Fig 24).

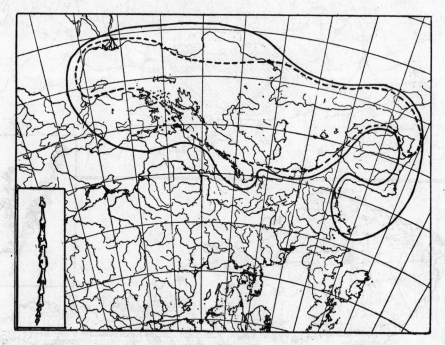

Figure 24. Distribution area of *Baldratia salicorniae* Kieffer (– – –) and its host plant, *Salicornia herbacea* L. (————) in the Mediterranean Subregion of the Palaearctic Region.

The majority of gall midges have a continuous distribution [e.g., the type of distribution of *Lasioptera rubi* (Schrank), *Dasineura rosaria* (H. Loew) and *D. salicis* (Schrank)] (Fig. 25). But, some have a disjunct distribution even though the host plant occupies a continuous area. For example, *Zeuxidiplosis giardi* (Kieffer) has such a distribution. Small territories, on which it occurs, lie about 1,000 km from each other (they are so-called hiates) (Fig. 26), but the distribution of the host *Hypericum perforatum* L. is continuous.

Several species of gall midges which are naturally distributed in other zoogeographical regions occur as well in the marginal parts of the Palaearctic Region. For example, *Contarinia sorghicola* (Coquillett) which occurs in all the tropics and subtropics of the world was found in southern France; it was probably imported to that area by man. Similarly, *Orseliola cynodontis*

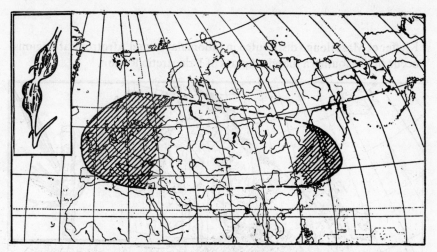

Figure 25. Distribution area of the gall midge *Lasioptera rubi* (Schrank) on host plant—various species of the genus *Rubus*, in the western and eastern part of the Palaearctic Region. From the central area of this region there are no data.

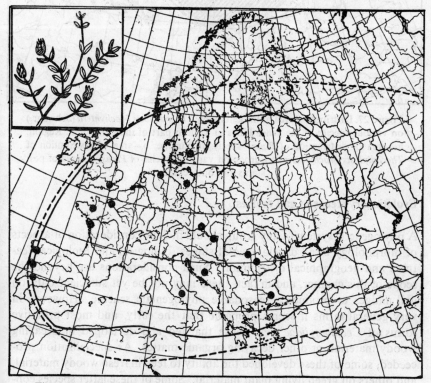

Figure 26. Disjuncted area of distribution of the gall midge *Zeuxidiplosis giardi* (Kieffer) (black circles) and continuous area of distribution of *Dasineura hyperici* (Bremi) (black line) and their host plant, *Hypericum perforatum* L. (- - -) in the Palaearctic Region.

Kieffer et Massalongo distributed in Indomalayan and Afrotropical Regions occur in the southern marginal parts of Palaearctic Region.

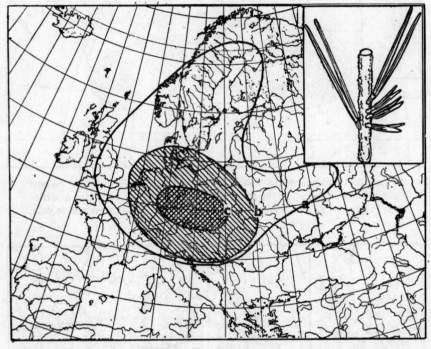

Figure 27. Distribution area of the gall midge *Thecodiplosis brachyntera* (Schwägr.) on various species of the genus Pinus in western part of the Palaearctic Region: a—area of distribution of *Pinus*—species in Europe; b—area of distribution of *Thecodiplosis brachyntera* in Europe on various species of *Pinus*; c—area of pest status and outbreaks.

Evolutionary trends in gall midges

A phylogenetic system of any group should include consideration of aspects of morphology of all developmental stages, biological and ecological features and geographical distribution. Such considerations are beyond the scope of this chapter, and in fact, there may not be yet sufficient data on gall midges to produce a reasonable phylogenetic scheme. However, in very general terms it seems probable that the early, and most primitive forms of *Cecidomyiidae* were those that fed and developed in decaying woody material, or other decaying organic matter. As specialization proceeded, some of these developed the ability to feed in fresh woody material, and others on green living plant material. Some of these latter species continued to specialize and were eventually were able to direct the growth of some part of the growing plant and thus cause galls to be formed. Some species of the gall formers probably became further specialized as inquilines while others, or perhaps some of the inquiline forms, adopted the zoophagous mode of life. If these assumptions are correct, it appears clear that the subfamily *Cecidomyiinae* is more advanced than either the *Lestremiinae* or *Porricondylinae* because a greater number of species in that group are of the more specialized types (Fig. 28).

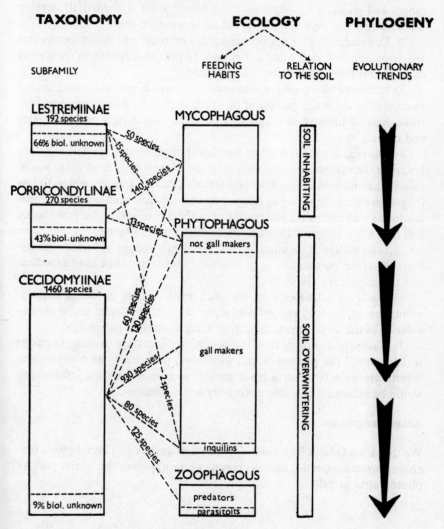

Figure 28. Evolutionary trends of gall midges according to different factors.

Problems in the study of gall midges

The current and future problems in the study of gall midges may be summarized as follows:

220

1) It is necessary to resolve the problems of generic, and species, level status of many gall midge species by thoroughly considering information on morphological characters, and their variation, on all developmental stages including larvae, pupae, and both sexes of adults. Details of life cycles are needed including embryonic and genetic information, host plant or prey range, and parasitiod complexes. This information is needed to develop usable diagnostic keys to many of the taxonomic groups.

2) Detailed biological studies are needed on those gall midge species that are damaging to forestry and agriculture so that suggestions to reduce the damage that these pests cause, can be made.

3) Increased efforts on the studies on zoophagous gall midges are needed to determine the potential role of these predacious forms as biological control-agents of forestry and agricultural pests, particularly against aphids and mites.

4) Distribution maps must be developed of gall midge species, and their hosts for the various zoogeographic regions of the world. Such maps would help in the understanding of the pest status of certain species, and to follow long-term fluctuations of gall midge populations. Information on such fluctuations should allow the forecasting of outbreaks of gall midge pest species and thus allow properly timed control efforts to be instituted.

5) Information is required on the role of gall midges in various ecosystems of nature, particularly as to how such insects might be used as indicators of ecosystem stability.

6) Study of all aspects of the cecidomyiid group, including biology, biotaxonomy, ecology and biogeography must be continued use in the elucidation of the phylogenetic process of the evolution of gall midges.

7) Lastly, and perhaps most important, it is necessary to continue efforts to understand the process of gall formation, particularly as to how these insects are able to control plant growth so precisely. Such information would be extremely valuable in forestry and agriculture.

Acknowledgements

We thank dr. Odette Rohfritsch for the photographs of Thecodiplosis brachyntera/sections/and Ladislav Havel who prepared the main part of photographs of galls.

REFERENCES*

1. Barnes, H.F. 1946–1956. Gall Midges of Economic Importance Vol. I–VII, London. 104, 160, 184, 165, 229, 270, 261 pp. respectively.
2. Buhr, H. 1964–1965. Bestimmungstabellen der Gallen (Zoo- und Phytocecidien an Pflanzen Mittel- und Nordeuropas.) Band 1+2. Jena. 1572 pp.

*Only fundamental works which include basic information and further references are given.

221

3. Docters van Leeuwen, W.M. 1982. Gallenboek (revised by A.A. Wiebes-Rijks and G. Houtman). Zutphen. 355 pp.
4. Edwards, F.W. 1938. On the British Lestremiinae, with notes on exotic species. *Proc. R. ent. Soc. Lond.* (B) 7 : 18-32, 102-108, 173-182, 199-210, 229-243, 253-265.
5. Felt, E.P. 1925. Key to Gall Midges. A resumé of studies I-VII, (Itonididae). *Bull. N.Y. St. Mus.* 257 : 1-239.
6. Gagné, R.J. 1968. Family Cecidomyiidae. In: A Catalog of the Diptera of the Americas South of the United States. Sao Paulo: Şecr. Agricultura, Dep. Zool. 62 pp.
7. Gagné, R.J. 1973. Family Cecidomyiidae. In M.D. Delfinado and D.D. Hardy: A Catalog of the Diptera of the Oriental Region. Vol. 1. Univ. Hawaii Press. 480-517.
8. Grover, P. 1970. Origin and classification of family Cecidomyiidae (Diptera: Nematocera). *Cecidol. Indica* 5 : 71-107.
9. Harris, K.M. 1966. Gall midge genera of economic importance (Diptera, Cecidomyiidae). Part I. *Trans. R. ent. Soc. Lond.* 118 (10) : 313-358.
10. Harris, K.M. 1967. A systematic revision and biological review of the Cecidomyiid predators (Diptera: Cecidomyiidae) on world Coccoidea (Hemiptera: Homoptera). *Trans. R. ent. Soc.* 119 (13): 401-494.
11. Harris, K.M. 1980. Family Cecidomyiidae. In: Crosskey R.W. (ed.) : Catalogue of the Diptera of the Afrotropical Region. London: British Mus. (Nat. Hist.): 238-251.
12. Hendel, F. 1937. Ordnung der Pterygogenea: Diptera-Fliegen. In: W. Kükenthal, Handbuch der Zoologie, Band IV, Berlin and Leipzig, 1729-1998.
13. Houard, C. 1908-1909. Les Zoocécidies des Plantes d'Europe et du Basin de la Mediterranée. Vol. 1+2, Paris, 1247 pp.
14. Kieffer, J.J. 1913. Family Cecidomyidae. Genera insectorum. *Fasc.* 152, 346 pp.
15. Kleesattel, W. 1979. Beiträge zu einer Revision der Lestremiinae (Diptera, Cecidomyiidae) unter besonderer Berücksichtigung ihrer Phylogenie. Dissertation, Stuttgart, 274 pp.
16. Mamaev, B.M. 1968. Evolution of gall midges. Leningrad. 236 pp. (in Russian)
17. Mamaev, B.M., Krivosheina, N.P. 1965. Larvae of gall midges, Diptera, Cecidomyiidae. Moscow. 276 pp. (in Russian).
18. Mani, M.S. 1946. Studies on Indian Itonididae (Cecidomyiidae: Diptera). VIII. Keys to the genera from the Oriental Region. *Indian J. ent.* 7 : 189-235.
19. Mani, M.S. 1964. *Ecology of Plant Galls.* The Hague: Dr. W. Junk Publishers. 434 pp.
20. Mani, M.S. 1973. *Plant Galls of India.* New Delhi: Macmillan India. 354 pp.
21. Mohn. E. 1955. Beiträge zur Systematik der Larven der Itonididae (=Cecidomyiidae, Diptera). 1. Teil: Porricondylinae und Itonidinae Mitteleuropas. *Zoologica*, 38, Heft 105, 247 pp.
22. Mohn, E. 1966-1971. 6. L. Cecidomyiidae=(Itonididae). In: Lindner, E.: Die Fliegen der Palaearktischen Region, Lief. 269. 248 pp.
23. Nijveldt, W. 1969. Gall midges of Economic Importance. Vol. VIII. 221 pp.
24. Panelius, S. 1965. A revision of the european gall midges of the subfamily Porricondylinae (Diptera: Itonididae). *Acta Zool. Fenn.* 113 : 157 pp.
25. Pritchard, A.E. 1960. A new classification of the paedogenetic gall midges formerly assigned to the subfamily Heteropezinae (Diptera: Cecidomyiidae). *Ent. Soc.. Amer. Ann.* 53 : 305-316.
26. Rohdendorf, B.B. 1964. Historical development of the Diptera. Moscow. 310 pp. (in Russian).
27. Roskam, J.C. 1977. Biosystematics of insects living in female birch catkins. I. Gall midges of the genus *Semudobia* Kieffer (Diptera, Cecidomyiidae). *Tijdschr. Ent.* 120 : 153-197.

28. Ross, R., Hedicke H. 1927. Die Pflanzengallen (Cecidien) Mittel-und Nordeuropas, ihre Erreger und Biologie und Bestimmungstabellen. Jena. 348 pp.

29. Rübsaamen, E.H., Hedicke, H. 1926–1939. Die Zoocecidien, durch Tiere erzeugte Pflanzengallen Deütschlands und ihre Bewohner. II. Band: H. Hedicke: Die Cecidomyiiden (Gallmucken) und ihre Cecidien. Unter Benutzung von Prof. Ew. H. Rübsaamen + bearbeitet. *Zoologica* 29 : 350 pp.

30. Skuhravá, M. 1973. Monographie der Gallmückengattung *Clinodiplosis* Kieffer, 1894. Studie ČSAV, Praha, Nr. 17, 80 pp.

31. Skuhravá, M. 1985. Family Cecidomyiidae. In: A Catalog of the Palaearctic Diptera. Budapest, Amsterdam (in Press).

32. Skuhravá, M., Skuhravý, V. 1960. Bejlomorky (Gallmücken). Praha. 260 pp. (In Czech with German summary).

33. Skuhravá, M., Skuhravý, V. 1973. Gallmücken und ihre Gallen auf Wildpflanzen. Die neue Brehm-Bücherei. Wittenburg. 117 pp.

34. Skuhravá, M., Skuhravy, V. 1973. Die Gallmücken (Cecidomyiidae, Diptera) des Schilfes (*Phragmites communis* Trin.). Studie ČSAV, Praha, Nr. 3, 150 pp.

35. Stone, A., et al. 1965. A Catalog of the Diptera of America north of Mexico. 1696 pp. Washington, D.C.

36. Sylvén, E., Carlbäcker, U. 1981. Morphometric studies on Oligotrophini adults (Diptera: Cecidomyiidae) including an attempt to correct for allometric deviations. *Ent. scand.*, Suppl. 15 : 185–210.

37. Udvardy, M.D.F. 1975. A classification of the biogeographical provinces of the World. IUCN Occasional Paper No. 18; Switzerland. 48 pp.

38. Yukawa, J. 1971. A revision of the Japanese Gall midges (Diptera: Cecidomyiidae). *Mem. fac. agric. Kagoshima Univ.* 8 : 1–203.

8. The Biology of Gall Wasps

R.R. Askew

Introduction and classification

Gall wasp biology is many-sided and complex. The gall wasps themselves are at the centre of an intricate web of mutual interrelationships that involve the galls they create, the host plants on which they feed, and a variety of other organisms which are more or less dependent upon them. Everywhere one finds the most delicate and subtle bonds linking the component species; the remarkable capacity of the gall wasp larva to regulate the growth of the host plant to produce a gall of often extraordinary complexity and the varied trophic relations of the animal species that unite them in an ordered and highly structured community. The focal organism of the community, the gall wasp, will be the chief consideration in this chapter but, to achieve a degree of completeness, attention must be given also to the inquilines and parasites that are dependent upon the galls produced.

Gall wasps belong to the family Cynipidae which is placed, with other families, in the superfamily Cynipoidea. Cynipoidea are parasitic Hymenoptera and all species except those in the subfamily Cynipinae behave as typical parasitoids or hyperparasitoids of other insects. The Cynipinae are the gall wasps and they alone induce gall formation and as larvae are phytophagous.

Classification of the Cynipoidea, as outlined by Quinlan (1979) and modified by Quinlan and Evenhuis (1980), is as follows:

Superfamily Cynipoidea
Family Ibaliidae Thomson 1862   Endoparasitoids of larvae of
           Siricidae (Hymenoptera)
Family Liopteridae Ashmead 1895   Biology unknown
  Subfamily Oberthuerellinae
   Kieffer 1903

223

Subfamily Liopterinae Ashmead 1905
Subfamily Mesocynipinae Hedicke &
   Kerrich 1940

Family Eucoilidae Thomson 1862   Endoparasitoids of Diptera

Family Figitidae Thomson 1862

   Subfamily Aspicerinae Kieffer 1910 Endoparasitoids of Syrphidae
                                      (Diptera)

   Subfamily Figitinae Thomson 1862   Endoparasitoids of Diptera
                                        (mainly Sarcophagidae)

   Subfamily Anacharitinae Thomson Endoparasitoids of Chrysopidae
     1862   and Hemerobiidae (Neuroptera)

   Subfamily Himalocynipinae
     Yoshimoto 1970

Family Cynipidae Hartig 1840

   Subfamily Austrocynipinae Riek 1971
   Subfamily Pycnostigmatinae
     Cameron 1905

   Subfamily Alloxystinae Hellén 1931 (= Allotriinae) Hyperparasitoids
                                        of Aphidae (Hemiptera)
                                      via Hymenoptera

   Subfamily Charipinae Dalla Torre & Endoparasitoids of Psylloidea
     Kieffer 1910   (Hemiptera)

   Subfamily Cynipinae Hartig 1840   Gall wasps

Placing the Cynipinae at the end of the list of cynipoid taxa acknowledges their status as one of the most specialized or 'advanced' groups in the superfamily. Telenga (1969) considers Cynipinae to be of relatively recent origin, having without doubt evolved from an entomophagous stock. Gall wasps may be thought of as parasites of plants rather than of insects. Evolutionary switching between entomophagy and phytophagy is not unusual among parasitic Hymenoptera, although the direction of the dietary change may sometimes be debatable. In the Chalcidoidea, for example, larval phytophagy occurs in several distantly related taxa and the habit must have evolved separately several times. The chalcid family Torymidae is basically entomophagous but most species of the genus *Megastigmus* have adopted a plant diet, although a few species have retained or, more likely, again reverted to an entomophagous habit and feed as larvae on gall wasp larvae. Also in the family Torymidae, some species of *Torymus* (subgenus *Syntomaspis*) develop as larvae phytophagously in fruits and the larva of *T. (S.) cyaneus* Walker, a parasite in oak galls, is known to feed initially on gall tissue and only later on the gall wasp larva (Askew, 1961a). In the Cynipinae there are some gall wasps whose larvae have lost the ability to induce gall formation and these larvae live obligatorily as inquilines in the galls of gall-forming Cynipinae. Here they are phytophagous but their presence may result in the death of the gall-maker and it is possible that in some cases the gall-maker is eaten by the inquiline. Inquilines may modify the structure of their host galls.

Distinction between inquiline and gall-making Cynipinae marks an important and obvious biological division of the subfamily. Hartig (1840) recognised this in organising the gall wasps into two groups, Inquilinae (inquilines) and Psenides (gall-producers) and Krombein et al. (1979) follow this division in cataloguing American species under Synerginae (inquilines) and Cynipinae (gall-makers). The separation of inquilines and gall-makers, however, is based upon their present larval feeding habits rather than upon their phylogeny. The inquilines do not easily form a discrete entity, distinguished by morphological characters of the adult insects, indicating that the inquiline habit is probably polyphyletic. Nevertheless, provided that it is borne in mind that the inquilines may have a number of origins, and until their phylogenetic relationships to other gall wasps are more clearly understood, it is useful to consider them together as a biological unit.

The gall-making Cynipinae may be separated into three groups so that a division of Cynipinae into four groups, first proposed as tribes by Kinsey (1920), is adopted here:

Aylax group (tribe Aylaxini)	Gall-makers on plants other than *Rosa* and *Quercus*
Synergus group (tribe Synergini)	Inquiline gall wasps
Diplolepis group (tribe Rhoditini)	Gall-makers on *Rosa*
Cynips group (tribe Cynipini)	Gall-makers on *Quercus* and allied Fagaceae

Aylaxini includes rather diverse elements with the genus *Pediaspis* fitting here particularly uneasily, and in Cynipini is *Neuroterus* which shares some morphological features with the Aylaxini. In spite of these imperfections, the above classification neatly segregates a number of biological characters including host plant relations.

Adult morphology

Gall wasps are of rather uniform structure and are readily recognisable. They possess all the features of Hymenoptera, but their general appearance is not at all wasp-like. The spherical gall chamber which they almost fully occupy during their immature stages seems to impose its stamp on adult body shape which is short and globular. Bright yellow, red or white pigmentation, found in some other groups of Hymenoptera, is absent from gall wasps which instead are mainly black or black and dull yellow to brown (reddish brown in *Xestophanes*, *Diplolepis* and some others). Metallic colours derived from cuticular structure are not displayed by gall wasps.

The head is hypognathous with biting mouthparts, an important function of the powerful mandibles being to gnaw an exit hole, often through hard and woody tissue, from the gall. Feeding requirements of adult gall wasps are little known; I have never found them feeding at flowers. Compound eyes, three ocelli and a pair of filiform antennae comprise the main sense organs. The antenna includes thirteen to fifteen segments, the males of a species generally having one more antennal segment than their females as is frequently the case among Hymenoptera. They are long but with little obvious modification, the scape being short and the flagellum comprising a series of similar segments without a differentiated club or clava. The third antennal segment is more or less enlarged and sometimes contorted in some males, notably of *Synergus* and *Neuroterus*. Antennae are employed by females to explore the surface of the host plant preparatory to oviposition, and they are well-provided with sensilla.

The thorax is short and deep, strongly arched dorsally with the short pronotum sloping upwards from the neck to join, without the formation of a posterior collar, the mesonotum which makes up the main part of the dorsal surface. Posteriorly the short metanotum and the propodeum (first abdominal segment fused with thorax) are very steeply inclined. The mesonotum is clearly differentiated into anterior mesoscutum and posterior scutellum, and the mesoscutum is divided by more or less complete notaulices (parapsidal furrows) and sometimes by a partial median groove. Anteriorly the scutellum is often provided with two pits or foveae, the form of which may be of taxonomic value. The type of sculpture on the mesonotum and mesopleuron is also used in taxonomy of the group. The thorax accomodates the flight muscles and the forewings of gall wasps are relatively large. The small hindwings are coupled to the forewings by a row of small hooks or hamuli on the costal margin of the hindwing. Female gall wasps on the host plant walk over its surface and do not fly a great deal. Their movements are often slow and deliberate, especially so in *Diastrophus*, more rapid in *Neuroterus* and the inquiline genera. Some species in the *Cynips* group (*Biorhiza, Trigonaspis*) are apterous or brachypterous, at least in one of their generations, and this clearly must restrict dispersion between host plants. Dispersion is of greatest importance to those species galling short-lived, scattered host plants or different species of host plant in alternating generations; aptery is thus confined to non-heteroecious oak gall wasps. When disturbed, gall wasps frequently seek refuge by dropping from the host plant and feigning death, rather than by taking flight. The venation of gall wasp forewings is very characteristic with an elongated, pointed radial cell which may or may not be completely enclosed at the costal margin. The radial cell is shorter and broader in the *Aylax* and *Synergus* groups than it is in the more specialized species.

The gaster (the abdomen posterior to the petiole) is strongly compressed laterally so that in dorsal view it is very narrow, but in females it is deep, housing the long ovipositor which is curved and concealed inside when not in use. The hypopygium on the ventral surface of the female gaster is well-developed and it terminates in a process or spine, the length and adornment of hairs on which provide characters of taxonomic value.

The ovipositor is composed of three parts, a pair of dorsal (or second) valves which are fused to form a single unit, a pair of ventral (or first) valves which are longitudinally grooved to receive complementary ridges on the dorsal valves, and a pair of sheaths (third valves) which are broader than the other valves which they enclose. During drilling, the dorsal and ventral valves acting together are worked into the host plant. They are equipped with cutting ridges at their apices and the ventral valves, sliding beneath the dorsal valves, are forced into the plant by the action of powerful muscles in the abdomen. The two ventral valves work alternately, first one and then the other being driven a short distance into the plant. During this process the dorsal valves function mainly as a support and guide for the ventral valves which are the actual organs of penetration. Sensilla on the valves act as proprioceptors during drilling, whilst others at the tips provide information on the substrate being drilled. The length of the ovipositor of a species is related to the depth at which the eggs are placed below the plant surface. The drill-hole, however, is seldom perpendicular to the surface, and when oviposition is into buds it often follows a curved track since it is started by the female gall wasp inserting the tip of its ovipositor under the edge of an outer bud scale.

Ritchie and Peters (1981) provide a recent account of the external morphology, and it nomenclature, of a gall wasp species (*Diplolepis rosae* L).

The four groups of gall wasps, listed at the end of the previous section, may be partly distinguished on morphological characters of the adult insects as follows:

1. Pronotum short, without median pits; hypopygial spine long or females apterous...2
— Pronotum longer, usually with median pits; hypopygial spine short; never apterous.........................*Aylax* and *Synergus* groups
2. Hypopygium ploughshare-shaped; mesopleuron with longitudinal furrow; tarsal claws simple...........................*Diplolepis* group
— Hypopygium not ploughshare-shaped; mesopleuron lacking longitudinal furrow; tarsal claws sometimes with a basal lobe or tooth............
....*Cynips* group

The immature stages

Gall wasps have a high reproductive potential and females of the larger species of *Andricus* may contain about one thousand eggs. Each egg is white with a smooth chorion and consists of a rather small body and a long pedicel which is slightly clavate at its free end. The pedicel is usually three to seven times as long as the egg body. After the apex of the ovipositor has penetrated the host plant to a location suitable for oviposition, eggs are worked down the egg canal between dorsal and ventral ovipositor valves by small movements of these valves. The body of the egg precedes its pedicel down the ovipositor and because it is of greater diameter than the egg canal, a part of its contents are temporarily diverted into the pedicel. Another function of the egg pedicel is to serve as a respiratory organ for the developing embryo. It extends up the drill-hole towards the plant surface and is longest in those species that deposit their eggs most deeply.

The gall wasp larva is short and stout, broadly rounded at both ends and more or less c-shaped. Thirteen body segments follow the head and the integument is smooth and devoid of setae. A pair of inconspicuous open spiracles is located on each of body segments two to ten inclusively. There is no trace of legs and the larva is very sluggish in its movements. The head bears a pair of very short antennae, but its most obvious appendages are the mandibles. These are strongly sclerotized and quadridentate. The larva feeds and grows but does not defaecate until immediately prior to pupation. This avoids contamination of the larval cell. When fully grown the larva pupates without any trace of a cocoon. The pupa is exarate. Pupation always takes place inside the gall, escape from which is effected by the adult insect; the larva does not make a partial exit from the gall before pupating as do some other groups of gallicolous larvae.

Gall formation

The real nature of the association between galls and the insects that they contain was misunderstood for a long time. There were some who believed that insects appeared inside galls as a result of spontaneous generation. Before it was realized that oak gall wasps have alternation of generations, with the production of totally different galls by the two generations of a species, it was very difficult to explain the apparent gap in the life cycle of a species that, for example, produces galls in spring. Adults emerge from these galls in early summer and soon die. Similar galls do not appear again until the following spring, in buds, leaves or catkins that have freshly grown. One interesting but erroneous theory was that the gall wasp egg was laid in the ground, taken up by the roots of the oak, and transported in the sap to the site of gall development.

Marcellus Malpighi, physician to Pope Innocent XII, published in 1675 a 28-page treatise entitled 'De gallis', and this was the first extended literary account of gall insects. Malpighi clearly established that galls are vegetable products whose formation is induced by insects. Our knowledge today is greater but still incomplete and an understanding of the precise biochemical interrelationship between gall wasp and plant, whereby each gall wasp species engenders its own characteristic and complex gall structure, remains fragmentary.

The female gall wasp uses its elongated ovipositor to penetrate more or less deeply into plant tissue wherein the eggs are deposited. Each species is precise in the location of its eggs, selecting with very few exceptions a single plant species and almost always a particular organ on that plant. Galls may be formed on roots, stems, fine branches, leaves, buds or flowering parts. It was originally believed that pricking of plant tissue by the gall wasp's ovipositor, or perhaps irritation caused by lubricating secretions produced by the female gall wasp during oviposition, induced gall formation. That these are not the causes of cecidogenesis by Cynipidae (although true for galls of Tenthredinidae (sawflies)) became clear from the works of Beijerinck (1883) and Adler (1881) who demonstrated that gall formation does not begin until after the gall wasp larva has emerged from its egg, and that gall development ceases upon death of the larva. It is a constituent of the larva's salivary secretions which acts upon the plant tissues in such a way as to stimulate hypertrophy and hyperplasy resulting in the production of the type of gall specific to each gall wasp species. The active substance in the larval secretion may well be an analogue of the plant's auxins, but no explanation can yet be offered for one of the most remarkable aspects of cynipid biology; the variety of galls produced by different gall wasp species on the same plant and sometimes the same structure. Much has yet to be learnt about the biochemical interaction between larva and plant.

Galls provide the cynipid larva with food and shelter. Initially they probably represent a defensive reaction by the plant, an attempt to isolate the gall-inducing organism, but the course of their subsequent development is controlled by the gall wasp larva in a manner beneficial to itself. Because it is modified plant growth that produces galls, the majority form in actively growing parts, and oak buds are the most frequent site for oviposition. Oviposition in buds does not necessarily mean, however, that a bud gall will result; galls on catkins and leaves may follow egg-laying on these structures when they were still enclosed by bud scales. The most elaborate galls are usually those on flowers and buds, the least complex are stem and root galls, an inverse relationship between gall organisation and the age of the plant structure that was first noted by Zweigelt (1931) in the case of aphid galls.

The predominance of galls of the more specialized Cynipini on leaves, buds and floral parts, and of the less specialized Aylaxini on stems and runners, is illustrated by the following data showing the distribution over plant structures of galls of British species of Cynipidae.

	roots runners	stems branches	leaves	buds	flowers fruits
Aylaxini	2	11	3	0	5
Rhoditini	0	2	4	0	0
Cynipini	2	4	16	26	17

Figures for Cynipini represent generations. Where a gall is found on more than one structure, the more usual location is given.

GALL STRUCTURE

Galls may be unilocular (monothalamous) or plurilocular (polythalamous). The former contain just a single larval chamber unless secondarily attacked by inquilines, but the latter contain many gall wasp larval chambers. Plurilocular galls result from the coalescence of gall tissue surrounding a number of larvae that have emerged from eggs laid in close proximity. Plurilocular galls are generally less specialized than unilocular galls and are the type most often produced by Aylaxini. Only relatively few Cynipini form plurilocular galls.

After oviposition, the egg is usually more or less embedded in plant tissue. Very soon after the larva hatches the plant tissue in its vicinity breaks down so that it sinks into a small pocket which forms the larval chamber. Cells surrounding the larval chamber proliferate and close over the external opening so that the larva is completely enclosed (Magnus, 1914). Less commonly the egg is laid subepidermally and the gall wasp larva is surrounded by plant tissue from the outset. Gall cells are mostly larger than normal cells and their nuclei may be about twice the usual diameter (Meyer, 1950). Surrounding the larval chamber is a layer of tissue, rich in nitrogenous compounds and termed the nutritive zone, upon which the larva browses.

The structure of galls peripheral to the nutritive zone is varied. In the simplest types, such as those of *Neuroterus aprilinus* (Giraud) ♂♀ and *N. quercusbaccarum* (L.) ♂♀, the nutritive zone is surrounded by thin-walled parenchyma cells. Such galls are soft with a high water content. In most galls, between the parenchyma and the nutritive zone, there is a layer of sclerenchyma which forms a protective shell about the nutritive zone and larval chamber. Sclerenchyma, nutritive layer and larval chamber constitute an inner gall. The outer gall may be composed of thin-walled parenchyma, as in the galls of *Andricus curvator* Hartig ♂♀ and the spangle galls of the agamic generations of some species of *Neuroterus*, or it may be composed of thick-walled parenchyma and of a harder, woody consistency, as in galls of *Cynips divisa* Hartig ŏ and *C. longiventris* Hartig ŏ. One of the most specialized galls is that of *A. kollari* (Hartig) ŏ in which eight concentric zones of different tissues have been described (Adler and Straton, 1894).

Galls may or may not break through the epidermis of the host plant. Those that do not, such as the stem galls of *Aulacidea* and *Phanacis* and the galls of *Andricus inflator* Hartig ♂♀ and *A. curvator* ♂♀ in oak twigs and leaf petioles, merge more or less imperceptibly into ungalled plant tissue. The gall of *A. inflator* often bears buds which later develop and old galls, which remain attached to the tree, sprout twigs. Galls that erupt through the epidermis are usually less persistent, since they are often attached to the host plant by a narrow stalk, and some are regularly deciduous. Spangle galls of *Neuroterus* do not complete their development unless they fall from the oak leaves in autumn; these galls are remarkable in continuing to enlarge after they have dropped to the ground. The gall of *A. ostreus* Hartig ŏ is also deciduous but it only partially breaks through the epidermis, two epidermal scales remaining on either side of the small gall growing on a leaf vein and suggesting its common name of oyster gall. The surfaces of galls that erupt may become differentiated into a secondary epidermis and this may carry normal epidermal structures, as for example on the oak leaf gall of *Cynips quercusfolii* L. ŏ whose surface sometimes bears stomata. Surfaces of galls are frequently hairy. Bedeguar galls of *Diplolepis rosae* (L.) are covered by a dense coat of branched processes like those of the normal rose epidermis but much longer, and the surface of galls of *D. nervosa* (Curtis) may be either smooth or equipped with a few thorn-like spikes. Many oak galls have a hairy surface; the plurilocular gall of *Andricus quercusramuli* (L.) ♂♀, formed on male catkins of oak, has a dense vestiture of long, white hairs and looks like a ball of cotton wool. Spangle galls of *Neuroterus quercus baccarum* (L.) ŏ (= *lenticularis* Olivier) and *N. numismalis* (Geoffroy in Fourcroy) ŏ bear respectively stellate and biramous hairs, such hairs being peculiar to the surfaces of these galls and not present elsewhere on the oak tree.

Galls provide the gall wasp larva with food, shelter and a degree of protection. Insectivorous birds and mammals may open galls to feed on the larvae but most mortality of gall wasps is probably inflicted by parasitic Hymenoptera, particularly Chalcidoidea. Many species of chalcid wasp use gall wasps as hosts, piercing the gall wall with their ovipositors in order to lay eggs in or on the bodies of gall wasp larvae. Gall wasp larvae control the development of their galls and many of the features of galls may be interpreted as devices, evolved by natural selection acting on the gall wasps, that enhance their protectiveness. Several galls have a high tannin content which renders them bitter-tasting and probably deters potential vertebrate predators from breaking them open. The probability of successful attack by chalcids is reduced in a number of ways. A hairy covering makes it difficult for chalcids to explore a gall surface with their antennae and to insert their ovipositors. A few galls, for example those of *Andricus quercuscalicis* (Burgsdorf) ŏ on acorns, are covered by a sticky secretion that, as well as making surface exploration by chalcids hazardous, attracts ants which appear to gather the secretion and probably also ward off attacking chalcids. Hard galls with thick-walled parenchyma are drilled into more slowly than soft galls, and the broken off ovipositors of chalcids (especially *Torymus* species) sometimes found embedded in woody galls testifies to the fact that chalcid attack can be thwarted. Gall hardness is the protective strategy adopted by *Cynips divisa* ŏ and *C. longiventris* ŏ whereas the allied *C. quer-*

cusfolii ŏ has a soft gall of thin-walled parenchyma and derives its protection from the great depth of parenchymatous tissue (Askew, 1961e). The layer of sclerenchyma surrounding the inner gall must also present something of a hindrance to penetration of the larval chamber by ovipositing chalcids, and this is probably especially true when the inner gall lies loosely in a cavity in the outer gall, as is the case in the gall of *Andricus curvator* ♂♀. Other galls have smaller cavities or atria in addition to the larval chamber and these may serve to foil chalcid attack by acting as decoys; a chalcid penetrating such an empty atrium with the tip of its ovipositor will receive no stimuli to elicit oviposition and it may leave the gall with the gall wasp larva remaining unharmed. *Cynips disticha* Hartig ŏ provides an example of a gall with a single atrium laying immediately above the larval chamber.

Notwithstanding these diverse protective measures, most types of galls suffer from heavy attack by parasites. Galls that appear to escape most lightly are those that form early in the season and develop very rapidly before most chalcid species are active. *Neuroterus aprilinus* ♂♀ has such a gall, the adult wasps mostly emerging in April only a few days after the galls first become visible. The alternate generation of this species, *N. aprilinus* ŏ (=*schlechtendali* Mayr), also develops rapidly in spring on male catkins which soon fall to the ground. The agamic generation emerges in August and September and females oviposit in buds, the eggs remaining dormant until the following spring. The total period spent feeding on the host oak tree in the entire life cycle of *N. aprilinus* is only about four weeks.

Chalcidoid parasites do not seem to search on the ground for fallen host galls so that galls are at risk from parasite attack only as long as they remain on the host plant. Deciduousness, or alternatively falling with the organ on which they develop, are therefore effective protective devices. In the case of the gall of *Andricus fecundator* (Hartig) ŏ, it is only the inner gall that is deciduous, it being extruded from the apex of the hypertrophied bud that forms the outer gall and which remains attached to the oak tree for a year or more.

Host plants

The association between gall wasps and their plant hosts is one of such delicacy that one might anticipate that the plants used as hosts by Cynipinae would all be closely allied or at least share some physiological property that made them susceptible to and suitable for gall wasp attack. The former is not the case and the latter is not immediately apparent. The numbers of Cynipinae associated with different plant genera in North America and Europe are displayed in Table 1. From this it can be seen that the proportions of species associated with different plant genera are very similar on the two continents. Oak (*Quercus*), attacked by almost all Cynipini, is the most heavily infested (88.2 per cent of American species and 75.3 per cent of European species) with *Rosa*, the sole genus exploited by Rhoditini, the next most often attacked (5.4 and 4.2 per cent of species in America and Europe respectively). It should be noted that the number of species given as occurring on oaks in America is probably rather too high, some species being likely to be listed twice under each of their alternate generations. It is the Aylaxini which have a remarkable diversity of plant hosts, most of these being in the family Compositae, but genera in Labiatae, Papaveraceae, Valerianaceae, Umbelliferae, Rosaceae and Aceraceae are also included in the host plant list. Unfortunately, too little is known of the history of plant colonization by Aylaxini to make it possible to account for the present-day distribution of the group. As understood here, Aylaxini is rather heterogeneous. Quinlan (1968) restricts inclusion in the tribe to those genera forming galls on 'herbaceous plants, *i.e.* Compositae, Labiatae, Papaveraceae, the aromatic plants with milky juice'. Such restriction excludes *Pediaspis* which galls *Acer* and differs from all other Cynipinae in having a heart-shaped impression on the scutellum, the genera attacking Rosaceae (*e.g. Xestophanes, Diastrophus*), and *Aylacopsis* on *Heracleum*.

Table 1. Numbers of species of gall-making Cynipidae associated with different plant genera in Europe (data modified from Dalla Torre and Kieffer, 1910) and in North America (from Krombein *et al.*, 1979).

Plant families and genera	Cynipid genera	European species	North American species
ACERACEAE			
Acer	*Pediaspis*	1	0
COMPOSITAE			
Ambrosia	*Aulacidea*	0	1?
Centaurea	*Isocolus*	5	0
	Phanacis	1	0
Chrysothamnus	*Antistrophus*	0	1
Crepis	*Phanacis*	1	0
Hieracium	*Aulacidea*	5[1]	0
Hypochaeris	*Phanacis*	1	1[2]
	Aulacidea	1	0
Lactuca	*Aulacidea*	0	4[3]
	Phanacis	1	0
Lapsana	*Phanacis*	1	0
Lygodesmia	*Antistrophus*	0	1
Microseris	*Antistrophus*	0	1
Picris	*Phanacis*	1	0
Prenanthes	*Aulacidea*	0	1
Scorzonera[a]	*Aulacidea*	1	0
Serratula	*Phanacis*	1	0
Silphium	*Antistrophus*	0	4

1	2	3	4
Smilax	*Diastrophus*	0	1?
Sonchus	*Aulacidea*	1	1[4]
	Phanacis	1	0
Taraxacum	*Phanacis (Gillettea)*	1	1[2]
Tragopogon	*Aulacidea*	1	0

FAGACEAE

Castanea	*Dryocosmus*	0	1[5]
Castanopsis	*Dryocosmus*	0	1
Lithocarpus	*Andricus*	0	1
Quercus	*Acraspis*	0	17
	Amphibolips	0	23
	Andricus	94	93
	Antron	0	10
	Aphelonyx	1	0
	Atrusca	0	15
	Bassettia	0	8
	Belonocnema	0	3
	Besbicus	0	8
	Biorhiza	2	0
	Callirhytis	7	104
	Chilaspis	1	0
	Cynips	9	0
	Disholcaspis	0	38[6]
	Dros	0	5
	Dryocosmus	5	16
	Eumayria	0	4
	Fioriella	1	0
	Heteroecus	0	8
	Holocynips	0	4
	Liodora	0	3
	Loxaulus	0	11
	Neuroterus	12	50
	Odontocynips	0	1
	Paracraspis	0	3
	Philonix	0	7
	Phylloteras	0	3
	Plagiotrochus	5	1
	Sphaeroteras	0	7
	Synophrus	3	0
	Trichoteras	0	7
	Trigonaspis	4	0
	Trisoleniella	0	3
	Xanthoteras	0	12
	Xystoteras	0	3
	Zopheroteras	0	6

LABIATAE

Glechoma	*Liposthenus*	1	1[2]
Nepeta	*Ayalx*	1	0

1	2	4	4
Phlomis	Panteliella	1	0
	Rhodus	1	0
Salvia	Aylax	1	0
PAPAVERACEAE			
Papaver	Aylax	2	0
	Phanacis	1	0
ROSACEAE			
Fragaria	Diastrophus	0	1
Potentilla	Diastrophus	1	4
	Gonaspis	0	1
	Xestophanes	3	0
Rosa	Diplolepis	7	29[7]
	Liebelia	1	0
Rubus	Aylax	0	1?
	Diastrophus	1	6
UMBELLIFERAE			
Heracleum	Aylacopsis	1	0
VALERIANACEAE			
Valerianella	Cecconia	1	0

[1]*A. hieracii* also recorded from *Solidago, Linaria*.
[2]same species introduced from Europe to N. America.
[3]nearctic '*Aulacidea*' probably in *Phanacis*.
[4]this species also listed under *Lactuca*.
[5]introduction from eastern palaearctic.
[6]alternate forms probably in other genera.
[7]includes two species introduced from palaearctic.

Nearly all species are host plant specific or at least restricted to attacking a single plant genus. There are a few apparent exceptions. *Aulacidea hieracii* (Bouché), in addition to galling several species of *Hieracium* (Folliot, 1964), is recorded as also having been reared from galls on *Linaria* and *Solidago* (Eady and Quinlan, 1963). *A. tumida* (Bassett) is similarly alleged to attack both *Lactuca* and *Sonchus* (Krombein et al., 1979). These cases perhaps require further study since some misidentification may be involved. In any event they are vastly outnumbered by the plant specific species. Finally, mention must be made of those oak gall wasps in the *Andricus kollari* group of species that are obligatorily attached to different species of *Quercus* in their agamic and sexual generations.

Geographical distribution

Cynipinae are found in most parts of the world although they are predominantly a group associated with north temperate regions. Over five hundred species are described from North America and about two hundred from

Europe (Table 1). These two continents are entomologically the most studied but it is unlikely that they will be displaced from their position as the foremost centres of gall wasp abundance when the faunas of other areas become better known. The Eastern palaearctic also has a large gall wasp fauna and many species are known from Japan and the Soviet Union. In contrast, few species have been described from South America, Africa, India and the oriental region, and the only endemic gall wasps in Australia are the inquiline Austrocynipinae (seven species). Cynipinae are, however, readily transported by human agency and two palaearctic species have become established in Australia, at least one in New Zealand, and a number in North America.

A notable case of introduction concerns *Andricus kollari* in Britain. This insect requires two species of oak for the completion of its life cycle, *Quercus robur* which is a native British tree and *Q. cerris*. *Q. cerris* was introduced to Britain as an ornamental plant in 1735. About a century later *A. kollari* appeared in Britain, perhaps intentionally introduced to the southwest of England in connection with the tanning industry (Marsden-Jones 1953), and it spread rapidly so that now galls of the agamic generation can be found wherever *Q. robur* grows. *Q. cerris* has a limited distribution in Scotland, but such are the powers of dispersal of the sexual generation (*A. kollari* ♂♀ (= *circulans* Mayr)) that they find and attack *Q. robur* many miles from the nearest tree of *Q. cerris*. Another species with a similar alternation of host oak trees is *A. quercuscalicis*, first recorded in England from Northamptonshire by Claridge (1962). It too has spread and galls of the agamic generation on acorns (knopper galls) have now been found over most of the southern half of England and Wales. The dispersive capabilities of sexual females of *A. quercuscalicis* (= *cerri* Beijerinck), emanating from small galls on *Q. cerris*, have been demonstrated recently by McGavin et al. (pers. comm.) who found that small trees of *Q. robur* with very few acorns are just as likely to be found and galled as very fruitful trees. *A. lignicola* (Hartig) is yet another species with similar host plant relations to *A. kollari* and *A. quercuscalicis* that recently appeared in Britain (first certainly recorded in 1973) (Hutchinson 1974, Quinlan 1974) and is now common from the south coast of England northwards to at least as far as Cheshire.

A. kollari, the first of the trio to colonise Britain, is now heavily parasitised by chalcids (Blair 1946, Askew 1961e) but the agamic generation galls of *A. lignicola* and *A. quercuscalicis* so far suffer very little parasitism in Britain (Askew 1982, Martin 1982), although on the European mainland they have a number of chalcid parasites (Fulmek, 1968). The paucity of parasitism in Britain is puzzling since the species of parasites that attack the galls in countries where they have been long established also occur in Britain as parasites of other gall wasp species. However, parasitism of the sexual generation galls of *A. quercuscalicis* by three *Mesopolobus* species has recently been reported (McGavin, pers. comm.) and these same there species attack the similar sexual generation galls of *A. kollari.*

Dryocosmus kuriphilus Yasumatsu, the chestnut gall wasp, oviposits in the vegetative buds of *Castanea* and its gall causes severe deformation of shoots to the extent of reducing fruit-formation and, in cases of very severe infestation, even killing the tree (Payne, 1978). Where chestnuts are grown commercially, *D. kuriphilus* can be a very serious pest. It is another gall wasp to significantly expand its geographical range in recent years and, because of its economic status, it has attracted considerable attention.

D. kuriphilus was described in 1951 from Japan having been discovered first in Okayama, Honshu in 1941 (Yasumatsu, 1951). It probably entered Japan from China (Murakami, 1980) and spread rapidly throughout the country reaching northern Honshu by 1961 and being found on Hokkaido in 1964 (Kamijo and Tate, 1975). In 1961 it was accidently introduced into Korea and in 1974 an infestation was discovered in Georgia, U.S.A. (Payne et al., 1975). The spread of the gall wasp has undoubtedly been augmented by human agency through the movement of infested twigs, but in addition it must possess considerable powers of natural dispersion. It reproduces by obligatory parthenogenesis and a single female could therefore establish a population. There is only one generation a year.

Parasitism of *D. kuriphilus* in Japan was initially at rather low levels (Miyashita et al., 1965) although it involved at least five species of chalcids. Miyashita and co-workers report wide regional variation in the levels of parasitism in Japan with a maximum of only twenty-eight per cent of gall wasps killed by chalcid parasites in 1964. Perhaps the parasite load on *D. kuriphilus* in Japan is increasing, for Kamijo and Tate (1975) report a maximum of forty-nine per cent parasite-inflicted mortality in 1973 in Hokkaido, Yasumatsu and Kamijo (1979) mention twelve chalcid species attacking the gall wasp, and Murakami (1980) records fifteen species. These are all species originally associated with Cynipini on *Quercus.*

Dryocosmus kuriphilus is a pest but *Diplolepis rosae* is a potentially beneficial insect. Its introduction to New Zealand has been recommended (Schröder, 1967) to attempt to control the rapid increase there of sweet briar (*Rosa rubiginosa*) which poses a serious weed problem.

The species mentioned above owe their range expansions both to man and to their own dispersive and host plant-locating abilities. The galling of isolated host plants bears testimony to the latter. The Aylaxini, which mostly attack early successional plant species, must be adapted to discover isolated, short-lived plants whose location changes sometimes annually.

Distribution of galls on host plants

It is not sufficient for a gall wasp to find its correct host plant species. The plant part that is attacked must be in a suitable physiological state to permit oviposition to occur and later to form a gall under the influence of the larval cecidogenic secretions. That all plants of a species are not equally susceptible to successful gall wasp attack can clearly be seen in the variation in gall species and numbers on different oak trees growing close together. Numbers of galls of different gall wasp species on forty-nine small oak trees planted in Wytham Wood near Oxford are shown in Table 2. The numbers are totals for two consecutive years, 1958 and 1959 (see legend to table). The trees were all *Quercus robur* planted in 1949 and they averaged about 2.5 m in height. The uneven distribution of the galls is at once apparent. No tree bore galls of all species and only tree 21 carried galls of twelve of the thirteen types listed. Tree 32 bore galls of only two species, tree 5 had only five galls found on it in the two years of study whilst in contrast 1,273 galls were found on tree 49. The distribution of each type of gall over the trees was similar in both years of study; a tree that had large numbers of a gall in 1958 usually carried large numbers of the same type of gall in 1959.

Table 2. Numbers of galls of the more numerous species of Cynipini on individual oak trees in Wytham Wood, Berkshire, England. All galls of each species found on the trees in 1958 and 1959 are recorded except for the spangle galls (*Neuroterus numismalis* ♂, *N. albipes* ♀ (=*laeviusculus*), *N. quercusbaccarum* ♀ (=*lenticularis*)) where numbers on samples of 200 leaves (100 in each year) are given.

Tree No.	C. divisa	C. longiventris	C. quercusfolii	A. kollari	A. fecundator	A. ostreus	A. curvator	N. numismalis	N.a. laeviusculus	N.q. lenticularis	N.n. vesicator	N. albipes	N. quercusbaccarum
1	2	3	4	5	6	7	8	9	10	11	12	13	14
1	2	3	1	0	0	35	0	0	25	2	8	64	1
2	102	4	0	0	0	89	0	5	54	10	4	52	14
2b	5	0	0	0	38	68	0	695	52	41	0	11	0
3	82	4	1	0	1	34	9	51	28	30	7	7	2
5	0	1	1	0	0	2	0	0	1	0	0	0	0
6	43	2	0	1	0	45	0	0	33	3	0	7	1
7	2	1	0	0	7	15	0	213	16	30	0	4	1
8	117	0	2	2	1	52	0	0	32	25	0	10	2
9	0	2	0	0	0	12	0	0	10	0	0	1	0
10	14	8	0	0	0	21	6	5	5	10	5	26	11
11	6	6	0	0	0	54	0	32	22	25	0	1	0
12	87	7	0	0	7	73	0	146	54	41	4	19	0
13	23	16	2	0	3	75	0	211	32	57	0	2	0
14	93	14	18	3	0	40	0	22	25	67	3	10	8
15	274	22	10	1	0	150	15	92	14	99	5	23	1
16	122	9	0	1	0	117	1	783	55	115	1	11	0
18	0	2	0	0	0	44	0	24	63	8	19	13	1
19	0	1	0	3	0	35	35	139	43	21	15	39	0

1	2	3	4	5	6	7	8	9	10	11	12	13	14
20	0	0	2	1	0	25	0	20	47	34	0	6	0
21	68	4	3	3	17	140	10	204	48	67	1	3	0
22	0	0	1	6	0	22	1	0	0	26	7	4	0
24	0	0	3	3	0	31	2	49	65	98	27	35	0
25	0	0	0	0	0	43	0	3	48	39	4	5	0
26	108	14	3	0	1	66	0	8	41	21	2	17	0
27	165	0	0	0	0	47	0	127	15	0	0	0	0
28	29	8	10	2	3	72	0	31	19	47	5	66	7
29	115	3	2	1	0	37	41	23	14	6	8	0	6
30	17	1	0	0	0	63	0	0	16	37	16	4	0
31	15	12	26	0	0	89	0	608	85	91	2	10	0
32	0	0	0	0	0	28	0	0	3	0	0	0	0
33	56	5	2	1	0	94	0	44	20	128	0	3	0
34	13	3	1	0	0	86	1	134	41	103	7	9	0
36	0	0	0	0	0	17	2	0	10	1	10	11	3
37	75	3	4	0	0	87	5	113	84	72	2	2	0
38	131	10	17	1	2	134	0	713	62	162	0	0	0
39	17	4	2	1	3	34	12	0	53	14	16	39	4
40	26	6	4	0	0	11	1	1	36	57	8	17	1
41	0	6	4	0	3	94	22	16	80	21	47	32	16
42	84	14	8	0	0	54	1	79	57	116	1	6	0
43	0	0	0	1	0	27	1	223	88	132	1	15	0
44	50	2	5	0	0	50	3	257	11	49	2	1	0
46	1	2	0	0	2	57	0	31	64	61	6	11	1
47	51	8	3	0	0	62	0	65	33	148	0	8	0
48	10	2	0	0	0	31	1	4	31	7	17	19	2
49	62	20	40	0	4	79	0	802	138	100	0	28	0
50	66	11	0	3	0	32	1	123	35	88	0	4	8
52	0	8	0	1	0	38	0	81	15	39	0	4	1
53	43	3	6	0	0	86	22	80	45	119	1	1	0
55	3	1	8	0	0	54	0	9	30	0	0	16	0

This indicates that gall wasps have preferences for particular individual plants of the host species. The trees studied were too young to produce flowers and so agamic females of those species requiring catkins for the formation of galls of the sexual generations were obliged to leave the plantation and seek mature oaks some distance away. Such a species is *Andricus fecundator* but the correlation between the distribution of galls in 1958 and in 1959 is remarkably high ($r = 0.77$, $P < 0.01$) showing that sexual females returned to the plantation and sought the same trees as the sexual females of the previous year. The alternative explanation that all trees were more evenly attacked and that survival was high on only a few is not favoured.

Correlations between the distributions of all gall types are presented in Table 3. A tendency for equivalent generations of allied species to have similar distributions is apparent, the distributions of the three species of *Cynips* and of the three species of *Neuroterus* in both generations being quite

Table 3. Correlations between the distributions of galls shown in Table 2. The enclosed values of _r_ are significant at the 1% level

	C. dietsa	C. longiventris	C. quercusfolii	A. ostreus	N.q. lenticularis	N. numismalis	N.a. laeviusculus	A. fecundator	N. albipes	N. quercusbaccarum	N. n. vesicator	A. curvator
C. longiventris	.52											
C. quercusfolii	.22	.62										
A. ostreus	.57	.48	.34									
N.q. lenticularis	.28	.44	.40	.54								
N. numismalis	.19	.35	.55	.47	.49							
N.a. laeviusculus	−.01	.26	.53	.36	.44	.53						
A. fecundator	−.07	−.08	−.02	.21	−.01	.42	.15					
N. albipes	−.07	−.10	.11	.03	−.16	−.10	.20	−.01				
N. quercusbaccarum	.05	.12	−.02	.01	−.24	−.24	−.01	−.07	.44			
N.n. vesicator	−.24	−.16	−.12	−.04	−.20	−.25	.18	−.09	.40	.44		
A. curvator	.16	−.03	−.05	.12	−.08	−.11	0	−.03	.10	.26	.38	
A. kollari	.05	−.03	.01	−.02	.11	−.06	−.12	0	.08	.06	.07	.16

similar. In fact, the distributions of all species forming leaf galls in summer are relatively highly correlated suggesting that the leaves of certain trees are especially suitable for gall formation by all species. In contrast, *A. kollari* ♂ and *A. fecundator* ♂, which are the only species surveyed to form summer bud galls, have distributions uncorrelated with each other and with almost every other species.

Distributions of galls of both generations of three species of *Neuroterus* have been studied in detail (Askew, 1962). The site of study was the plantation of young oak trees in Wytham Wood referred to previously. *N. albipes* (Schenck), *N. quercusbaccarum* and *N. numismalis* are positively associated in both generations and the distribution of each gall type was similar in the two years (1958, 1959) of study. The data on which each tree first came into leaf in spring was recorded and spangle galls of the agamic generations of *N. numismalis* and *N. quercusbaccarum* (=lenticularis) were found in greatest numbers in samples of leaves from those trees that had a late bud-burst. Their alternate sexual generations, however, were most abundant on the earlier opening trees. Numbers of galls of *N. albipes* in both generations were independent of the date of leaf-opening. This differing association between date of leaf-opening and the two generations of *N. numismalis* and *N. quercusbaccarum* was reflected in negative correlations between the distributions of agamic and sexual generation galls of both species, whilst the two generations of *N. albipes* were weakly positively correlated. Agamic generation galls often occur at such high densities on leaves that interspecific interference is likely; the tendency for spangle galls of all three species to be concentrated on the same trees increases this possibility. However, when the distributions of spangle galls are investigated with respect to their position on individual trees and leaves, differences between *N. albipes* (=*laeviusculus* Schenck), *N. quercusbaccarum* (=*lenticularis*) and *N. numismalis* become apparent. On the young trees in the plantation, *numismalis* galls were most numerous on peripheral foliage and especially towards the tops of trees, *laeviusculus* was concentrated on foliage nearer the ground and closer to the tree trunks, and *lenticularis* galls tended to predominate in an intermediate situation. Ejlersen (1978), studying the distribution of the same species on young trees in Denmark found a rather different distribution with *numismalis* again predominating at the top and periphery of a tree but this same pattern being shown also by *lenticularis*, and *laeviusculus* was apparently distributed evenly.

The zonational pattern found by Askew (1962) was paralleled on individual leaves with *numismalis* occurring mostly towards the leaf apex, *lenticularis* in the region somewhat behind the apex, and *laeviusculus* on the more basal parts of the leaf. Ejlersen (1978), employing a full statistical analysis, confirms this distributional pattern on the leaves of young trees although he found that *lenticularis* was displaced from an apical situation only when it occurred on the same leaves as *numismalis*. The distribution of spangle galls on oak leaves has been studied over many years as a class exercise at Manchester University and in Fig. 1 is reproduced the combined data obtained from very large numbers of galled leaves collected in Cheshire, England from mature *Quercus robur* in autumn, and also the author's original data from young trees in Wytham Wood. The zonation of spangle galls is less sharply defined on mature trees than on young trees and *lenticularis*, the dominant type of gall on mature trees, is shifted more apically on the leaves. The pattern of distribution of spangle galls is clearly related to the timing of the life cycles of the gall wasps (Askew, 1962). Sexual females of *N. albipes* are the first of the three species to oviposit in spring and leaf buds on young trees have only just opened at this time. Eggs are perhaps distributed evenly over the leaves but, because of later maximum leaf growth and expansion towards the apex, galls of *laeviusculus* occur at greatest densities basipetally. Sexual females of *N. quercusbaccarum* appear after those of *N. albipes* and slightly before those of *N. numismalis*. Both these species apparently select growing parts of the leaves for oviposition (Hough, 1953) and hence galls of the agamic generations have an apical or centroapical distribution which is most apparent when oviposition is into older leaves (Fig. 1), mature oak

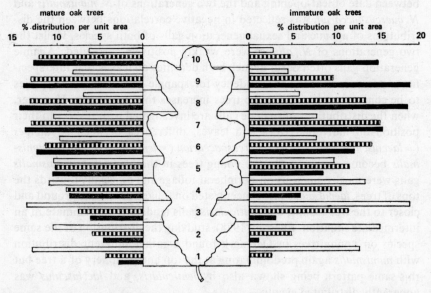

Figure 1. The percentage distribution of spangle galls of three species of *Neuroterus* per unit area in ten horizontal divisions of leaves of mature (left) and immature (right) *Quercus robur*. Data for immature trees are in Askew (1962) and are based on the distributions of 613 *N. albipes* ♂ (=*laeviusculus*) galls, 1,221 *N. quercusbaccarum* ♂ (=*lenticularis*) galls and 3,672, *N. numismalis* ♂ galls. The data for mature oaks are based on 1,520 *laeviusculus* galls, 113,664 *lenticularis* galls and 54,575 *numismalis* galls. The distributions are represented by open bars (*laeviusculus*), banded bars (*lenticularis*) and solid bars (*numismalis*) and are corrected for the differing areas of the leaf zones.

trees coming into leaf in advance of young trees. As Ejlersen (1978) concludes, "the timing of life cycles...in relation to leaf development seems to have reduced the effect of competition between the three species concerned".

A fourth species of *Neuroterus* to form agamic spangle galls is *N. tricolor* (Hartig) (=*fumipennis* Hartig) and the position of these galls on *Q. robur* differs from that of the three species discussed above in that they alone develop on the secondary flush of summer leaves (lammas leaves).

In conclusion, individual plant characteristics determine the extent to which they are galled by different species so that plants of the same species may present an entire spectrum of susceptibility to a gall wasp species. Further, as is seen in the case of *Neuroterus*, subtle differences in chronology between gall wasp species and between gall wasps and their host plants can result in a zonation of galls which effectively reduces interference between them. The relationship between a gall wasp and its plant host is even more complex than might at first appear involving plant species suitability, physiological suitability of an individual plant and plant part suitability, the situation as seen today being the product of evolutionary interaction between not only plants and gall wasps but also between gall wasps themselves.

Life cycles and reproduction

Haplodiploidy, males developing by arrhenotokous parthenogenesis from unfertilized eggs hence being haploid and females developing from fertilized eggs hence being diploid, is the normal mode of reproduction in Hymenoptera. In Cynipinae thelytokous parthenogenesis is also prevalent; that is the development of diploid females from eggs that have not been fertilized by sperm. The most complex reproductive cycle is seen in most of the gall wasps that attack oak. Here there is an alternation of generations (or heterogony), a sexual generation of males and females alternating with a so-called agamic generation comprising only females. Such a cycle involves two types of female (and may hence be termed heterogyny) which differ considerably, not only in their modes of reproduction, but also in their morphology and in the typ es of gall that their offspring produce. In a few species there is even an obligatory alternation of host plants involving different species of *Quercus*.

Alternation of generations in Cynipinae was discovered first in America by Bassett (1873) and Riley (1873) and then, independently, by Adler (1877) working in Germany. The two female forms of a heterogonous species differ so much that they were usually originally placed in different genera. After alternation of generations was discovered, subsequent work by Beijerinck, Mayr, Kieffer, Niblett, Folliot and others demonstrated it to be of almost universal occurrence amongst oak gall wasps. In gall wasps attacking other plant hosts it is exceedingly rare.

Folliot (1964) worked in France with several gall wasp species, both

Aylaxini and Cynipini, and his careful research has shed a great deal of light upon gall wasp reproductive biology. The following account is largely based on Folliot's studies.

The Aylaxini, regarded as the least specialized of Cynipinae, may be divided into three reproductive groups, bisexual, unisexual and heterogonous species.

1) BISEXUAL SPECIES

In bisexual species both sexes regularly occur, although females usually outnumber males (Table 4). After mating, females lay fertilized eggs which give rise to female offspring and unfertilised eggs which produce male offspring. Virgin females will oviposit under experimental conditions and the insects emerging from galls so produced are all, except for occasional developmental aberrations, males. Such males are fully functional and able to fertilies females, but unmated females probably oviposit only rarely in nature. In spermatogenesis the first meiotic division is abortive so that haploid males produce haploid sperm. Oogenesis is normal and results in haploid ova. Reproduction in bisexual Aylaxini is therefore typical of Hymenoptera in general and may be represented as:

Examples of bisexual Aylaxini are *Diastrophus rubi* (Bouché), *Aulacidea hieracii* (Bouché), *Aylax minor* Hartig, *Xestophanes potentillae* (Retzius) and *X. brevitarsis* (Thomson). *D. rubi* is a species that has been considered (Beijerinck, 1883) to be moving in the direction of thelytoky, that is towards the group of species next described, because males in some populations are said to be scarce although in others they comprise about thirty per cent of the population (Table 4).

Table 4. Percentage males in species of Aylaxini, reared from galls collected in France (Folliot, 1964).

	Number of gall wasps reared	Percentage males
Aylax minor	1761	48
Xestophanes potentillae	1481	40
Diastrophus rubi	890	31
Aulacidea hieracii	979	30

Bisexual Aylaxini are typically univoltine, adults emerging in early summer and producing galls of the next generation in which the larvae overwinter and complete their development the following year. An exception to this pattern is found in certain populations (e.g. in western France) of *X. potentillae* which are bivoltine, the early summer generation of adults giving rise to a second adult generation later in the same year. Galls of these bivoltine *X. potentillae* are greenish-brown in colour and formed on the creeping stolons and leaf petioles of the host plant, *Potentilla reptans*. Univoltine *X. potentillae* develop either in galls similar to these or in rather different galls, dark brown in colour and found on the roots of *P. reptans*.

2) UNISEXUAL SPECIES

In some Aylaxini males are absent, or at most very rare, and females reproduce by thelytokous parthenogenesis to produce all female offspring. During oogenesis a maturation division fails to take place so that the diploid females lay diploid, female-producing eggs:

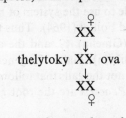

Species in this group are *Phanacis hypochoeridis* (Kieffer), *P. lampsanae* (Perris), *Aulacidea pilosellae* (Kieffer), *A. subterminalis* Niblett and *Liposthenus latreillei* (Kieffer). Males are occasionally produced by most species but their maximum frequency, as reported by Folliot, is about only six per cent (*P. hypochoeridis*).

As in the bisexual Aylaxini, the unisexual species also are normally univoltine, taking a year to complete their development. However *A. pilosellae*, at least in western France, is bivoltine.

3) HETEROGONOUS SPECIES

Heterogonous species are those that exhibit alternation of an agamic (unisexual) generation with a sexual generation. Heterogony is a phenomenon associated with the more specialized gall wasps (Cynipini) and its occurrence in Aylaxini is apparently restricted to one species, *Pediaspis aceris* (Gmelin).

P. aceris is a gall-maker on certain species of *Acer*. Galls that house the sexual generation are found on the undersurfaces of leaves, sometimes on the petioles, in May and June. They are globular, five or six millimetres in diameter, smooth, unilocular and green to reddish in colour. The larval chamber is very large. Adults emerge in June, mate, and inseminated females oviposit in the roots of *Acer*, preferentially selecting rootlets one millimetre or less in diameter. Several eggs are usually laid in a row in the same rootlet and the resulting galls tend to coalesce to form an irregular mass, although individual galls also occur. These galls house the agamic generation which consists only of females. For the most part they form in the autumn following oviposition by the sexual females but, according to Folliot (1964), some eggs diapause for up to two years. Duration of larval development is also variable but most individuals complete their metamorphosis in two years. The agamic females emerge in March, drill with their ovipositors into the buds of *Acer*, and lay their eggs on the embryonic leaves. These eggs develop parthenogenetically and galls of the sexual generation appear three or four weeks later.

The agamic generation was named *P. sorbi* (Tischbein) because it was originally thought to represent a distinct species. Nearly all heterogonous gall wasps have for the same reason been given two names. According to the rules of nomenclature only the senior name (in this case *P. aceris*) should be retained, but since the names of both generations are so extensively used in the literature it seems preferable to use the system of nomenclature applied by Eady and Quinlan (1963) and Folliot (1964). Thus the sexual generation is designated *Pediaspis aceris* (Gmelin) ♂♀, and the agamic generation *P. aceris* (Gmelin) ♂ (=*sorbi* Tischbein). Galls of a generation are the galls in which the generation develops, not the galls that follow its oviposition. Thus the agamic generation galls of *P. aceris* are the root galls, the sexual generation galls are the leaf galls.

Unfertilized eggs laid by the agamic generation of *P. aceris* produce the sexual generation. Agamic females give rise to sexual males and females, but an individual agamic female lays eggs that are all, or almost all, of one sex. There are, therefore, two types of agamic female, androphores that lay only male eggs and gynephores that lay only female eggs. By a penetrating series of experiments, Folliot (1964) was able to demonstrate that individual fertilized sexual females produced in their agamic offspring both androphores and gynephores. In this respect the life cycle of *Pediaspis* is less complex than that of most of the Cynipini.

The five types of life cycle found in Aylaxini are illustrated in Figure 2. Rhoditini, the tribe that galls exclusively species of *Rosa* and comprises *Diplolepis* (=*Rhodites*) species, usually has a unisexual, univoltine life cycle like that of *Liposthenus*, etc. In Europe males of *Diplolepis* are generally rare and the species must reproduce by thelytokous parthenogenesis. The only exception to this is *D. kiefferi* (Loiselle) in which males are almost as numerous as females. In *D. rosae* (L.), the incidence of males appears to vary geographically, more northerly populations including a higher percentage of males than occurs in populations further south (Askew, 1960). Even so, the maximum representation of males recorded, from north-east England, is only 4.2 per cent and their reproductive contribution to the population is uncertain. In Asia, males of *Diplolepis* are said to be as fre-

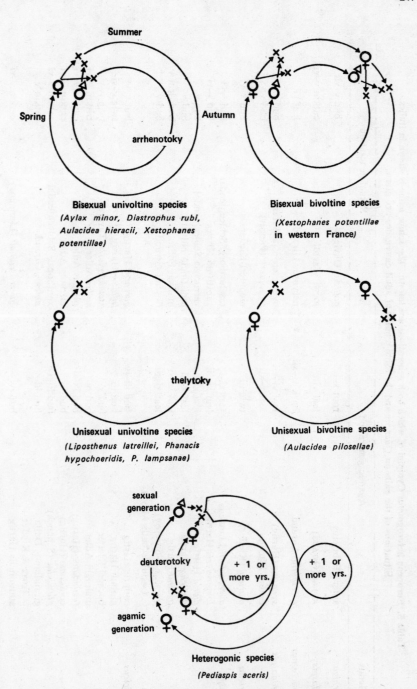

Summer

Spring

Autumn

arrhenotoky

Bisexual univoltine species
*(Aylax minor, Diastrophus rubi,
Aulacidea hieracii, Xestophanes
potentillae)*

Bisexual bivoltine species

*(Xestophanes potentillae
in western France)*

thelytoky

Unisexual univoltine species
*(Liposthenus latreillei, Phanacis
hypochoeridis, P. lampsanae)*

Unisexual bivoltine species
(Aulacidea pilosellae)

sexual
generation

deuterotoky

+ 1 or
more yrs.

+ 1 or
more yrs.

agamic
generation

Heterogonic species

(Pediaspis aceris)

Figure 2. Diagrammatic representations of the different life cycles of species
of Aylaxini (modified after Folliot, 1964).

Table 5. European heterogonous Cynipini of which both generations are known. The senior name is shown in italics. Situations of the galls on *Quercus* and the usual months of adult emergence are given.

Genus	Agamic generation		Sexual generation	
Biorhiza	aptera Fabricius	root x–i	*pallida* (Olivier)	bud v–vii
Chilaspis	*nitida* (Giraud)	leaf viii	loewi Wachtl	catkin v–vi
Dryocosmus	cerriphilus Giraud	bark xi–xii	nervosus Giraud	leaf vi
Trigonaspis	*synaspis* (Hartig)	leaf vi–vii	megapteropsis Wriese	bud vi–vii
	renum Hartig	leaf x–xii	*megaptera* (Panzer)	bud vi–vii
Fioriella	mariani (Kieffer)	twig iii	meunieri Kieffer	bud v–vi
Cynips	*quercusfolii* Linnaeus	leaf xi–i	taschenbergi Schlechtendal	bud iv–v
	longiventris Hartig	leaf xi–iii	substituta Kinsey	bud v–vi
	divisa Hartig	leaf xi–i	verrucosa Schlechtendal	bud v
	disticha Hartig	leaf x–xi	indistincta Niblett	bud v
	agama Hartig	leaf ix–xi	mailleti Folliot	bud, leaf v
	quercus (Fourcroy) (=pubescentis Mayr)	leaf xi–xii	flosculi Giraud	bud v
Andricus	collaris Hartig	bud iii–iv	*curvator* Hartig	leaf v–vi
	globuli Hartig	bud iii	*inflator* Hartig	twig v–vi
	bocagei Kieffer	bud x	*pseudoinflator* Tavares	twig v–vi
	malpighi Adler	bud iv	nudus Adler	catkin v–vi
	fecundator (Hartig)	bud ix–iii	pilosus Adler	catkin v–vi
	callidoma (Hartig)	bud iii	cirratus Adler	catkin v–vi
	glandulae (Schenck)	bud iii–iv	xanthopsis Schlechtendal	catkin v–vi
	giraudianus Dalla Torre & Kieffer	bud iii–v	*amenti* Giraud	catkin v
	autumnalis Hartig	bud iv	*quercusramuli* (Linnaeus)	catkin v–vi

urnaeformis Fonscolombe	leaf iv	sufflator Mayr	leaf vi
alniensis Folliot	bud iii–iv	rupellensis Folliot	catkin v
quercuscorticis (L.)	bark v–vii	gemmatus Adler	bud vi–viii
sieboldi Hartig	bark iii–iv	testaceipes Hartig	leaf viii
rhizomae (Hartig)	bark iii–iv	nodifex Kieffer	leaf viii
occidentalis Folliot	bark iii–iv	poissoni Folliot	bud viii
(described as subspecies of testaceipes)			
quercusradicis (Fab.)	bark, root iv	trilineatus Hartig	twig vii–ix
solitarius (Fonscolombe)	bud ix	occultus Tschek	catkin iv
ostreus Hartig	leaf x–xii	furunculus Beijerinck	bud iv–vi
quadrilineatus Hartig	catkin iii	kiefferi Pigeot	catkin v
albopunctatus (Schlechtendal)	bud iii	barbotini Folliot	catkin v
infectoria Hartig	bud ?	burgundus Giraud	bud v
kollari (Hartig)	bud ix–x	circulans Mayr	bud iv–vi
corruptrix	bud viii	larshemi Docters van	bud iv–v
(Schlechtendal)		Leeuwen & Dekhuijzen-Maasland	
lignicola (Hartig)		vanheurni Docters van	
		Leeuwen & Dekhuijzen-Maasland	
quercuscalicis (Burgsdorf)	bud v–vi	cerri Beijerinck	bud iii–v
	acorn ii–iv		catkin v
Neuroterus			
fumipennis Hartig	leaf iv–v	tricolor (Hartig)	leaf vi–vii
lenticularis Olivier	leaf ii–iv	quercusbaccarum (Linnaeus)	leaf, catkin v–vi
laeviusculus Schenck	leaf ii–iii	albipes (Schenck)	leaf v–vi
numismalis (Fourcroy)	leaf iii	vesicator Schlechtendal	leaf v–vii
schlechtendali Mayr	catkin vii	aprilinus (Giraud)	bud iv–v
saliens (Kollar)	leaf iv	glandiformis Giraud	acorn v–vi

quent as females (Kuznetzov-Ugamskij in Vandel, 1931). This demonstrates the biological variation, even within a species and often geographically based, that is frequently encountered in gall wasps.

Alternation of generations in Cynipini

Most Cynipini have a heterogonous life cycle, although this may be of varying complexity. Galls of the sexual generation usually form in spring or early summer, they develop quickly and males and females of the sexual generation soon emerge. After mating and oviposition, galls of the agamic generation develop slowly, the larvae overwintering to produce agamic females the following spring, or in some cases one, two or more years later. Alternating forms of European species are set out in Table 5.

Biorhiza pallida (Olivier) has a rather unusual phenology. Agamic females emerge from root galls on oak in November and December. They are relatively large and wingless and they crawl up the trunk to oviposit in oak buds, their heads directed towards the base of the bud as in most species laying eggs in buds. Oviposition may continue for more than an hour during which time more than one hundred eggs may be laid. Larvae emerging from these eggs the following spring induce the formation of large, plurilocular oak apple galls, which are the galls of the sexual generation. Some of these galls fall to the ground, others remain attached to the twigs. Adults of the sexual generation emerge in June, males usually a day or so before females. In Britain sexual females are almost always fully winged like their males, but in Germany they may be macropterous, brachypterous or apterous (Weidner, 1960). After mating, the sexual females oviposit in oak roots two to four millimetres in diameter and sometimes at a depth of a metre below the soil surface. Galls of the agamic generation *B. pallida* ŏ (=*aptera* F.) develop during the following year, at the end of which most of the agamic females emerge although a minority do not complete their development until the next year, some thirty months after oviposition. All gall wasps emerging from a single sexual generation oak apple are most certainly the offspring of a single agamic female. Usually, but not always, all emergents from a single sexual gall are of one sex, indicating that the agamic females are, as in *Pediaspis aceris* described above, either androphores or gynephores. However a proportion of oak apples does produce both sexes (Table 6). Folliot (1964) also found that the sexual generation progeny of individual agamic females, reared experimentally, most often included both sexes (Table 7). Therefore by no means all agamic females of *B. pallida* are strict gynephores or androphores. A variable proportion may be described as gynandrophores and these determine the sex of the following sexual generation. In those that are strict gynephores and androphores, as also in *Pediaspis aceris*, the determination of the sex of the sexual generation is pushed back to the preceding sexual generation

(Fig. 3). Both these types of life cycle are described by Folliot as simple compared with the life cycle of species of *Andricus* and *Neuroterus*. In these genera it is usual to find not only that the agamic generation of females is composed of strict androphores and gynephores, but also that one sexual

Table 6. The sexual composition of broods of *Biorhiza pallida* emerging from individual plurilocular galls of the sexual generation.

| Country | Numbers of galls producing: | | | Source |
	males only	females only	both sexes	
England	13	15	3	Folliot 1964
England	0	0	'all galls'	Niblett 1941
England	7	6	2	Askew unpublished
France	6	13	6	Folliot 1964

Table 7. The sex of the progeny of individual agamic female *Biorhiza pallida* (Folliot, 1964).

Agamic female	Male progeny	Female progeny
A	64	121
B	74	12
C	22	251
D	48	2
E	132	67
F	0	63
G	0	432

generation female will produce in her progeny either agamic androphores or gynephores but not usually both. Thus there is here even earlier determination of sex of the sexual generation; at the agamic generation one-and-a-half life cycles previously. These three types of life cycle are illustrated diagrammatically in figure 3.

Total conformity to a particular life cycle type is, however, seldom encountered, intraspecific variation often becoming manifest at various stages. Variation in *B. pallida* has already been described. *Andricus quercusradicis* (F.) has a life cycle basically of the third type although the progeny of one sexual female (*A. quercusradicis* ♀ (= *trilineatus* Hartig) emerging from a swollen leaf petiole), which develop in a large, plurilocular root gall, most often (in twelve out of twenty-two cases examined by Folliot) includes both androphores and gynephores with one type predominating. Further, an agamic androphore will occasionally produce a few female offspring together with many males.

Doncaster (1910, 1911, 1916) made some pioneer life cycle studies of gall wasps, and his work was later supplemented by that of Dodds (1939).

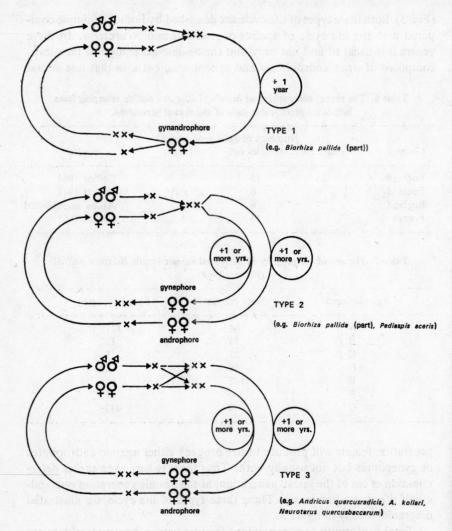

Figure 3. The three types of life cycle exhibited by heterogonous gall wasps, with progressively earlier establishment of the sex of individuals of the sexual generations.

Both studied *Neuroterus quercusbaccarum* which has an advanced type three life cycle. Sexual females lay eggs that are apparently always homogeneous, all agamic offspring of an individual being either androphores or gynephores and never both. Basically similar observations were made by Folliot (1964) with *Andricus kollari* and his results are reproduced in Table 8. From these it can be seen that four of the six sexual females used in the series of experiments produced agamic offspring that were all gynephores. The remaining two sexual females had twenty-six descendants, sixteen of which were absolute androphores and ten were predominantly androphores although producing also a small number of female progeny.

A. kollari is one of a small group of heterogonous gall wasps that, as a further elaboration of its life cycle, is heteroecous. Heteroecy is the obligate alternation of host plants and it occurs in a few allied species of *Andricus*.

Table 8. The sex of the progeny of agamic female *Andricus kollari* descended from individual sexual females (Folliot, 1964).

Sexual female	Number of agamic descendants	Numbers of each sex in the progeny of each agamic female	
		males	females
A	4	0	9
		0	1
		0	9
		0	2
B	3	0	14
		0	5
		0	11
C	2	0	7
		0	1
D	2	0	1
		0	50
E	3	3	0
		8	0
		7	1
F	23	18	0
		3	0
		5	0
		20	0
		12	0
		6	0
		53	0
		5	0
		5	0
		10	0
		7	0
		2	0
		3	0
		15	0
		2	1
		6	1
		2	3
		37	7
		16	5
		1	1
		6	2
		7	1
		1	1

Sexual females of *A. kollari* (=*circulans*) oviposit in May or June in buds of *Quercus robur* (also *Q. petraea* and *Q. pubescens*). They prefer buds that are swollen but not yet burst, and for this reason coppiced oak, in which bud burst is usually somewhat delayed, is often very heavily attacked. Agamic galls of *A. kollari*, the familiar globular, woody marble galls, mature in August and agamic females mostly emerge in September although some larvae overwinter and do not emerge until the following May or June. Agamic females oviposit in buds of *Quercus cerris*, depositing eggs between the embryonic leaves. Development during winter is slow and the sexual galls, which are small, thin-walled, ellipsoid structures, do not become visible among the bud scales until March or April. The alternation of host trees was first established by Beijerinck (1902) who reared the sexual generation on *Q. cerris* from sleeved agamic females obtained from galls on *Q. robur*, and Marsden-Jones (1953) obtained agamic marble galls on *Q. robur* from sleeved sexual females reared from galls on *Q. cerris*.

Parthenogenesis in the life cycle of advanced heterogonous species is limited to the agamic generation. Doncaster (1916) recorded oviposition by virgin sexual female *N. quercusbaccarum* but no galls ensued, and Folliot (1964) made similar observations on sexual female *A. kollari* and *A. lignicola*. In less specialized heterogonous species, however, sexual females may retain some capacity for thelytoky. Folliot reared numbers of agamic female *P. aceris* (=*sorbi*) from root galls produced by virgin sexual generation *P. aceris*, and he also succeeded in getting one agamic *Biorhiza pallida* (=*aptera*) gall after sleeving ten virgin sexual *B. pallida* on oak rootlets. Therefore, in *P. aceris* mating is not essential for completion of the life cycle but in *B. pallida* the ability of the sexual generation to reproduce thelytokously is much reduced. The importance of males in the life cycles of heterogonous Cynipini is evident from the numbers that are produced. Of 19,578 sexual *B. pallida* reared by Folliot, forty-five per cent were males, and 9,298 sexual *A. radicis* (=*trilineatus*) included forty-three per cent males.

Doncaster, following his researches on *N. quercusbaccarum*, was able to explain some of the cytological processes involved in sex determination in heterogonous gall wasps. During oogenesis in agamic females, eggs either fail to undergo a maturation division, in which case diploid sexual females result, or they do undergo a reduction division, with ejection of a polar body, to produce haploid sexual males. In sexual females all developing eggs undergo a double maturation division followed by reduction so that the ova are haploid. In sexual males, which are haploid at least in the germinal tissue, spermatogenesis involves a first abortive meiotic division and a second division that yields two haploid spermatids. Doncaster did not claim to completely explain gall wasp sex determination, but his studies led others to evolve a theory which was generally accepted until questioned by Folliot (1964).

In heterogonous species with a perfect type three life cycle (Fig. 3), some sexual females have offspring which are all androphores and others have offspring which are all gynephores. The classical theory suggests that these two types of sexual females are generated during oogenesis in agamic gynephores (Fig. 4). Folliot argues, however, that it is difficult to believe that agamic gynephores can generate two types of eggs since oogenesis proceeds without a maturation division and apparently provides no opportunity for diversification. On the contrary, spermatogenesis in agamic androphores involves elimination of a polar body and could, in theory, engender the necessary diversity. Folliot suggests that there are two types of sexual males produced by agamic androphores and that these form spermatozoa with differing sexual potential. Sexual females could then be all alike and mate once only, as is usual in many parasitic Hymenoptera, with one or other type of male. Their progeny would then be either agamic androphores or agamic gynephores (Fig. 4).

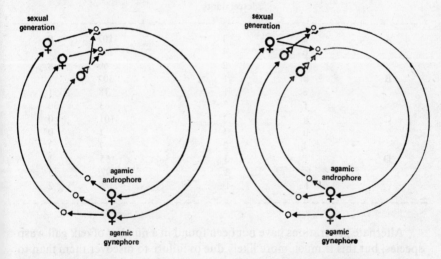

Figure 4. Illustration of the two theories concerning sex determination in heterogonous gall wasps (after Folliot, 1964). The classical theory (left) involves two types of sexual females. Folliot's alternative theory (right) involves two types of sexual males.

To test the validity of the theory of differing sexual males, Folliot reared the progeny of individual males to the third generation (next sexual generation). If these were all of the same sex, it would be strong evidence in support of his theory. A. kollari was selected as the experimental species since its complete life cycle does not normally extend over more than one year, and because males readily copulate with a succession of females. The results (Table 9) show that each of the four males used sired agamic androphores. A small number of sexual females was also produced but Folliot considers this to be within the normal limits of variation and that the results corroborate his theory. Sex is therefore determined by the production of genetically differing males, illustrated diagrammatically as follows:

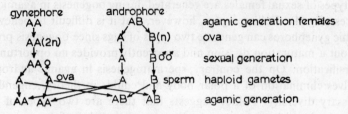

Table 9. Sex of the descendants of four male *Andricus kollari* (Folliot, 1964).

Male	x	Sexual ♀ →Number of agamic ♀♀→ produced that left descendants	Sexual generation offspring ♂♂	Sexual generation offspring ♀♀
A	a	1	107	0
	b	4	178	3
	c	2	66	8
B	d	1	302	1
	e	1	18	1
	f	7	83	30
C	g	2	101	0
	h	1	1	0
	i	2	7	3
D	j	3	55	1
	k	3	97	5
	l	1	2	2

Alternate generations have not been found in a number of oak gall wasp species, but this is much more likely due to failure to discover them than to the loss of heterogony. Three species of *Andricus* in the well-studied European fauna, *A. seminationis* (Giraud), *A. quadrilineatus* Hartig and *A. albopunctatus* (Schlechtendal), were for a long time thought to exist only as agamic forms and to have lost their sexual generations. Folliot (1964) shows however that both *A. albopunctatus* and *A. quadrilineatus* have sexual forms (respectively, *barbotini* Folliot and *kiefferi* Pigeot), although *A. quadrilineatus* is so far as is known unique in that the progeny of agamic females can produce either agamic or sexual galls. Thus in *A. quadrilineatus* heterogony is not obligatory and the sexual generation is possibly superfluous. The occurrence of galls of agamic *A. kollari* in regions where no *Quercus cerris*, the host plant of the sexual generation, seems to grow has caused speculation that heterogony may be facultative also in *A. kollari*, but despite repeated attempts agamic females have not been shown to be capable of directly producing agamic marble galls. Agamic females will prick buds of *Q. robur*, and even lay eggs in them, but no galls result (Marsden-Jones, 1953).

Summarizing the diverse life cycles of gall wasps is difficult. Even the relatively unspecialized Aylaxini range from being bisexual to thelytokous, univoltine or bivoltine, and *Pediaspis* has become heterogonous. Heterogony is possible only on long-lived host plants with an extended growing season, and its occurrence in *Pediaspis* is linked with its association with the tree *Acer*. Other Aylaxini attack herbaceous plants, but amongst these are found examples of bivoltine species and also, in *Xestophanes potentillae*, the behavioural plasticity to attack different host plant organs, both requirements of heterogony. The Rhoditini have become largely thelytokous and this habit probably developed after their radiation on *Rosa* since at least one species, *Diplolepis kiefferi*, remains bisexual. The evolution of heterogony by Cynipini on *Quercus* is associated with a reduction in and eventual loss of the capacity of sexual females to reproduce thelytokously, and for agamic females to change from deuterotoky (as still found in some *Biorhiza pallida*) to thelytoky and arrhenotoky. As a final step, arrhenotokous parthenogenesis may be omitted so that no males are produced in the life cycle and reproduction is entirely thelytokous (as in *Andricus quadrilineatus*).

What are the selective pressures leading to the development of the various life cycles of Cynipinae? Thelytoky and bivoltinism increase the reproductive potential of a species, although thelytoky will reduce genetic variability and bivoltinism is possible only when the host plant has organs that are actively growing for most of the year. Among the Aylaxini, species of *Xestophanes* galling *Potentilla* are bivoltine but bisexual, *Aulacidea pilosellae* is both bivoltine and thelytokous, and only *Pediaspis aceris* is heterogonous, preserving genetic variability in the sexual generation whilst profiting from the increased reproductive potential of thelytoky in the agamic generation. Establishment of Cynipini upon *Quercus*, very long-lived plants with actively growing structures present over much of the year, allowed the full expression of heterogony, the sexual generation tending to develop in galls early in the year with an often reduced parasite load, and the structurally usually more protective galls of the agamic generation carrying the species through autumn and winter. This is, of course, a generalization; there are many exceptions. A suggestion as to why galls of the two generations of a heterogonous species are so very different is given later in this paper.

Inquiline gall wasps

Many groups of Hymenoptera include species that exploit the larval food supply of their allies. Well-known examples are the bees, wasps and ants that usurp the nests of social species, kill the foundress queen, and leave their own progeny to be raised by the host workers. Solitary bees and wasps are also exploited by allied species which deposit their own eggs in cells constructed and provisioned by the host for its own offspring. This type of relationship between two species, in which the benefits are entirely uni-lateral, is termed inquilinism. The inquiline larva may kill the host larva directly or it may destroy it by developing more rapidly and consuming all available food. The relationship is not a casual one, the inquiline being obligatorily dependent upon its host to which it is often very closely taxo-nomically allied. Inquilinism is more common among aculeate than para-sitic Hymenoptera, but in the latter are inquiline (or cleptoparasitic) Ich-neumonidae such as *Pseudorhyssa* which takes possession of the woodwasp host of *Rhyssella* (Askew, 1971), and inquiline Cynipidae whose larvae feed in galls induced by other gall wasps.

Inquiline gall wasps are mostly included within the tribe Synergini. As previously remarked, this is a somewhat heterogeneous assemblage in the Cynipinae, indicating that the inquiline habit has arisen within gall wasps more than once. The Synergini are certainly derived from cecidogenic ancestors. Adler (1881) instances the frequent occurrence of agamic galls of *Andricus curvator* (= *collaris* Hartig) among bud scales of *A. fecundator* galls as possibly indicating how the inquiline habit may arise. The enlarged bud housing a developing *A. fecundator* gall is apparently especially attrac-tive to ovipositing *A. curvator* so that galls of the two species often develop side by side.

The largest genus of inquiline gall wasps is *Synergus* and its species are associated with Cynipini on *Quercus,* as are the holarctic genera *Ceroptres* and *Saphonecrus* and the nearctic genus *Euceroptres.* Species of the holarc-tic genus *Periclistus* are all inquilines of Rhoditini on *Rosa* whilst in America species of *Synophromorpha* have also been reared from galls of *Diplolepis* as well as from galls of *Diastrophus* on *Rubus.* Whilst galls of Cynipini and Rhoditini harbour very many species of inquiline gall wasps, galls of Ayla-xini are almost free from their attack. The only inquilines known to regularly oviposit in galls on plants other than *Quercus* and *Rosa* are *Synophromor-pha rubi* Weld, which inquilines galls of *Diastrophus cuscutaeformis* (Osten Sacken), and *S. terricola* Weld which attacks the more or less subterranean galls of *D. bassetti* (Beutenmueller) and *D. radicum* Bassett, all on *Rubus.* Ross (1951) records a *Synergus* species from galls of *Phanacis hypochoeridis* on *Hypochaeris* but this can be no more than a casual association. Records of *Synergus ruficornis* Hartig being reared from galls of *Diplolepis* (Morley 1931, Blair 1943) are almost certainly erroneous and refer to *Periclistus brandti* Ratzeburg (Eady and Quinlan, 1963). *Synergus clandestinus* Eady inhabits stunted acorns and was originally thought to be free-living, not associated with a gall wasp host, but Wiebes-Rijks (1980) has shown that it is an inquiline, like all of its congeners, attacking galls of *Andricus legitimus* Wiebes-Rijks.

Synergini lay their eggs in developing host galls. Often these are deposited inside the larval chamber and the rapid growth of the inquiline larvae destroys the gall-making larva. The inquiline larva does not appear to consume its host, subsisting entirely on gall tissue. Being smaller than the host gall wasps, several inquilines often develop together in a single gall. Thin walls of gall tissue separate the inquilines which are thus capable of slightly modifying the structure of a host gall, although no Synergini can induce the formation of a new gall. Some inquiline gall wasps do not kill their hosts and both inquilines and gall wasp may emerge from a single gall. This situation arises when the host gall has an extensive parenchyma peripheral to the larval chamber providing space for the inquiline cells. Both of these two types of inquiline larvae may be seen in galls of *Andricus kollari* in which *Synergus reinhardi* Mayr occludes the larval chamber and destroys the gall-making larva and *S. umbraculus* Olivier occupies peripheral cells in the parenchyma. Galls inquilined by *S. umbraculus* may also produce adult *S. reinhardi* or *A. kollari*.

Inquiline gall wasps are virtually host plant specific, although most species have broad host gall ranges. Strict host specificity is exceptional but many species are predominantly associated with a particular type of gall. Univoltine and bivoltine species both occur in temperate regions; *Periclistus* species, like their *Diplolepis* hosts, are usually univoltine, and bivoltinism is found in some of the *Synergus* species that attack the bivoltine Cynipini. Interestingly, in parallel with their hosts, bivoltine *Synergus* may exhibit seasonal or generational dimorphism, individuals of the two generations differing in several structural characters (Ross 1951, Eady 1952). The two generations of bivoltine *Synergus* species do not necessarily attack both generations of the same host species.

Another group of gall wasps, quite different from the Synergini, is also composed of inquilines. This is the Austrocynipinae, a subfamily created for the endemic Australian genus *Thrasorus*. The seven known species are inquilines in galls on acacias and eucalypts. These galls are formed not by gall wasps but by chalcids belonging to the pteromalid subfamily Brachyscelidiphaginae. One of the Cynipinae, *Myrtopsen mimosae* Weld, is similarly an inquiline in chalcid galls of *Tanaostigmodes* on *Mimosa* in Arizona. These gall wasps, in being trophically dependent on chalcids, represent a fascinating departure from the more usual relationship in which chalcids are parasites of gall wasps.

In addition to inquiline gall wasps, cynipid galls may provide accomodation also for inquiline gall midges (Cecidomyiidae). *Clinodiplosis galliperda* (Loew) lives as a larva in spangle galls of *Neuroterus* species. Other types of insect to feed on the tissues of living galls include the weevil *Curculio* (=*Balaninus*) *villosus* Fabricius whose larvae are frequently to be found in oak apples of *Biorhiza pallida*.

Inquilines are obligatorily associated with galls but a variety of other organisms use the structure of galls, more or less facultatively, after the gall wasps, inquilines and their parasites have emerged. The empty larval chamber of an old gall, still often attached to the host plant, provides a ready-made shelter for other organisms that gain admittance through the exit hole. Old galls are frequently found to contain solitary bees and wasps, larvae of Lepidoptera and Neuroptera, Thysanoptera, Hemiptera, Psocoptera, spiders and, less commonly, sawflies and pseudoscorpions. These facultative associates of gall wasps may be termed successori (Mani, 1964).

Gall wasp parasites and communities

If a few mature galls are collected and kept in a container, one cannot fail to be impressed by the variety of insects that emerge from them. As well as gall wasps and inquilines, there will in all probability also be a host of their parasites. The parasites (or parasitoids) of Cynipinae are parasitic Hymenoptera mainly of the superfamily Chalcidoidea (chalcids) although Ichneumonidae may also be reared from a few types of gall. Chalcids dominate the parasite fauna and they are represented by several families. The high family diversity of parasites in galls of Cynipinae is in striking contrast to the parasite faunas of other endophytic host groups. The parasites of leaf-mining Lepidoptera, for example, belong almost entirely to the chalcid family Eulophidae and to the ichneumonoid family Braconidae. The following families and genera include species known to be regular parasites in galls of palaearctic Cynipinae and those genera that seem to be exclusively parasitic in galls of Cynipinae are marked with an asterisk. Gall wasp tribes with which species in the genera are associated are appended.

Ichneumonoidea		
Ichneumonidae	*Orthopelma	Rhoditini
Chalcidoidea		
Eurytomidae	Eurytoma	all tribes
	Sycophila (=Eudecatoma)	mainly Aylaxini, Cynipin
Torymidae	Torymus (Torymus)	all tribes
	Torymus (Syntomaspis)	Cynipini
	Megastigmus	Cynipini
	Liodontomerus (=Lochites)	Aylaxini
	*Glyphomerus (=Oligosthenus)	Aylaxini, Rhoditini
Ormyridae	Ormyrus	Aylaxini, Cynipini
Pteromalidae	*Ormocerus	Cynipini
	Cyrtoptyx	Cynipini
	*Phaenocytus	Aylaxini
	Pteromalus (=Habrocytus)	Aylaxini, Rhoditini
	Mesopolobus	mainly Cynipini
	*Caenacis	Rhoditini, Cynipini
	*Cecidostiba	Aylaxini (Pediaspis), Cynipini
	*Hobbya	Cynipini

	Cecidolampa	Aylaxini (*Pediaspis*)
	Arthrolytus	Cynipini
Eupelmidae	*Macroneura* (=*Eupelmella*)	Aylaxini, Rhoditini
	Eupelmus	all tribes
Encyrtidae	*Cynipencyrtus*	Cynipini (known only from Japan)
Eulophidae	*Aulogymnus* (*Olynx*)	mainly Cynipini
	Aulogymnus (*Aulogymnus*)	Aylaxini (*Pediaspis*)
	Dichatomus	Aylaxini (*Pediaspis*)
	Pediobius	Cynipini
	Tetrastichus	all tribes

Several other genera are reported in the literature (e.g. Fulmek, 1968) as parasites in galls of Cynipidae but these, for the most part, may be dismissed as cases of misidentification or casual inhabitants perhaps parasitic upon other insects using galls facultatively.

Nearly all species of parasites in galls of Cynipinae are found only in this situation. Only two eupelmids, *Eupelmus urozonus* Dalman and *Macroneura vesicularis* (Retzius), are regularly recorded from a wider range of hosts. Further, more species have more or less restricted host ranges, sometimes being known from only one type of gall, and only a few species include in their host gall ranges gall-makers on more than one plant genus. Thus oak galls, rose galls and galls of Aylaxini each have distinctive parasite faunas with only occasional overlapping; *Sycophila biguttata* (Swederus), *Mesopolobus jucundus* (Walker) and *Aulogymnus skianeuros* (Ratzeburg) are predominantly oak gall parasites but all exceptionally attack galls of *Diplolepis rosae* (L.). *Eupelmus urozonus* is the only species that regularly attacks the inhabitants of galls of both Rhoditini and Cynipini. Parasites in galls of Aylaxini are frequently plant specific and hence, since several plant species each support just one species of gall wasp, they are often also gall specific.

A parasite reared from a gall is by no means necessarily a parasite of the gall-maker. The biologies of the inhabitants of galls are very varied. Some species are endophagous but most are ectophagous. Most feed on other larvae or pupae in the galls but a few, such as *Eurytoma brunniventris* Ratzeburg, *Torymus* (*Syntomaspis*) *cyaneus* Walker and *Dichatomus acerinus* Förster, subsist at least partially on gall tissue. Some feed only on the gall-maker whilst some will consume any species, gall-maker, inquiline or parasite. A capacity to modify gall structure is employed by some species. *Aulogymnus euedoreschus* (Walker) makes its host gall harder and more resistant to penetration than normal (Askew, 1961f), but how this is achieved is not known. *D. acerinus*, which lives gregariously in galls of *Pediaspis aceris* ♂♀ and appears to be phytophagous, divides the original larval chamber so that each parasite larva occupies a separate compartment, rather like some *Synergus* (Askew, 1963).

The trophic interrelationships of gall inhabitants are not difficult to work out. A gall is a closed community and all insects that inhabit a larval chamber leave traces of their occupancy. Egg shells, exuvia and mandibles may all be found in the debris in a cell and these, together with direct observations of feeding, enable food webs to be constructed. The immature stages of a number of parasite genera are described in Claridge and Askew (1960) (*Eurytoma*), Askew (1961c) (*Eupelmus*), (1961d) (*Mesopolobus*), (1961a) (*Torymus* sg *Syntomaspis*), (1961f) (*Aulogymnus*), (1963) (*Aulogymnus, Dichatomus*), (1965) (*Torymus*) and (1966) (*Megastigmus*). The following key will assist in the identification of larvae (and some eggs) of most gall inhabitants.

1. Larval mandibles quadridentate.................................2
— Larval mandibles with one or two teeth..........................4
2. Body with short setae and tapering posteriorly. Head with a median frontal pit. Anterior edge of labrum crenate.....................
....*Megastigmus*
— Body without setae, broadly rounded posteriorly. Other characters disagreeing (Cynipinae)..3
3. Larva in stunted gall or with larval chamber modified, sometimes divided into compartments. Often several in a single gall...........
....Synergini
— Larva in unmodified gall. Solitary unless the gall is normally plurilocular.................................Aylaxini, Rhoditini, Cynipini
4. Larval mandibles bidentate. Egg with long pedicel (Eurytomidae) 5
— Larval mandibles unidentate. Egg often lacking a pedicel..........6
5. Ectophagous. Ventral body setae almost as long as body segments. Mandibles strong and conspicuous. Egg sometimes with spiny chorion .... *Eurytoma*
— Endophagous. Body setae short. Mandibles weak.........*Sycophila*
6. Larva with numerous body setae, the longest about as long as the body segments..7
— Larva with body setae absent or much shorter than the body segments 9
7. Anterior edge of clypeus serrated. Egg with long pedicel and usually bound against wall of larval chamber by a mat of silken threads ....*Eupelmus*
— Anterior edge of clypeus entire. Egg without a long pedicel and lying free in larval chamber..8
8. Head with two large, reniform frontal pits..............*Glyphomerus*
— Head without frontal pits...............................*Torymus*
9. Larva with short body setae. Antennae relatively conspicuous, borne on frontal prominences...10
— Larva without body setae. Antennae inconspicuous (Eulophidae)....11

10. Head with four pairs of long setae, almost as long as distance separating
    antennae..Ormyridae
— Head with very short setae.........................Pteromalidae
11. Endophagous............................*Pediobius*, ? *Tetrastichus*
— Ectophagous..12
12. Ectophagous and solitary.............................*Aulogymnus*
— Phytophagous and gregarious.........................*Dichatomus*

Congeneric parasite species have similar trophic habits and this allows
a generalized food web to be constructed for the genera of Hymenoptera
inhabiting oak galls (Fig. 5). In oak galls, only a few parasite and inquiline

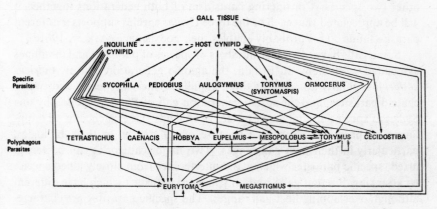

Figure 5. Trophic interrelationships of genera of cynipid oak gall inhabitants in
Britain. Arrows point to the consumer.

species are gall specific, the majority attacking a limited range of galls. The
same parasite species are not usually found in galls of alternate generations
of the same gall wasp species, but food webs in galls of corresponding
generations of closely allied gall wasps are often similar. The factors that
mainly determine the parasite complement of a gall are its structure, its
position on the tree and its season of growth (Askew, 1961e). Galls that
resemble each other in these respects, even though they may be formed by
quite unrelated gall wasps, have similar parasite faunas. Parasites of gall
wasps inflict considerable mortality and if each type of gall was to be attacked by the same species of parasites in the same proportions, gall wasps
might well be in competition with each other to avoid parasite attack. The
fact that no two species of gall wasps have identical galls ensures that competition of this sort is reduced and it permits the coexistence on oak of a
large number of similar species, all of which derive nourishment as larvae
from the tree. It is gall variety that plays a very large part in allowing many
otherwise similar species to live together. The formation of different types
of galls by the alternate generations of most species of oak gall wasps which

attract different parasite complements further diminishes competition. In Europe there are three common species of the genus *Cynips*: *C. quercusfolii*, *C. divisa* and *C. longiventris*. Galls of the agamic generations of *C. divisa* and *C. longiventris* are similar in form, position and their seasons of growth whilst the galls of *C. quercusfolii* have a different form. Thus, in the agamic generation, *C. quercusfolii* is attacked by a rather different parasite fauna to the one attacking *C. divisa* and *C. longiventris*. In galls of the sexual generations, however, the situation is reversed. Here *C. longiventris* and *C. quercusfolii* have very similar galls, and hence identical parasite faunas, whilst the sexual generation gall of *C. divisa* differs both in structure and position on the tree and it is apparently not attacked by the parasites of the other two species. Considering parasitism of both generations together, it will be appreciated that each of the three *Cynips* species supports a different parasite fauna. This probably enables their coexistence (Askew, 1961e).

Some galls have very large numbers of species of parasites and inquilines associated with them. In Britain, the agamic generation gall of *Andricus kollari* is regularly attacked by three inquiline *Synergus* species and twelve chalcid parasites, and the sexual generation gall of *Biorhiza pallida* by one *Synergus* species and nineteen species of chalcids. Coexistence of all inquilines and parasites is possible because of differences in their biologies, particularly in their host gall ranges. Two major categories can be recognized, specific parasites which avoid interference from closely allied species by attacking different galls, and polyphagous parasites which have different although overlapping host gall ranges. The specific parasites are characterized by generally attacking only the gall-maker, having a narrow host gall range, attacking the gall at an early stage of its development, having a short flight period synchronized with the periods when its host galls are in a susceptible state, and by having a relatively high reproductive potential producing many eggs and having a strongly female-biassed sex ratio (some species are thelytokous). The polyphagous parasites have a contrasting set of characteristics (Askew, 1975).

The composition of gall communities remain remarkably constant over wide geographical areas. An illustration of this is provided by rearings from galls of *Diplolepis rosae* from different European countries (Table 10). Although the community may include at least fourteen species, it is dominated by just six: the gall-maker and its inquiline *Periclistus*, the ichneumonid parasite *Orthopelma* and three species of chalcid parasites, *Torymus bedeguaris*, *Glyphomerus stigma* and *Pteromalus bedeguaris*. These species constitute over ninety per cent of the sample from every country. There are some regional differences in the relative abundance of species in the community and some of the rarer species may be absent from certain areas, but the overall conclusion is that 'the complex of inhabitants of the gall of *D. rosae* is very much the same all over Europe' (Schröder, 1967).

Relative constancy of community composition permits the meaningful

Table 10. Percentage representations of the constituent species in the *Diplolepis rosae* community in European countries. Data from Schröder (1967) (columns 3–8) and Askew (unpublished) (columns 1 and 2). Column 5 gives the mean of three sets of percentages from different regions in Switzerland.

	Britain	France	France	Spain	Switzerland	Germany	Austria	Czechoslov
Diplolepis rosae (L.)	25.0	42.4	13.3	35.0	24.7	9.7	12.4	25.9
Periclistus brandti (Ratzeburg)	20.0	7.0	10.9	8.6	7.5	10.0	23.2	17.9
Orthopelma mediator (Thunberg)	36.7	25.9	36.4	1.1	15.1	33.1	38.0	30.5
Glyphomerus stigma (F.)	1.2	8.6	9.3	28.2	17.3	28.4	4.1	11.0
Torymus bedeguaris (L.)	6.8	4.2	22.0	7.1	21.9	4.9	4.8	4.4
Pteromalus bedeguaris Thomson	5.8	8.9	6.6	17.2	11.3	9.8	8.5	5.9
Caenacis inflexa (Ratzeburg)	2.6	1.9	0.5	–	0.1	1.2	5.3	1.2
Eurytoma rosae Nees	1.2	0.3	0.5	1.6	0.4	1.8	2.2	0.9
Eupelmus urozonus Dalman	0.5	0.4	0.4	0.6	1.1	0.9	0.9	1.6
Torymus rubi (Schrank)	0.2	–	0.1	0.6	0.6	0.2	0.6	0.7
Tetrastichus sp.	<0.1	0.1	–	–	–	–	–	–
Mesopolobus jucundus (Walker)	<0.1	–	–	–	–	–	–	–
Macroneura vesicularis (Retzius)	–	0.2	–	–	–	–	–	–
Sycophila biguttata (Swederus)	<0.1	–	–	–	–	–	–	–
Species diversity (H')	1.63	1.44	1.69	1.60	1.82	1.73	1.77	1.76

application of diversity indices to the communities (Askew, 1980). Species diversities of gall communities, excluding the host gall-makers, for which adequate data (rearings of at least one hundred parasites and inquilines) are available are shown in Table 11 (further details in Askew, 1980). The

Table 11. Shannon-Weaver indices of species diversity of gall communities (excluding gall-makers). Data pertain to Great Britain except those for *Pediaspis aceris* which are French.

Aylaxini
Xestophanes brevitarsis (Thomson) 1.12
Liposthenus latreillei (Kieffer) 0.65
Aulacidea tragopogonis (Thomson) 0.86
A. hieracii (Bouché) 0.40
Phanacis hypochoeridis (Kieffer) 0.68
Diastrophus rubi (Bouché) 0.29
Pediaspis aceris (Gmelin) 1.05
Rhoditini
Diplolepis rosae (L.) 1.42

Cynipini
Cynips divisa Hartig ŏ 1.84
C. quercusfolii L. ŏ 1.34
C. longiventris Hartig ŏ 1.75
Neuroterus albipes (Schenck) ♂♀ 1.59
N. quercusbaccarum L. ♂♀ 1.78
N. numismalis (Geoffroy) ♂♀ 1.64
N. numismalis ŏ 0.36
Andricus ostreus (Hartig) ŏ 1.53
A. kollari (Hartig) ŏ 1.93
A. quercusradicis (F.) ŏ 1.29
A. curvator Hartig ♂♀ 2.30
Biorhiza pallida (Olivier) ♂♀ 2.08

species diversity of oak gall (Cynipini) communities is generally higher than that of gall communities on other plants and galls of Aylaxini usually support communities of relatively low species diversity. The numbers of species of specific parasites and inquilines in all galls are small, usually four or fewer, and it is the large numbers of polyphagous parasites in oak galls that account for their highly diverse communities. Polyphagous parasites have been able to become so well-established in oak galls because the wide variety of galls on oak provides sufficient scope for the development by the parasites of differing host gall ranges (Askew, 1980).

Gall communities are not only specifically very diverse but also generic and family diversities are high. This suggests that gall communities have evolved over a prolonged period. A community probably initially consists of polyphagous parasites and only after a lengthy association with a host do some of these, as a result of interspecific competition, become specific parasites. New polyphagous parasites are probably continually being added to communities and replacing some of those that were previously established. Replacement of specific parasites is likely to occur at a much slower rate since a specific parasite must be more finely tuned to the demands imposed upon it by a single host species. The differential rate of change between specific and polyphagous parasites could explain the taxonomic isolation in the communities of many of the specific parasites. The only regular ichneumonid inhabitant of galls in Europe is *Orthopelma mediator*, a specific parasite in galls of *Diplolepis rosae*. Eulophidae are not commonly associated with cynipid galls and the species of three eulophid genera, *Dichatomus*, *Aulogymnus* and *Pediobius*, all behave as specific parasites. The only genus of miscogasterine Pteromalidae in gall communities is *Ormocerus* and both of its known species are specific parasites (Askew, 1961b).

The sequence of events envisaged in the development of a gall community, parasite colonization, radiation, specialization and extinction, are reminiscent of the 'taxon cycle' proposed by Ricklefs and Cox (1972) to explain the restriction of endemic species of island birds to mature ecosystems.

Whichever branch of cynipid biology one elects to investigate, involvement with several aspects of the life of these fascinating insects almost inevitably follows. Evolution by natural selection has moulded gall wasps into a successful and highly specialized group, living in intricate relationship with their host plants and with the variety of inquilines and parasites dependent upon them. The galls that they produce manifest the interplay between the various organisms. They are produced by the plant and therefore have constraints of plant morphogenesis imposed upon them, but their form is controlled to a large extent by the gall wasp to provide itself with a suitable larval environment. One aspect of the suitability of a gall is the control it exercises over the community of inquilines and parasites that attack its contents but some of these gall inhabitants may themselves

(A) Plurilocular galls of *Phanacis hypochoeridis* (Kieffer) on *Hypochaeris*.

(B) An old plurilocular gall of *Diastrophus rubi* (Bouche) on Rubus; some of the larval chambers have been opened by birds.

(C) Plurilocular bedeguar gall of *Diplolepis rosae* (L.) on *Rosa*.

(D) An agamic female *Neuroterus quercusbaccarum* (L.) (=*lenticularis* Olivier) ovipositing in an oak bud.

(E) Unilocular galls of *Diplolepis nervosa* (Curtis) on *Rosa*.

(F) Sexual generation currant galls of *Neuroterus quercusbaccarum*.

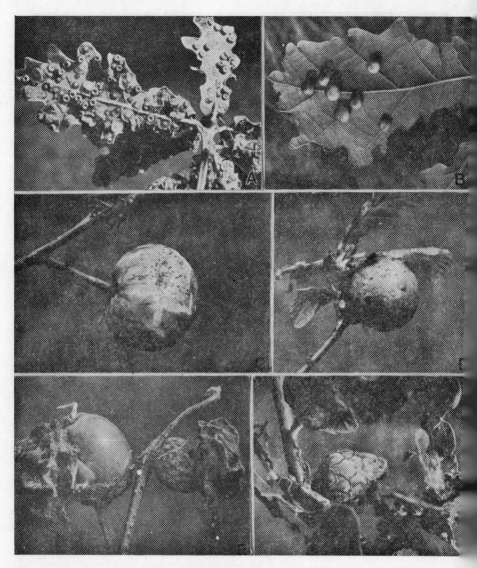

(A) Agamic generation spangle galls of *Neuroterus quercusbaccarum* (=*lenticularis*) and *N. numismalis* (Geoffroy in Fourcroy).

(B) Pea galls of the agamic generation of *Cynips divisa* Hartig on the underside of an oak leaf.

(C) Plurilocular oak apple gall of *Biorhiza pallida* (Olivier).

(D) Unilocular marble gall of *Andricus kollari* (Hartig).

(E) Knopper galls of *Andricus quercuscalicis* (Burgsdorf) on acorns.

(F) Artichoke gall of *Andricus fecundator* (Hartig) formed in a greatly enlarged oak bud.

be able to modify gall structure. In this chapter some aspects of gall wasp biology have been considered in depth, others have been passed over briefly. To some extent this reflects the incomplete state of present knowledge and one hopes it urges others to contribute to a fuller understanding of the complex machinery of cynipid biology.

REFERENCES

1. Adler, H. 1877. Beitrag sur Naturgeschichte der Cynipiden. *Dt. ent. Z.* 21 : 209–248.
2. Adler, H. 1881. Uber den Generationswechsel der Eichengallwespen. *Z. wiss. Zool.* 35 : 151–246.
3. Adler, H., Straton, C.R. 1894. *Alternating Generations. A Study of Oak Galls and Gall Flies.* Oxford: Clarendon Press. 198 pp.
4. Askew, R.R. 1960. Some observations on *Diplolepis rosae* (L.) (Hym., Cynipidae) and its parasites. *Entomologist's mon. Mag.* 95 : 191–192.
5. Askew, R.R. 1961a. On the palaearctic species of *Syntomaspis* Förster (Hym., Chalcidoidea, Torymidae). *Entomologist's mon. Mag.* 96 (1960) : 184–191.
6. Askew, R.R. 1961b. *Ormocerus latus* Walker and *O. vernalis* Walker (Hym., Pteromalidae), parasites in cynipid oak galls. *Entomologist,* 94 : 193–195.
7. Askew, R.R. 1961c. *Eupelmus urozonus* Dalman (Hym., Chalcidoidea) as a parasite in cynipid oak galls. *Entomologist* 94 : 196–201.
8. Askew, R.R. 1961d. A study of the biology of species of the genus *Mesopolobus* Westwood (Hymenoptera: Pteromalidae) associated with cynipid galls on oak. *Trans. R. ent. Soc. Lond.* 113 : 155–173.
9. Askew, R.R. 1961e. On the biology of the inhabitants of oak galls of Cynipidae (Hymenoptera) in Britain. *Trans. Soc. Br. Ent.* 14 : 237–268.
10. Askew, R.R. 1961f. The biology of the British species of the genus *Olynx* Förster (Hymenoptera: Eulophidae), with a note on seasonal colour forms in the Chalcidoidea. *Proc. R. ent. Soc. Lond.* (A) 36 : 103–112.
11. Askew, R.R. 1962. The distribution of galls of *Neuroterus* (Hym.: Cynipidae) on oak. *J. Anim. Ecol.* 31 : 439–455.
12. Askew, R.R. 1963. Some Chalcidoidea (Hymenoptera) reared from galls of *Pediaspis aceris* (Förster) (Hym., Cynipidae). *Entomologist* 96 : 273–276.
13. Askew, R.R. 1965. The biology of the British species of the genus *Torymus* Dalman (Hymenoptera: Torymidae) associated with galls of Cynipidae (Hymenoptera) on oak, with special reference to alternation of forms. *Trans. Soc. Br. Ent.* 16 : 217–232.
14. Askew, R.R. 1966. Observations on the British species of *Megastigmus* Dalman (Hym., Torymidae) which inhabit cynipid oak galls. *Entomologist* 99 : 124–128.
15. Askew, R.R. 1971. *Parasitic Insects.* London: Heinemann Educational Books: 316 pp.
16. Askew, R.R. 1975. The organisation of chalcid-dominated parasitoid communities centred upon endophytic hosts. In *Evolutionary Strategies of Parasitic Insects and Mites,* Ed. P.W. Price, pp. 130–153. New York & London: Plenum Press.
17. Askew, R.R. 1980. The diversity of insect communities in leafmines and plant galls. *J. Anim. Ecol.* 49 : 817–829.
18. Askew, R.R. 1982. *Andricus lignicola* (Hartig) (Hym., Cynipidae), another gall wasp with an expanding British distribution. *Entomologist's mon. Mag.* 118 : 116.
19. Bassett, H.F. 1873. On the habits of certain gall insects of the genus *Cynips. Can. Ent.* 5 : 91–94.

270

20. Beijerinck, M.W. 1883. *Beobachtungen uber die ersten Entwicklungsphasen einiger Cynipidengallen*. Amsterdam: J. Müller. 198 pp.
21. Beijerinck, M.W. 1902. Uber die sexuelle Generation von *Cynips Kollari*. *Marcellia* 1 : 13–20.
22. Blair, K.G. 1943. On the rose bedeguar gall and its inhabitants. *Entomologist's mon. Mag.* 79 : 231–233.
23. Blair, K.G. 1946. On the economy of the oak marble gall (*Cynips kollari* Hartig). *Proc. S. Lond. ent. Nat. Hist. Soc.* 1945–1946 : 79–83.
24. Claridge, M.F. 1962. *Andricus quercus-calicis* (Burgsdorff) in Britain (Hym., Cynipidae). *Entomologist* 95 : 60–61.
25. Claridge, M.F., Askew, R.R. 1960. Sibling species in the *Eurytoma rosae* group (Hym., Eurytomidae). *Entomophaga* 5 : 141–153.
26. Dalla Torre, C.W. von, Kieffer, J.J. 1910. Cynipidae. *Das Tierreich* 24 : 1–891.
27. Dodds, K.S. 1939. Oogenesis in *Neuroterus Baccarum* L. *Genetica* 21 : 177–193.
28. Doncaster, L. 1910. Gametogenesis of the gall-fly *Neuroterus lenticularis* (*Spathegaster baccarum*). Part I. *Proc. roy. Soc. B.* 82 : 88–113.
29. Doncaster, L. 1911. Gametogenesis of the gall-fly *Neuroterus lenticularis* (*Spathegaster baccarum*). Part II. *Proc. roy Soc. B.* 83 : 476–489.
30. Doncaster, L. 1916. Gametogenesis and sex-determination in the gall-fly, *Neuroterus lenticularis* (*Spathegaster baccarum*). Part III. *Proc. roy. Soc. B.* 89 : 183–200.
31. Eady, R.D. 1952. A revision of section I (Mayr, 1872) of the genus *Synergus* (Hym. Cynipidae) in Britain, with a species new to science. *Trans. Soc. Br. Ent.* 11 : 141–152.
32. Eady, R.D., Quinlan, J. 1963. Hymenoptera Cynipoidea. Key to families and subfamilies and Cynipinae (including galls). *Handbk Ident. Br. Insects* 8 (la) : 1–81.
33. Ejlersen, A. 1978. The spatial distribution of spangle galls (*Neuroterus* spp.) on oak (Hymenoptera, Cynipidae). *Ent. Meddr.* 46 : 19–25.
34. Folliot, R. 1964. Contribution a l'etude de la biologie des cynipides gallicoles (Hymenopteres, Cynipoidea). *Ann. Sci. Nat. Zool.* 12 ser. 6 : 407–564.
35. Fulmek, L. 1968. Parasitinsekten der Insektengallen Europas. *Beitr. Ent.* 18 : 719–952.
36. Hartig, T. 1840. Uber die Familien der Gallwespen. *Z. Ent. Germar* 2 : 176–209.
37. Hough, J.S. 1953. Studies on the common spangle gall of oak. III. The importance of the stage in laminar extension of the host leaf. *New Phytol.* 52 : 229–237.
38. Hutchinson, M.M. 1974. *Andricus lignicolus* (Hartig) (Hym., Cynipidae) in S.E. England: a species new to Britain. *Entomologist's Rec.* 86 : 158–159.
39. Kamijo, K., Tate, K. 1975. Distribution of chestnut gall-wasp, *Dryocosmus kuriphilus* Yasumatsu and its infestation to chestnuts in Hokkaido. *Bull. Hokkaido Forest Exp. Stn.* no. 13 : 27–35.
40. Kinsey, A.C. 1920. Phylogeny of cynipid genera and biological characteristics. *Bull. Am. Mus. Nat. Hist.* 42 : 357–402.
41. Krombein, K.V., Hurd, P.D., Smith, D.R., Burks, B.D. 1979. *Catalog of Hymenoptera in America North of Mexico. Vol. 1. Symphyta and Apocrita (Parasitica)*. Washington: Smithsonian Institution Press. 1198 pp.
42 Magnus, W. 1914. *Die Entstehung der Pflanzengallen verursacht durch Hymenopteren*. Vienna. 160 pp.
43. Mani, M.S. 1964. *Ecology of Plant Galls*. The Hague: Junk, 434 pp.
44. Marsden-Jones, E.M. 1953. A study of the life cycle of *Adleria kollari* Hartig, the marble or Devonshire gall. *Trans. R. ent. Soc. Lond.* 104 : 195–222.
45. Martin, M.H. 1982. Notes on the biology of *Andricus quercuscalicis* (Burgsdorf) (Hymenoptera: Cynipidae), the inducer of knopper galls on the acorns of *Quercus robur* L. *Entomologist's mon. Mag.* 118 : 121–123.

46. Meyer, J. 1950. Gigantisme nucleolaire et cecidogenesis. *C.R. Acad. Sci. Paris* 2131 : 1333–1335.
47. Miyashita, K., Ito. Y., Nakamura, K., Nakamura, M., Kondo, M. 1965. Population dynamics of the chestnut gall-wasp, *Dryocosmus kuriphilus* Yasumatsu (Hymenoptera: Cynipidae). III. Five year observation on population fluctuations. *Jap.J. appl. Ent. Zool.* 9 : 42–52.
48. Morley, C. 1931. A synopsis of the British Hymenoptera; family Cynipidae. *Entomologist* 64 : 248–249.
49. Murakami, Y. 1980. The parasitoids of *Dryocosmus kuriphilus* in Japan and the introduction of a promising natural enemy from China. *Abstracts XVI Int. Congr. Entomology. Kyoto* : 363.
50. Niblett, M. 1941. Notes on the cynipid genera *Cynips, Biorhiza* and *Megaptera* (Sic). *Entomologist* 74 : 153–157.
51. Payne, J.A. 1978. Oriental chestnut gall wasp: new nut pest in North America. *Proc. Am. Chestnut Symp.* 1978 : 86–88.
52. Payne, J.A., Menke, A.S., Schroeder, P.M. 1975. *Dryocosmus kuriphilus* Yasumatsu, (Hymenoptera: Cynipidae), an oriental chestnut gall wasp in North America. *U.S. Dep. Agric. Coop. Econ. Insect Rep.* 25 : 903–905.
53. Quinlan, J. 1968. Cynipinae (Hymenoptera) occurring on *Phlomis. Trans. R. ent. Soc. Lond.* 120 : 275–286.
54. Quinlan, J. 1974. On the occurrence of *Andricus lignicola* (Hartig) (Hym., Cynipidae) in Britain. *Entomologists, Gaz.* 25 : 293–296.
55. Quinlan, J. 1979. A revisionary classification of the Cynipoidea (Hymenoptera) of the Ethiopian zoogeographical region. *Bull. Br. Mus. nat. Hist. Entomology* 39 : 85–133.
56. Quinlan, J., Evenhuis, H.H. 1980. Status of the subfamily names Charipinae and Alloxystinae (Hymenoptera: Cynipidae). *Systematic Entomology* 5 : 427–430.
57. Ricklefs, R.E., Cox, G.W. 1972. Taxon cycles in the West Indian avifauna. *Amer. Nat.* 106 : 195–219.
58. Riley, C.V. 1873. Controlling sex in butterflies. *Amer. Nat.* 7 : 513–521.
59. Ritchie, A.J., Peters, T.M. 1981. The external morphology of *Diplolepis rosae* (Hymenoptera: Cynipidae, Cynipinae). *Ann. ent. Soc. Am.* 74 : 191–199.
60. Ross, J. 1951. A study of Some British species of *Synergus* (Hym. Cynipoidea). *Trans. Soc. Br. Ent.* 11 : 81–96.
61. Schröder, D. 1967. *Diplolepis* (=*Rhodites*) *rosae* (L.) (Hym : Cynipidae) and a review of its parasite complex in Europe. *Commonw. Inst. Biol. Control Tech. Bull.* 9 : 93–131.
62. Telenga, N.A. 1969. *Origin and Evolution of Parasitism in Hymenoptera Parasitica and Development of their Fauna in the USSR.* Jerusalem: Israel Program for Scientific Translations Ltd. 112 pp.
63. Vandel, A. 1931. *La Parthenogenese.* Paris: Encycl. Scientifique.
64. Weidner, H. 1960. Die Cynipidengallen des westlichen Norddeutschlands und ihre Bewohner. *Abh. naturwiss. Ver. Bremen* 35 : 477–548.
65. Wiebes-Rijks, A.A. 1980. The identity of the gall-wasp causing stunted acorns (Hymenoptera, Cynipidae). *Neth. J. Zool.* 30 : 243–253.
66. Yasumatsu, K. 1951. A new *Dryocosmus* injurious to chestnut trees in Japan (Hym., Cynipidae). *Mushi* 22 : 89–93.
67. Yasumatsu, K., Kamijo, K. 1979. Chalcidoid parasites of *Dryocosmus kuriphilus* Yasumatsu (Cynipidae) in Japan, with descriptions of five new species (Hymenoptera). *Esakia* 14 : 93–111.
68. Zweigelt, F. 1931. Blattlausgallen. Histogenetische und biologische Studien an *Tetraneura-* und *Schizoneura-*gallen. Die Blattlausgallen im Dienste prinzipieller Gallenforschung. *Z. angew. Ent.* 27 : 1–684.

9. Chalcids and Sawflies Associated with Plant Galls

T.C. Narendran

Introduction

Intricate relationships exist among many hymenopterous insects and various angiosperms, with many species inducing galls, and others being only gall *locatari* such as inquilines or parasites of gall formers. A species that is an inquiline in one gall system may be a gall former on another plant species. The interrelationships between cecidicoles and cecidozoans vary in different galls, and it is not always easy to assign a given species precisely to any of the above-mentioned groups.

While most of the chalcids are entomophagous, some are also phytophagous. Entomephagous chalcids are among the most important agents in many biological control programmes against pests of agricultural crops since they appear relatively more successful than taxa such as Tachinidae, Ichneumonidae, Braconidae and Proctotrupoidea. Phytophagy in chalcids probably has evolved on more than one occasion and occurs most frequently in association with gall-forming habits. The evolutionary significance of this aspect remains unknown (Gordh, 1979), and very little is known about the zoogeography of chalcids associated with plant galls, though most of the chalcids associated with plant galls are tropical and subtropical in distribution.

The habits and host relationships of sawflies are highly varied and diverse. While many are phytophagous, a few, such as members of the family Orussidae are entomophagous being parasitic in wood-boring insects. Among the phytophagous forms, many feeding on foliage, and some inducing galls, while a few others are leaf/petiole miners or stem borers or leaf rollers. The zoogeography of gall sawflies is also poorly known: although the sawflies are widely distributed, interestingly enough they have not been found in the Pacific including Hawaii.

The Chalcids (Hymenoptera : Chalcidoidea)

Chalcidoidea is one of the most important superfamilies of the order Hymenoptera, varying greatly in their form, habits and host relationships and have a wide distribution. Based on an analysis of various characters of each group, the author has recognised 16 families under the Chalcidoidea; viz., Agaonidae, Chalcididae, Leucospidae, Torymidae, Eurytomidae, Eucharitidae, Pteromalidae, Eulophidae, Tetracampidae, Elasmidae, Aphelinidae, Signiphoridae, Eupelmidae, Encyrtidae, Mymaridae and Trichogrammatidae. More than 2000 valid genera and 16000 valid species of chalcids are known so far (Noyes, 1978).

Among the chalcids it is difficult to separate the obligatory gall inhabitants and those which have an indirect association with galls. All the chalcids having an indirect as well as direct association with plant galls are discussed here. Among the 16 families of Chalcidoidea, the Agaonidae, Torymidae, Eurytomidae, Pteromalidae, Eulophidae, Encyrtidae, Aphelinidae, Eupelmidae, Trichogrammatidae, Mymaridae, and Chalcididae are treated here.

Key to the Families of Chalcids Associated with Plant Galls

1. Tarsi only with 3 segments. Fore wing with postmarginal vein absent; wing disc usually with setae in radiating rows. Antennae very short; in female with maximum 2 distinct funicular segments between pedicel and club.........*Trichogrammatidae*
= Tarsi with 4 or 5 segments (except in some apterous forms, mainly of fig insects). Antennae mostly with more funicular segments. Setae on wing disc arranged differently...2
2. Hind wing narrowed into a stalk at base; wing frequently with long fringe. Antennal toruli widely separated from each other. Frons with transverse sulcus above antennae, which extends along the inner margins of eyes as vertical sulci. Tarsi with 4 or 5 segments. Body usually shorter than 1 mm...............*Mymaridae*
= Hind wing not narrowly stalked, broader in winged forms; fringe usually short. Antennal toruli generally close to each other or if widely separated then other characters different. Most often longer than 1 mm...............................3
3. Body mostly less than 1 mm in length; always non-metallic. Each antenna with maximum 8 segments distinct (rarely with 9 distinct segments). Fore wing with postmarginal absent; stigmal extremely short, often very indistinct; marginal extremely long. Notaulices complete. Axillae broadly separated in the middle region with their anterior margins more or less projected forward. Abdomen broadly sessile ....*Aphelinidae*
= Body with other characters; if rarely showing some resemblance (as in some members of Eulophidae) then axillae strongly advanced............................4
4. Hind femur greatly swollen, its ventral margin with a row of irregular or comb like teeth. Thorax with very coarse and usually dense punctures. Body always without metallic colouration. Cerci reduced to tubercles. Ovipositor mostly not exerted and in any case never turned upwards over dorsum......................*Chalcididae*
= Hind femur not greatly swollen or if so then each cercus in the form of an elongate

segment and puncturation of thorax usually not very coarse and then usually with metallic colouration...5

5. Middle tibial spur strongly developed. Large and covex mesopleura without vertical lines..6
 = Middle tibial spur not strongly developed. Mesopleura impressed often grooved..7

6. Coxae of middle legs attached nearer to fore coxa in front of the middle of mesopleura. Antenna mostly with 6 or less than 6 funicular segments, very rarely with more than 6 segments. Mesoscutum convex. Notaulices usually absent and rarely like a line. Fore wing with marginal vein short or rudimentary or absent. Prepectus not large and not conspicuous in dorsal view.......................*Encyrtidae*
 = Middle coxae situated not closer to fore coxae, situated either closer to hind coxae (Eupelminae) or placed about midway between fore and hind coxae (Tanostigminae). Prepectus in some cases (Tanostigmina) large and convex and very conspicuous in dorsal view. Notaulices usually present. Antenna in most cases with 7 funicular segments. Marginal vein usually long.............................*Eupelmidae*

7. Hind coxa usually three or more than three times larger than fore coxa. Pronotum large; notaulices complete. Marginal vein long; stigmal very short or capitate. Antenna 13 segmented...*Torymidae*
 = Hind coxa only a little larger than fore coxa. Other characters partly different....8

8. Pronotum large, often as broad as mesoscutum, dorsally rectangular. Body usually non-metallic in colour, often black or yellowish. Notaulices complete. Abdomen convex. Postmarginal, marginal and stigmal well developed. Thorax dorsally often with coarse punctures..*Eurytomidae*
 = Pronotum not rectangular dorsally but narrowed dorsally atleast in the middle or body with metallic colour and thorax then reticulate, abdomen often shrunken, other characters partly different...9

9. Thorax without conspicuous sculpture, only with tiny piliferous pits, almost smooth. Females winged with distinct stigmal vein (rarely veins reduced or indistinct). Head usually prognathous, elongate and flattened often with a median longitudinal furrow; mandible with a proximal appendage on the ventral side of the head in females. Scape swollen. First flagellar segment tooth-like in appearance. Fore tibia extremely short. Males apterous. Body yellowish in males with gaster turned forwards beneath the body. Eyes, ocelli, antennae, and tarsi reduced....................*Agaonidae*
 = Thorax with distinct sculptures or if rarely smooth then other features different, particularly mandibles in females without proximal appendage................10

10. Tarsi 4 segmented. Antenna often with less than 11 segments and rarely with 11 segments but never with more than 11 segments. Postmarginal sometimes reduced or absent. Fore tibial spur usually short and straight. Abdomen distinctly constricted at its junction with propodeum............................*Eulophidae*
 = Tarsi 5 segmented. Antenna usually with 11–13 segments. Postmarginal usually well developed; marginal relatively long. Fore tibial spur distinct and bent..*Pteromalidae*

1) AGAONIDAE

The agaonids are commonly known as 'fig insects', since they are all captrifiers living within the floral receptacles of figs. They are extremely specialized in their morphology as well as in their biology. The Agaonidae comprises two subfamilies viz., Agaoninae and Blastophaginae (Wiebes, 1982a). The *unplaced* subfamily Sycoecinae is closely related to Agaonidae and 'probably is a plesiomorphic (less specialised) group, pointing perhaps to a common origin of both groups in the pteromaloid complex' (Boucek *et al.*, 1981).

The secretion from the poison glands of the female agaonids is considered to be responsible for inducing parthenogenetic development of endosperm of *Ficus* leading to gall-formation. Three kinds of flowers viz., the male flowers, the female flowers with short styles, and the female flowers with long styles, occur within the fig hypanthodia. All these tree types exist in some species of *Ficus* which are monoecious (Ex.: *Ficus thonningii* Blume). In the dioecious figs (Ex.: *F. hispida* Linn.) the female flowers with long styles and female flowers with short styles occur in separate trees. In many species, the male flowers are located nearer the ostiole, while in other species they occur scattered among the pistillate flowers. The agaonids usually develop in the female flowers with short styles which are known as the 'gall flowers'. This is because the agaonids have a short ovipositor which can reach the oviposition site only through the short styles of the female flowers. The female flowers with long styles usually develop into seeds. In *Ficus carica* Linn., self-pollination is impossible because the pistillate flowers turn receptive several weeks before the staminate flowers, and this behaviour is effectively coordinated with the development of *Blastophaga psenes* (Linn.) (Free, 1970). Galil and Eisikowitch (1968a) recognised the following phases in the developmental period of the figs:

a) The Pre-female phase

From the time of appearance of syconial buds to the maturation of female flowers. The ostiole will be usually closed by overlapping scales at the entrance.

b) The female phase

From the time the female flowers are receptive for pollination and oviposition lasting about 72 hours. The ostiolar scales are withdrawn slightly, and narrow spaces appear between them.

c) The interfloral phase

Commences when the pollination and oviposition phases end, and it lasts till the maturation of male flowers, simultaneous with the emergence of the agaonids from the galls.

d) The male phase

The male flowers mature, and the emerged female agaonids search ripe anthers and emerge out of syconia carrying the pollen. The females will carry the pollen and will transfer them to other female flowers of the female phase.

e) The post-floral phase

After the exit of the female wasps from the syconia, post-floral changes take place.

On the ripening of the gall-fig, the male insects emerge first by cutting through the wall of the galls; they search with the help of antennae for galls containing females of their species, which may be generally in the same locality. An unknown factor in the females appears responsible for the attraction of males as they continue to attempt to open a female gall even when female has been removed, as in the agaonid *Elisabethiella stuckenbergi* (Grandi) (Boucek *et al.*, 1981). After locating a gall containing a female, the male makes a minute hole in the gall tissue with its mandibles and inserts its tubular gaster, and then mates. In some agaonids the male bites an opening into the 'female-gall' and slips in for mating. Males of some species of agaonids introduce their heads into the galls attempting to know the presence of the females with their antennae. In *Blastophaga psenes* about ten times as many females as males are produced, and therefore each male mates probably several times. Soon after mating the winged female gets out and leaves the gall-fig through the ostiole (Free, 1970). The males of the agaonid *Elisabethiella* make several pores in the receptacle and the females leave through these (Boucek *et al.*, 1981). Many of the agaonid females usually remain within the syconium for sometime, after emergence from the galls. They usually leave the fig in large numbers, particularly during early hours. Before the females leave the fig, the ostiolar opening becomes wide, and the scales become loose so that the female can crawl out through the ostiole. The males of *Blastophaga quadraticeps* Mayr make a few exit holes in the region of the ostiolar scales in case the scales do not themselves become loose in this process (Galil and Eisikowitch, 1968a). In *Platyscapa* also the male makes exit holes. The emergence of adult agaonids coincides with the anther dehiscence, and when the females leave the syconium they are covered or loaded with pollen from the male flowers nearer the ostiole. In certain agaonids such as *E. stuckenbergii* and *Ceratosolen arabicus* Mayr, specialised structures such as corbiculae occur within which they carry pollen, whereas in some other species like *B. psenes* no such structure is known. Great differences occur in the pollination mechanics of *F. carica* and in *F. sycomorus* which are associated with different behavioural patterns of their specific pollen vectors, viz., *B. psenes* and *C. arabicus* respectively (Galil and Meiri, 1981a). While *B. psenes* becomes passively coated with pollen when it emerges out of the fig (topocentric pollination), *C. arabicus* load the pollen into their corbiculae by themselves and help in the dehiscence of anthers. One species of *Platyscapa* with the brush of setae on the fore coxae, shovels the pollen into the pocket before it carries them to the fig. Immediately upon entering the young fig through the narrow ostiole, the female starts ovipositing in the pistil and towards the end of oviposition, the pollen is taken out with the tarsus and the flowers are pollinated (Wiebes, 1981a and 1982b). In *Elisabethiella*, when the female crosses the anther it is stimulated to gather pollen in the corbiculae (Boucek *et al.*, 1981). After crawling out of the receptacles, the female may sit on the fig to clean its body and may fly-off in search of tender figs of the same tree or of another tree of the same species. On reaching a developing syconium it searches the ostiolar opening with its antennae. The ostiole of such tender figs is surrounded by thick bracts, and the entrance to the receptacle is very much limited to the female agaonids which are adapted to crawl through the narrow spaces among the bracts. The head, antennae, legs, wings and the body of the female are all so modified that they facilitate the easy entry of the female into the syconium. The tailing segment of the antennae and the wings may get detached as the female enters the syconium. Upon reaching the interior of the receptacle the female begins to oviposit in the short stsled flowers and thus pollinates the flowers. During the process of oviposition the female injects a secretion into the ovary, which initiates the gall-development. The gall provides suitable conditions for feeding and the development of the immature stage of the aganoid (Longo, 1909; Grandi, 1929).

2) TORYMIDAE

This family was known in the past as Callimomidae. The family includes more than 100 genera and more than 1050 valid species. Seven subfamilies are recognised *viz.*, Sycophaginae, Sycorictinae, Toryminae; Megastigminae, Thaumatoryminae, Monodontomerinae and Ormyrinae. The taxonomic position of the subfamily Sycophaginae is uncertain and in the past it has been regarded as group of the Agaonidae (Prinsloo, 1980). This subfamily appears closer to Torymidae and hence has been placed by Boucek *et al.* (1981) as a special subfamily of Torymidae. The systematics of Sycorictinae is also problematic, yet there are almost as many reasons to include Sycorictinae under Pteromalidae as are in favour of inclusion under Torymidae (Boucek *et al.*, 1981). Hence this subfamily has been included only tentatively under Torymidae, and all the subfamilies of Torymidae contain species associated with plant galls.

Studying various behavioural aspects of *Sycophaga*, *Apocrypta* and *Eukoebelea* (subfamily Sycophaginae), Galil and Eisikowitch (1968b) stated that Idarninae (according to them Sycophaginae and Idarninae are two different subfamilies; the former belonging to Agaonidae and the latter belonging to Torymidae) developed only in ovaries previously occupied by *Ceratosolen* (Agaonidae) or *Sycophaga*; the Idarninae do not interfere with seed-setting in pollinated flowers which do not harbour either *Ceratosolen* or *Sycophaga* and the males of Idarninae do not bore exit holes, so that whenever there are only a few *Ceratosolen* males inside the syconium, the female idarnins remain entraped. The females of *Idarnes testacea* (Mayr) *Apocrypta westwoodi* Grandi and *Parkoebelea stratheni* (Joseph) oviposit as they move their gaster in antero-dorsal and postero-ventral directions till the ovipositor penetrates the wall of the fig-syconium. After puncturing, the ovipositor is thrust deep into the fig tissue. The ovipositor sheaths are not involved in the piercing process. The telescopic arrangement of the basal segments of the gaster and the rhythmic contraction of the gaster considerably helps *Apocrypta* to penetrate the ovipositor into the wall of the fig. Inhabitation by *Sycophaga sycomori* L., has a decisive effect on the subsequent development of the syconia and flowers of *Ficus sycomorus* L. Oviposition by *S. sycomori* induces sclerification of the pericarp and proliferation of the nucellus which serves as food for the larva of *S. sycomori*. Inhabitation by *Sycophaga* is indispensible for the maturation of male flowers (Galil *et al.*, 1970).

The biology of some members of the subfamily Sycorictinae has been studied by Mayer (1882), Lichtenstein (1919) and Joseph (1953a, 1958). Mating in *Philotrypesis caricae* (L.), generally, occurs while the females are still enclosed in the fig galls as in the case of most agaonids. Joseph (1958) has studied the biology, behaviour, postembryonic development and morphology of *Philotrypesis caricae* on a comparison with those of the agaonid *B. psenes*. He (1953a) also studied the biology, evolution, distribution and taxonomy of *Sycoscapteridea indica* (Joseph), but however did not state whether the female of *S. indica* really oviposited in ovaries of *Ficus* in which the agaonids oviposited earlier. The other important genera associated with the subfamily Sycorictinae, are *Sycoscapter*, *Sycorictes*, *Arachonia Sycorycteridea* and *Watshamiella*.

Most of the species of Toryminae are ectoparasitic upon cecidogenus insects of the families Cecidomyiidae and Cynipidae (Peck, 1963; Grissel, 1973a, 1976; Burks, 1979). A few species of the genus *Torymus* are parasitic upon gall-inducing Psyllidae, Eurytomidae, and Tephritidae. Askew (1965) described the immature stages of *Torymus* and discussed the biology of *T. auratus* (Fourcroy), *T. cingulatus* Nees, *T. nigricornis* Boheman, *T. erucarum* (Schrank), *T. nobilis* Boheman, *T. amoenus* Boheman and *T. pleuralis* Thompson. According to him, *T. auratus*, *T. cingulatus* and *T. nigricornis* has two generations a year. In *T. auratus* the overwintering generation is parasitic on *Neuroterus* (Cynipidae), while the autumn generation is parasitic on *Cynips*, *Andricus*, *Biorrhiza* (Cynipidae) and *Neuroterus* as well. The overwintering generations of *T. cingulatus* and *T. nigricornis* emerge in spring from the galls of *Andricus*, *Cynips* and *Biorrhiza* induced on leaves and buds, while the autumn generations are parasitic in the galls of *Biorrhiza pallida* (Oliver). The difference in the length of ovipositor in *T. auratus* between the two generations is remarkable; the insects of the first or overwintering generation usually have a longer ovipositor than that of the second generation. Differences in colour between the individuals of the two generations are also marked as in *T. auratus*, *T. cingulatus* and *T. nigricornis*. *Syntomaspis* of Toryminae include species which inhabit cynipid-induced galls on oak in their immature stages. *S. cyanea* lays its eggs singly in young galls of *Cynips divisa* and *C. longiventris* (Askew 1961a). Additional information on the biology of *S. apicalis* (Walker) (a parasite of *Biorrhiza pallida* (Oliver) of oak galls), *S. luzulina* (Forster (from the galls of *Cynips quercus-folli* (L.) and *C. longiventris* Htg.), *S. notata* (Walker) (a parasite of *Andricus curvator* Htg.) and *S. cerri* Mayr (from galls of *A. singularis* Mayr) is provided by Askew (1961a). Apart from *Torymus* and

Syntomaspis, the subfamily Toryminae includes genera such as *Diomorus*, *Lioterphus*, *Callimomus*, *Odopoia*, etc., which all comprise species parasitic on other gall formers (Nikol'skaya, 1952; Boucek, 1964; Peck, 1963; Risbeck, 1952; Boucek, 1978; Kamijo, 1963, 1979, 1982).

In the subfamily Thaumatoryminae, *Thaumatorymus notanisoides* Ferriere & Novicky is reported from the cynipid-induced galls of *Hypochoeris* (Boucek, 1964, 1978).

In the subfamily Megastgminae, many are cecidogenous, while some live as inquilines or parasitoids of other gall-forming insects or live as phytophagous insects on the seeds or plant shoots. Information on the biology of *Megastigmus dorsalis* (Fabricius) and *M. stigmatizans* (Fabricius), (Askew, 1966), and *M. brevivalus* (Girault) (under the synonym, *Epimegastigmus* Girault (Noble, 1938) parasitic on gall-wasps is available. *M. brevivalus* is a solitary parasitoid of the larvae of citrus gall chalcid *Eurytoma fellis* Girault in Australia. The female of *M. brevivalus* oviposits on any of the immature stages of the host. The egg of this parasite has a stalk, which usually bends over the egg after deposition. The length of the egg is almost the length of the egg of its host. The incubation period of *E. fellis* is variable ranging from 13 to 30 days and the incubation period of the egg of the parasite is correspondingly longer to be 34–39 days so that eclosion of the egg of the parasite does not take place until the eclosion of the egg of the host. The larval development of the host extends over about 10 months. Out of these 10 months, the first stage larva takes about eight months, during the period the parasite also remains in its first instar larval stage, freely floating in the haemolymph of its host. During the spring, when the host attains larval maturity, the parasite is still in its first instar stage but the parasite speeds up its development and consumes the entire body-contents of the host before the host is able to pupate. After the completion of feeding activities, the mature final instar larva emerges out of the skin of the host and pupates in the gall tissue. The two features viz., the deposition of the egg in the host egg and the synchronization of embryonic and larval development with that of the host are quite distinct in the biology of this torymid *M. brevivalus*, (Clausen, 1940). Such a synchronization in the development of the parasite and in the host occurs in Icheneumonidae, Braconidae and Platygasteridae of similar habits. The extension of the first larval instar is common to all species having similar oviposition habits, but the protracted period of incubation of *M. brevivalus* is a particular adaptation to ensure that the egg does not hatch until the host has attained the active first instar stage (Clausen, 1940).

Monodontomerinae comprises species which are parastic particularly on, Diptera, Hymenoptera, Lepidoptera and Coleoptera, and some species of this group are directly or indirectly associated with plant galls *Liodontomerus papaveris* (Foerst), *L. mayri* (Wachtl.), *Dimeromicrus smithi* (Schread), parasitic on gall-formers (Nikolskaya, 1952), and *Pseudotorymus lazulellus*

(Ashmead) associated with gall-systems (Grissel, 1979) are good examples. Detailed information on the biology of species of Ormyrinae is still lacking. Their role within the gall system is not yet clearly understood. However it is believed that many of the ormyrids may be entomophagous on other gall-insects rather than feeding on plant tissues. Several species of *Ormyrus* have been reported from plant galls from different regions of the world: *O. subconicus* Boucek, *O. flavipes* Boucek and *O. watshami* Boucek from the galled figs of *Ficus burkei* (Boucek *et al.*, 1981); *O. diffenis* Fons. parasitic on the gall-wasp *Xestophanes potentiallae* Ratz. (Nikol'skaya, 1952). Besides the reports of Fullaway (1912) describing *O. distinctus* from the galls of strawberry *Quercus dumosa* from California, and Ashmead (1885-1887) reporting several species of ormyrids obtained from the cynipid galls, Peck (1951, 1963) and Burks (1958, 1979) have given long lists of ormyrids of the American Continent, most of them being parasitic on gall forming cynipids. *O. orientalis* Walker, known earlier from Sri Lanka (Walker, 1871), is parasitic on a cecidogenus trypetid infesting a composite and is widely distributed in mediterannean and central European regions (Boucek, 1977). Girault (1920a) reported *O. australiensis* and *O. langlandi* (both described by Girault) from Australia. Hedqvist (1968) described a new species *O. philippinensis* from Philipines. *O. sculpyilis* Crosby is parasitic on an unknown agromyzid that induces galls on *Fluegea obovata* (L) (Crosby, 1909). Washburn & Cornel (1981) have reported *O. brunneipes* Provancher parasitic on *Andricus pattonii* (Basset) that induces galls on *O. stellata*.

3) EURYTOMIDAE

This family comprises more than 75 genera and more than 1070 species throughout the world. They are widely distributed all over the world. Burks (1971) classified eurytomids into eight subfamilies, viz. Rileyinae, Harmolitinae, Philoleminae, Eudecatominae, Aximinae, Heimbrinae, Prodecatominae and Eurytominae. Subba Rao (1978) considered Eurytomidae is best divided into two subfamilies viz. Rileyinae and Eurytominae, the former comprising Rileyini, Heimbrini and Buresiini as tribes and the latter comprising all the genera listed by Burks (1971) under Eurytominae, Eudecatominae, Prodecatominae, Harmolitinae and Aximinae. As pointed out by Subba Rao (1978) I am also of the opinion that no good reason appears obvious to split the subfamily Eurytominae into several tribes, and therefore the genera could ultimately be placed in two subfamilies like Eurytominae and Rileyinae without any tribes at all in Eurytominae. At this juncture it must be mentioned that there are some differences of opinion regarding the synonymy involved with the names *Tetramesa* and *Harmolita*. These names relate to one genus of Eurytomidae associated with plant galls. Boucek (1964), Claridge (1958, 1961), and Peck (1963) treat *Harmolita* as a synonym of *Tetramesa*, while Burks (1971) expressed some doubts regarding this synonymy and has stated that Gahan's notes made over 40 years earlier, need not be changed. According to these notes of Gahan *Harmolita* and *Tetramesa* differ in shape of abdomen, in nature of propodeum and in antennae. However, a critical analysis of this problem shows that the whole conglomerate of species cannot be subdivided on good characters and therefore *Tetramesa* should be regarded as a genus by priority, and *Harmolita* as a junior synonym of *Tetramesa*.

The family Eurytomidae like Torymidae has diverse host relationships. While some are phytophagous, forming galls or feeding seeds, quite a few are parasitic on gall-making insects associated with galls. Some are also found to be inquilinous in the galls. Among the gall inducers, *Tetramesa* is perhaps the most important genus of this family, whose species form galls in the stems or develop within the cereal stalks.

A number of species of *Tetramesa* have been reported from the plant galls by many authors like Nikol'skaya (1952), Boucek (1977), Peck (1963), Burks (1979), and Roskam (1982). The information on the parasitic habits of the family is limited mostly to *Eurytoma*. This genus has different types of host preferences; besides some being external parasites of the immature stages of hymenopterans, particularly gall-cynipids, others parasitise larvae of Lepidoptera, Diptera and Coleoptera inhabiting plant galls, tunnels, etc., (Clausen, 1940). Many species of *Eurytoma* are exoparasitic on the larvae of gall making trypetid flies as well. *Eurytoma parva* Philips is an interesting species partly parasitic on gall-inducing insect and partly feeding on the gall tissue. Initially the eurytomid larva attacks and consumes the first or second instar larva of its gall-forming eurytomid host *Tetramesa tritici* (Fitch) in stems of wheat. When the host larva is completely consumed, the parasite shifts its feeding on to the plant sap. The female *E. parva* oviposits through the wall of the stem by inserting its ovipositor into the cavity of the stem occupied by the young gall making *T. tritici* larva. The egg is laid usually close to the host, and rarely outside the plant-host tissue. Incubation period lasts for four or five days. Soon after emergence the parasitic larva consumes its host larva, and later the plant sap (Philips, 1927). A considerable amount of laceration of plant tissue during the course of feeding by *E. parva* has been noticed (Clausen, 1940) which is probably more than that would have been done by the larvae of *T. tritici*. Another eurytomid species viz., *E. curta* Walker is parasitic on the larval stages of the gall forming trypetid *Euribia jaceana* Her., in England (Varley, 1937). Several other species of *Eurytoma* are reported from the galls of plants either as parasites or as inquilines (Nikol'skaya, 1952; Burks, 1979, Boucek *et al.*, 1981). *Eurytoma bugbeei* Grissel, reared from thick shelled monothalamous galls harbouring not only *Disholcaspis* larva, but also several larvae of other species of the cynipid *Synergus*, is probably parasitic on one or more species of *Synergus* rather than on the gall-former *Disholcaspis* larva itself (Grissel, 1973b).

The genus *Sycophila* contains mostly species associated with figs in which they are probaby inquilines. Several species of this genus have been described erroneously under the generic name *Decatoma*. The name *Decatoma* was erected by Spinola and this was later recognized as junior synonym of *Eurytoma*. *Eudecatoma* is another latest synonym of *Sycophila* (Boucek, 1974; Boucek *et al.*, 1981). The neotropical genus *Calorileya* has one of its species recorded from the galls of *Mayrellus*, and another reported to make galls on grasses (Burks, 1971). Several other genera of this family such as *Phylloxeroxenus, Nikanoria, Prodecatoma, Rileya* and *Tenuipetiolus* include species associated with plant galls. According to Van Staden *et al.* (1977) an undescribed species of *Eurytoma* make true galls on the leaves on *Erythrina latissima* E. Mey. Agarwall and Jain (1981) have reportedly studied leaf and shoot galls of *Emblica officinalis* Gaertn., formed by an unknown chalcid belonging to Eurytomidae.

4) PTEROMALIDAE

As the largest family of Chalcidoidea, Pteromalidae comprises more than 560 genera and more than 2800 species. Much inconsistency exists in the classification of Pteromalidae. Some have given family status to perilampids, eucharitids, splangiids, ormyrids, etc., while others treated these groups as subfamilies of Pteromalidae. The following subfamilies are recognised: Spalangiinae, Miscogasterinae, Cercocephalinae, Diaparinae, Ceinae, Cleonyminae, Trydyminae. Colotrechninae, Pteromalinae, Epichrysomalinae, Otitesellinae, Asaphinae, Brachyscelidiphaginae, Eutrichosomalinae, Macromesinae, Eunotinae, Panstenoninae and Perilampinae. Perhaps the best taxonomic treatment on the family Pteromalidae so far published is that of Graham (1969).

A good account of many gall inhabiting pteromalids has been provided by Gahan and Ferriere (1947), dealing with the taxonomy of several genera supplemented with the data on the biology of these insects. The larvae of *Mesopolobus* are polyphagus occurring in a wide range of galls, evidently parasitising other cecidiocoles or gall-makers. From every gall chamber usually one adult *Mesopolobus* will emerge. *Ormocerus latus* Walker and *O. vernalis* Walker are parasitic on the cynipids inducing galls on oak (Askew, 1961e). There are several other genera like *Dinarmus, Caenacis, Homoporus, Cicidostiba, Arthrolytus, Lariophagus, Chlocytus, Cyrtoptyx, Hobbya, Halticoptera, Spaniopus* and *Rhizomalus* associated with cynipid induced galls on oaks. Two species of Pteromalidae viz., *Zatropis albiclavatus* (Girault) and *Habrocytus* species are associated with the galls on *Chrysothamnus* sp. in Mosco (Wangberg, 1977). Kamijo (1981a) has reported six species viz., *Pteromalus apantelophagus* (Crawford), *Cecidostiba semifascia* (Walker), *C. fushica* Kamijo, *Caenacis peroni* Kamijo, *Mesopolobus yasumatsui* Kamijo and *Arthrolytus usubai* Kamijo from cynipid galls on Oak and Chestnut in Japan. The females of the subfamily Epichrysomalinae induce the gall on the female florets of figs in a similar way as do the females of the agaonids but the Epichrysomalinae oviposit from the exterior of the hypanthodium (Boucek *et al.*, 1981). They are not pollinators and attack only small sized figs with florets within the reach of their coiled ovipositors. The association between *Ficus microcarpa* L. and the fig wasp *Odontofroggatia galili* Wiebes (Galil and Copland, 1981) appears interesting. The figs occupied by the wasps do not drop at the usual time but ripen fully remaining attached to the branch. The nucellar tissue proliferates and the carpellary tissue develops into a gall providing nourishment to the wasp-larva. *O. galili* can therefore develop independently even in the absence of legitimate pollen vectors of *F. microcarpa*, resulting in parthenocarpic fruits, developing in large quantities. Unlike most fig wasps, the male *Odontofroggatia* has well-developed wings, and mating of these wasps take place outside the fig. During oviposition the ovipositor is gradually pierced into the fig tissue by rotary movement and the telescoping of the proximal end of the inner ovipositor plate independently of the gater tip. Often weak and old females of *Odontofroggatia* get fixed to the fig as a result of oviposition, because the muscle activating the ovipositor sheath and stylets of old females are weak. Hence the ovipositor sheath slips in this direction, a little beyond the stylets, within the syconium and thus prevents the withdrawal of the ovipositor from the syconium.

The subfamily Otitesellinae comprises species which are associated with the galls of *Ficus burkei* (Miquel). The females are supposed to oviposit from within the receptacles, since they have a short ovipositor (Boucek *et al.*, 1981). However, *Otitesella* of *F. burkei* does not enter the receptacle for oviposition (Boucek *et al.*, 1981). Some of the other important genera associated with fig receptacle galls are *Micranisa, Walkerella, Grandiana, Sycobia, Sycotetra, Camarothorax* and *Pembertonia*. Quite a few like *Watshamia, Anogmoides, Systasis, Pseudocatolaccus, Gastrancistrus* of this family are associated with gall-forming cecidomyiids. For example, the female *Gastrancistrus hamillus* Walker oviposits in larvae of *Masalongia betulifolia* Harris (Cecidomyiidae) and the parasitic adult emerges in spring from the cocoon of its host in the soil, after having overwintered. It is likely that the development of *G. hamillus* does not proceed beyond the egg or first instar larva, while the host larvae remains in its gall. Parasitised midge larva remains active until it falls to the ground (Askew and Ruse, 1974). According to Yukawa *et al.* (1981) *Pseudocatolaccus syatamabae* Ishii is parasitic in the galls of *Aspondylia* spp. in Japan. Another pteromalid *Ormocerus* sp. is parasitic in the galls of Cynipids, (Askew, 1980). Pteromalids such as *Capellia orneus* (Walker) parasitise lepidopterans developing in resin galls, and genera like *Habrocytus, Lonchetron*, etc., have species associated with the galls induced by sawflies. An undetermined species of *Gastrancistrus* account for the reduction of populations of *Contarinia coloradensis* Felt by 7 per cent, which causes galls on Ponderosa pine (Brewer and Johnson, 1977). Among the gall makers, *Asparagobius braunsi* Mayr is restricted to South Africa, and forms galls on the stems of *Asparagus stricta* (Prinsloo, 1980). The genus *Hemadas* includes *H. nubilipennis* (Ashmead) which causes hard woody galls on the stems of *Vaccinium* in the eastern part of North America (Gahan and Ferriere, 1947). The genera *Semiotellus* and *Spaniopus* contain species associated with plant galls (Kamijo, 1977, 1981b).

5) EULOPHIDAE

This is a large family consisting about 325 genera and about 3000 species. Most of the members of this family are economically important being parasites of pests and are used as biological control agents. The family Eulophidae is recognised into five subfamilies, viz., Tetrastichinae, Entedontenae, Eulophinae, Elachertinae, and Euderinae by Boucek (1964), and into three subfamilies viz., Eulophinae, Entedontinae and Elasminae by Burks (1979). Burks (1979) has included Elachertinae as tribe Elachertini under Eulophinae. Euderinae and Tetrastichinae were recognised into tribes and included under Entedontinae, Elasmidae treated as a subfamily of Eulophidae. An analysis of the various characters of these groups shows that the inclusion of elasmids under eulophids is acceptable but yet, appears problematic. Hence one recognises Eulophidae as by Boucek (1964).

Some differences of opinion exist regarding the validity of certain generic names of this family, and particularly regarding the synonymy of *Tetrastichus* and *Sphenolepis* (Burks, 1979). Similar problem occurs relating to the synonymy of *Pediobius* and *Micropterus*. A case to recognise *Microterus* is very weak because of the already published information and the only name for this genus that could be retained on the basis of usage is *Pleurotropis* of Foerster (1856), used over a century.

In the subfamily Tetrastichinae, the genus *Tetrastichus* comprises several species which are associated as parasites on cecidomyiids or other gall-making or gall inhabiting insects. Askew and Ruse (1974) gave an account of the biosystematics of some species of *Tetrastichus* developing in the galls of *Massalongia* (Cecidomyiidae) on the leaves of birch (*Betula*). The female of *T. brevinervis* (Zetterstedt) oviposits and thus paralyses the larva of *Massalongia betulifolia* Harris. The parasite overwinters as a fully grown larva inside the host and usually pupates while being still enclosed by the remnant skin of the host. Superparasitism is often encountered in *T. brevinervis*. Another species, *T. ventricosus* (Graham), is parasitic on not only its principal larval hosts of *Massalongia rubra* (Kieffer), but also hyperparasitic on the larvae of *T. alveatus* (Graham) and on *T. citrinellus* (Graham) in the galls induced by *M. rubra*. *T. alveatus* is associated with the galls of *M. rubra* in which it parasitises the cecidomyiid larva. It is bivoltine and is apparently thelytokus (Askew and Ruse, 1974). Domenichini (1966) records this species from cynipid *Diplolepis mayri* Schlechtendal from Czechoslovakia and Morocco. *T. xanthosoma* Graham is a common ectoparasite in the galls of *M. betulifolia* and less frequently in the galls of *M. rubra* (Graham, 1974). This species is partly bivoltine. Askew and Ruse (1974) found *T. citrinellus* as a frequent ectoparasite of the larvae of *M. rubra*, or quite often as a secondary parasite through *T. ventricosus* or less commonly through its own species or *T. alveatus*. *T. citrinellus* is univoltine. The larvae of *T. cecidobroter* Gordh and Hawkins are phytophagous within the cecidomyiid galls in *Atriplex* spp. (Gordh and Hawkins, 1982), while *T. ardisiae* Ishii (Ishii, 1931), *T. cecidophagus* (Wangberg, 1977) and T. thripophonus Narayanan (Ananthakrishnan & Swaminathan, 1977) are known from diverse insect-induced galls as being parasitic on the gall-maker. There are several other genera in this subfamily (Tetrastichinae) which are associated with plant galls. The genus *Thymus* comprises species such as *T. albocinctus* (Ashmead) which are found in the cynipid galls of oak. Several species of *Galeospsomyia* have been recorded from the galls of Cynipidae and Cecidomyiidae from the American continents. The genera *Paragaleopsomyia, Galeopsomopsis, Crataepus,* and *Aprostocetus* contain species reared from the galls of Cecidomyiids.

The subfamily Entedontinae consists of several large genera. Askew (1962b) studied the biology of some gall inhabiting species of *Pediobius* viz., *P. lysis* (Walker) and *P. clita* (Walker). The two species are parasites in the spangle galls of *Neuroterus* in England. Both species have a single generation a year, and are specific parasites of cynipid larvae, attacking young galls; these eulophids search their host by tapping on the upper surfaces of the leaves of oak with their antennae. During this behaviour they usually take an extremely indirect searching path running sideways frequently (Askew, 1963b). *P. plagiotrochi* (Erdos) has two generations in a year and differs from *P. clita* and *P. lysis*. Another species of *P. aphidiphagus* (Ashmead), is parasitic on the tineid larva inhabiting the galls of cynipids on oak (Peck, 1963; Burks, 1979). *Crataepiella* has two species associated with *Lipara* galls on *Phragmites* (Boucek, 1964). While most of the species of the genus *Chrysocharis* are parasitic on leaf miners, some species such as *C. viridis* Provancher attack gall-cynipids on Lactuca in Canada (Ashmead, 1894). There are several other genera of this subfamily such as *Achrysocharis, Horismenus, Eprhopalotus, Omphale* and *Chrysonotomyia*, which include species that are parasites or inquilines in the galls on several plants.

The subfamily Euderinae has some species associated with plant galls. *Euderus crawfordi* from the cynipid galls (Yoshimoto, 1971), *E. saperdae* Miller from the galls formed by the beetle *Sapedra moesta* Le Conte (Yoshimoto, 1971) are two examples. An unidentified species of *Euderus* was reared from the galls of *Contarinia coloradensis* Felt (Brewer and Johnson, 1977). *Astichus auratus* Ashmead has been reported to have been reared from the cynipid galls (Burks, 1979).

In the subfamily Elachertinae, *Aulogymnus* is one of the major genera associated with plant galls. According to Boucek (1964) *Olynx* is a synonym of *Aulogymnus*, but Kamijo (1976) treats both as distinct genera and according to Kamijo (1976) *O. japonicus* (Ashmead) is parasitic in cynipid galls on *Quercus*. *Cirrospilus vittatus* Walker has often been reported from the galls of *Massalongia betulifolia* on birch leaves. *Dichatomus acerinus* (Giraud) has been reported from the galls of *Pediasapsis aceris* Gmelin by Boucek (1964), Nikol'skaya, (1952) & Bronner, (1981).

While the majority members of the subfamily Eulophinae are not associated with plant galls, the genus *Pauahiana* contains a species viz., *P. swezeyi* Yoshimoto which has been reported by Yoshimoto (1965) from *Trioza* gall on *Metrosideros*.

6) Encyrtidae

This is an extensive family of Chalcidoidea. It consists of more than 500 valid genera and over 2700 valid species. Many of the species of this family are economically important. Aphelinidae and Signiphoridae are more closely related to Encyrtidae than any of the families of Chalcidoidea. Hence Gordh (1979), and Burks (1979) included these two families as subfamilies under Encyrtidae. However, this is not acceptable for various taxonomical reasons, and the classifications of Prinsloo (1980) and Peck (1963) relating to this family have been followed here. Division of Encyrtidae into various subfamilies has been a matter of dispute. According to Hoffer (1964) and Prinsloo (1980) the family Encyrtidae contains three subfamilies viz., Encyrtinae, Arrhenophaginae and Antheminae. Tryapitzin (1973a, b) divided the family into two large subfamilies, viz., Tetracneminae and Encyrtinae based mainly on the presence or absence of paratergites. Noyes (1980) followed this classification eventhough he stated that the presence or absence of paratergite "may not be a reflection of natural grouping within the Encyrtidae since the genera with paratergites are almost all parasites of Pseudococcidae and these structures may have been evolved more than once as an adaptation to this form of parasitism; however some genera of Encyrtidae, lacking paratergites are also parasites of Pseudococcidae."

The number of genera and species associated with plant galls are relatively few in this family. Most of these species are primary endoparasites of insects, particularly coccids. Some are endoparasitoids of arachnids, while many encyrtids live as hyperparasites. A few species are polyembryonic mostly in lepidopteran larvae and rarely in aculeate hymenoptera. Some of the important encyrtids associated with plant galls are mentioned here. Several species such as *Pseudencyrtus clavellatus* (Dalman), *Pseudencyrtoides cupressi* Gordh and Tryapitzin, *Paraenasomyia orro* Girault, *Coccidoctonus* sp. etc. have been reported from the galls of Cecidomyiidae (Tachikawa, 1981; Hoffer, 1964; Nikol'skaya, 1952). *Mayrencyrtus claphyra* (Walker) (under the genus name *Liothorax*) is parasitic in the gall of *Andricus foecundatrix* Hart (Nikol'skaya, 1952). *Cynipencyrtus flavus* Ishii is parasitic in the galls of cynipids in Japan. The genus *Tineophoctonus* contains species which are parasites of gall inhabiting Teneidae, Cynipidae, and larvae of Anobiidae and Cerambycidae (Noyes, 1980). *Caenocerus puncticollis* Thompson is regarded from the cynipid galls on oak (Boucek, 1977). *Trechnites secundus* (Girault) (Mani, 1938) and *Psyllaephagus pachypsyllae* (Howard) (Peck, 1963) were recorded from the galls of Psyllids. There are some more genera of encyrtids reported as parasites of Psyllidae and Cecidomyiidae, but their association status with plant galls is not clear.

7) APHELINIDAE

This family comprises 44 genera (Hayat, 1983) and more than 800 valid species. Many species of this family are used as successful biological control agents. Aphelinidae is transitional between Eulophidae and Trichogrammatidae (Boucek, 1964). Gordh (1979) and Burks (1979) consider Aphelinidae as a subfamily of Encyrtidae while a few others treat aphelinids under Eulophidae. Many others have given aphelinids the family rank. An analysis of the various characters suggest that Aphelinids should be considered as a Family.

Most members of this family are not associated with plant galls. However, some records indicate that some species are parasites of some gall-forming aphelinids. For instance, a recent report from India (Bhagat, 1982) shows that *Populus* gall aphids are parasitised by an aphelinid parasitoid *Protoaphelinus nikolskajae* (Jashnosh) in Kashmir (India); *Aphelinus mali* Haldman is almost checking the population of the aphid pest *Eriosoma lanigerum* (Hausmann) which causes galls on the shoot axil of apple in Kashmir. These hymenopteran parasitoids are stated to invade the gall-making aphids through holes in different galls. There are some more records which show that some of the aphelinids parasitise cecidomyiids and psyllids but their exact nature of relationships with the gall making forms are not known.

8) EUPELMIDAE

This is one of the widely distributed families of Chalcidoidea. This family comprises more than 70 valid genera and about 740 valid species. In the past this family was included within Encyrtidae. According to Burks (1979) eupelmids and encyrtids diverged separately from the evolutionary stem of chalcids at a remote time in the development of the superfamily Chalcidoidea. The family Eupelmidae comprises three subfamilies viz., Eupelminae, Calosotinae and Tanaostigminae (Hedqvist, 1970; Burks, 1979). However some chalcidologists have given family rank to the last group (Tanaostigmatidae). The classification of Hedqvist (1970) and Burks (1979) has been followed here since this classification has been found to be more reasonable than any other concerning this family.

Though several species are reported from the galls of plants and some are supposed to be gall inducers, not much detailed information is available on the biology of these eupelmids except in the case of some species of the genus *Eupelmus*. A good account of the biology of *Eupelmus urozonus* Dalman found parasitic on the cynipid galls of oak in England is available (Askew, 1961d). This species attacks both bud and leaf galls induced by cynipids and probably any gall that it is able to locate. When a female of this species finds a gall, she 'drums' on the gall with its antennae and latter drills the gall with its ovipositor. The female drills a gall five or six times usually (a drill lasting 10 to 26 minutes) but only on that occasion of drilling when it takes about 26 minutes to drill a gall, can she pierce the gall and lay an egg. Unlike many other chalcids, the female never paralyses the host larva which may be an eurytomid or a torymid or a pteromalid. The female binds her eggs to the wall of the gall chamber nearer to the point of issue of its ovipositor, weaving fine strands of the silky substance over it. The male stands on the front of the female's thorax and vibrates his antennae and middle legs in the air. The tips of the female's antennae are caught inside the elliptical beating movement of the male, and finally the female lays her antennae upon the male's frons. Eventually the male moves backwards down to the female's body and then the female raises her gaster. Simultaneous with this the male quickly darts around the side of the female's gaster and copulation occurs with the male's head turned towards the female's posterior. There are several records of various species of eupelmids as reared from various plant galls (Nikol'skaya, 1952; Peck, 1963; Boucek, 1970; 1977; Burks, 1979; Hedqvist, 1970; Wangberg, 1975). Among the various genera of Eupelmidae, the genus *Eupelmus* includes a majority of species associated with plant galls. Some of the other important genera which comprise species associated with plant galls are *Brasema, Neanastatus, Anastatus, Calosota, Macroneura* (= *Eupelmella*), *Tanaostigmodes* etc.

9) TRICHOGRAMMATIDAE

Most of the members of this family are economically important being parasite of many pests of cultivated crops. They are widely distributed. The family comprises 70 genera and 438 valid species (Noyes, 1978). The trichogrammatids show close affinities with eulophids and aphelinids. They might have diverged from the eulophid stem at some remote time in the evolution of Chalcidoidea (Burks, 1979).

All members of this family are parasites in the eggs of other insects. Knowledge about their association with plant galls is scanty. This family has very little indirect association with plant galls. However, there are some records which show that some species of this family having associations with plant galls. *Thoreauia compressiventris* Girault, *T. gargantua* (Girault) and *T. gemma* (Girault) are reported from coccid-induced galls on *Eucalyptus* in Australia (Girault, 1916, 1920b, 1923; Doutt and Viggiani, 1968).

10) MYMARIDAE

Like the trichogrammatids, the members of this family are usually of minute size and fragile body. This is an extensive family which comprises more than 100 genera with about 1200 valid species. Many earlier workers considered this family independent or placed under Proctotrupoidea because of the presence of the unusual type of antennae. However, the structure of the species of this family convinces that they are true chalcids. The thoracic structure indicates that they have been derived from the same stem that produced the Eulophidae (Burks, 1979). The family comprises two subfamilies, Alaptinae and Mymarinae.

As in the case of Trichogrammatidae, all members of Mymaridae are egg parasites, and these insects too have a very indirect association with the plant galls. *Narayanella pilipes* (Subba Rao) has been obtained from the galls of *Lagestroemia flos-reginae* (Subba Rao, 1976). Smith (1970) has reported some unidentified mymarids parasitic in the galls of the sawflies of the genus *Euura*. (However, Dr. Smith in a communication to the author (1982) informed that unfortunately a fair number of incidental cecidomyiid galls were associated with stem galls of *Euura*, and hence the records of the hyperparasites and egg parasites are open to question.) According to Nikol'-skaya (1952) *Camptotera papaveris* Foerster was reared from the gall wasp *Aulax papaveris* Mayr in poppy heads.

11) CHALCIDIDAE

This is one of the taxonomically difficult families of Chalcidoidea. The main reason for the difficulty in studying these chalcids is that most species of this family are found comparatively less frequently than most of the members of other families of Chalcidoidea. There is much confusion regarding the systematics of many genera of this family except in the cases of genera viz. *Brachymeria*, *Dirhinus*, *Epitranus* and *Smicromorpha* (Narendran, 1973; 1980; Boucek and Narendran, 1981; Boucek, 1982a). The family comprises six subfamilies namely Brachymeriinae, Haltichellinae, Epitraninae, Dirhininae, Chalcidinae and Smicromorphinae (Schmitz, 1946; Narendran, 1980).

There is a recent record of a species viz. *Hockeria tamaricis* Boucek parasitic on a gall-making lepidopteran *Amblypalpis olivierella* Ragonot associated with Tamarix (Boucek, 1982b). There are some other reports (viz., Ashmead, 1887; Peck, 1951, 1963) according to which *Haltichella onatas* (Walker) is said to be associated with the galls of *Xanthoteras politum* (Cynipidae). However one doubts the authenticity of these reports on *H. onatas* since these have been excluded by Burks (1979).

The Sawflies (Hymenoptera : Symphyta)

Members of the suborder Symphyta are commonly known as the sawflies and the horntails. Various other names such as Chalastogastra, Sessiliventria, Phyllophaga, and Xylophaga, were given to these insects by early workers. The Symphyta comprises two series, the Orthandria and Strophandria. While the former includes three superfamilies, viz., Megalodontoidea, Siricoidea and Cephoidea, the latter contains only one superfamily Tenthredinoidea. Most of the sawflies associated with plant galls are gall makers. These are mainly confined to Tenthredinidae (of Tenthredinoidea) and Xyelidae (of Megalodontoidea).

1) Tenthredinidae

This includes the typical sawflies. This is the largest family of sawflies, with over 5000 species widely distributed in the world. Many members of this family are injurious to cultivated crops. The classification of the family Tenthredinidae into different subfamilies has always been controversial. According to a recent classification by Smith (1979) the family Tenthredinidae is divided into eight subfamilies viz., Selandriinae, Dolerinae, Susaninae, Nematinae, Heterarthinae, Belennocampinae, Allantinae and Tenthredininae. Most of the information on the biology of this family provided below is based on the articles of Smith (1970, 1968), Benson (1954, 1950), Carleton (1938) and Caltagirone (1964).

In the subfamily Nematinae there are three main genera viz. *Phyllocolpa*, *Pontania* and *Euura* which are associated with plant galls. Smith (1979) reported the record of an unplaced species viz., *Nematus inguilinus* Walsh 'bread from cecidomyiid galls of *S. rhodoides* Walsh'. Members of *Phyllocolpa*, *Pontania*, and *Euura* induce galls on Salicaceae. When the female oviposits, the accessory glands secrete a colleterial fluid into the tissue of the plant. It is this fluid that is mainly responsible for the subsequent development of a gall on the plant (Hovanitz, 1959; McCalla *et al.*, 1962). As a result of the develpment of these galls, the stem may get distorted or the surface of the leaf may get reduced or the branches may become weak and consequently break at the site of the gall. However these adverse affects are rarely found serious.

The gall-inducing nematine sawflies are relatively fragile insects. They usually have a short life span of two weeks in nature. The food of the adults is mostly pollen and nectar. They also feed on stamens and parts of fruits and not leaves. According to Smith (1970) "emergence of the imago is synchronized with daylight, and most leave their pupal chambers in early morning. However, they may remain hidden in the immediate vicinity and will not attempt to fly until the temperature rises later in the day. Since humidity is low at midday, the first flights may not be made until late afternoon. The optimal conditions for sawfly swarming would be at mid-afternoon of a still, warm day (28–30 °C) after a morning rain."

Phyllocolpa make open galls by deforming leaf margins of *Populus* and *Salix*. Members of this genus are not host-specific. The female usually oviposits first along one of the margins of the developing leaf and then along the other. The cataplasma commences immediately after oviposition and the margins of the leaf curl on themselves until the edges touch the under-surface of the leaf. The growth of the gall completes before the emergence of first instar larva. As a general rule only one larva occurs on the edge of the leaf. The mature larva pupates in the soil. According to Smith (1970) in California, most of the members of *Phyllocolpa* emerge without diapause in the laboratory but in nature they overwinter as prepupae. Large popula-tions usually occur for various species of the genus. Adults gather in groups at dusk and fly to the leaves where they mate and oviposit.

In the case of *Pontania*, each species makes a distinct, closed gall on the leaf of a particular species or of a particular species-group of willow. Gener-ally six types of galls recognisable into Types I–VI (Smith, 1970) are produc-ed by members of this genus. However according to Smith (personal com-munication in 1983) these six types (which tend to overlap in some regions) apply only to North American material as far as presently known, and other types are present elsewhere (Benson, 1954). Type I galls are with irregular walls and wringled surfaces. These galls form somewhat parallel rows between the leaf margin and midrib of the leaf. The Type II galls are glob-ular and thin-walled. They have slightly wrinkled surfaces and they protrude through the upper as well as the lower surfaces of the leaf. Type III galls are with slightly irregular and moderately thick walls. They are somewhat oval in shape and with smooth surfaces. Type IV galls hang from the ventral surface of the midrib of the leaf. They are with smooth and glossy surface. The galls may be ovoid or conical in shape. Type V galls also hang from the undersurface of the leaf midrib. They are round in shape. Their surfaces are smooth. Each of this gall is usually stalked. Type VI galls are large and with thick walls. They protrude through the upper as well as lower surfaces of leaf. The galls may be oval or rounded. The surfaces of the galls are glab-rous or pubescent. Apart from the abovementioned six types of galls, inter-mediates between these types also occur occasionally. There is substantial histological variation between the galls of the various types, but it is not clear how much of the difference reflects the structure of the willow and what is wrought by the sawfly (Smith, 1970). *Pontania* galls usually attained their maximum growth only under the additional stimulus caused by the larval feeding (Caltagirone, 1964; McCalla *et al.*, 1962).

Species of the subgenus *Gemmura* of the genus *Euura* induce closed bud galls on *Salix*. Members of the subgenus *Euura* form stem galls and petiole galls. The petiole gall makers of *Euura*, pupate in the soil as in the case of *Gemmura*. In some of these cases up to four larvae live in a single gall. The most common gall-making sawflies belong to the group of stem galling *Euura*. Based on the biology and morphology, Smith (1970) divided the stem galling euurans into two groups. In one group the gonopophyses of adults are only moderately long and larvae of most species of this group make exit holes in the side of the gall. They make galls on the exiguoid willows. In the other group the adults have long gonopohyses and typically the larvae do not make exit holes. They make galls on lesiolepoid willows. Galls of this group are usually more bulbous than those of the first group. When disturbed, most of the nematine gall makers freeze in position as observed by Narendran (1975) in the case of chalcids, whereas some members of *Phyllcolpa* and a few of *Euura* become truculent. There are a number of natural enemies which reduce the population of gall-making nematines. Among vertebrates, birds devour all stages and can defect 'galls that are aborts, are parasitised or contain dead immatures' (Smith, 1970). Among insect enemies, eurytomids, pteromalids, eulophids, mymarids, ichneumonids, proctotropoids, braconids, sphecoids, curculionids and cosmopterygids are important. Some species of mites of Anoetidae and Pyemotidae are also reported as the enemies of these gall makers (Smith, 1970).

Host-controlled diapause is an interesting phenomenon net with in some of the members of *Euura*. In the case of the stem-galling euurans, pupation takes place inside the gall. Dormant host trees will have diapausing prepupae while adjacent blooming host trees will have non-diapausing pupae and emerging adults. Another interesting phenomenon which occurs among euurans galling shoot axes is that the adults will not emerge from their respective galls until the cuticle of the gall is completely hardened. Certain species or clones or individuals of host-plants are more susceptible to the attack of *Euura* and *Pontania* than some other species or clones or individuals of host-plants. This difference in susceptibility could be either an adaptation of a few sawflies to a particular host clone or a gradual sensitization of the willow to repeated attack (Smith, 1970).

In the subfamily Blennocampinae two genera viz., *Blennogeneris* and *Ardis* make bud galls. The larva of *B. spissipes* (Cresson) induce terminal bud gall on *Symphoricarpos albus* (L.) (Smith, 1969). Ross (1932) published a short account of the biology of *B. spissipes*. The adults of this species are the first group of sawflies to appear in the spring. Usually one larva is met with in each gall. The head of the final instar larva is covered with several slender hairs. The labrum of this larva is shallowly emarginate. The genus *Ardis* is a small holarctic genus which comprises species which make bud galls.

2) XYELIDAE

This is a small and primitive family of sawflies. While most of its members are associated with conifers, members of the tribe Macroxyelini are associated with deciduous trees. The Xyelidae consists of two subfamilies viz., Xyelinae and Macroxyelinae. The former comprises two tribes, Xyelini and Pleuroneurini. The latter Macroxyelinae also comprises two tribes, Xyeleciini and Macroxyelini.

The larva of *Xyela gallicaulis* Smith form galls on shoots of *Pinus echinata*, *P. elliotti* and *P. taeda*. According to Smith (personal communication in 1982) the genus *Pleuroneura* also contains species that form bud galls. Members of this genus are associated with *Abies*. They oviposit in the buds of the host and the larvae feed in the developing buds or galls or new shoots. Webb and Forbes (1951) gave notes on the biology of *Pleuroneura bruneicornis* Rohwer which inhabit *Abies balsamea* and *Abies concolor*.

BIOECOLOGY

Among the Chalcidoidea, Agaonidae is perhaps the only family fully associated with plant galls. The family Torymidae contains many members which are directly or indirectly associated with plant galls. The Eurytomidae ranks next and Sycophila, Ficomila, Syceurytoma and Eurytoma are the principal genera of this family which includes both gall inducing and gall-associated species. In the family Pteromalidae, the subfamilies Otitesellinae and Epichrysomallinae are fully associated with plant galls (fig galls). Since phytophagy is relatively uncommon (except in a few cases) in Eulophidae, most members are parasitic on other inquilines or gall formers. Eupelmidae contains several species of plant galls. The family Mymaridae is not probably associated with plant galls and the reports that a few species of this family as associated with plant galls can be accepted only with reservation as it is quite possible that these reported mymarids might have laid eggs into the plant tissue which happen to form a wall of the gall (personal communication, Boucek, 1982). The same conclusion may also be applicable to the family Trichogrammatidae since they too are egg parasites like mymarids. The report of a species viz., *Hockeria tamaricis* Boucek as associated with plant galls (Boucek, 1982b) is perhaps the only reliable record which connects the family Chalcididae to plant galls.

Relatively not much detailed information is available on the economic importance of gall chalcids. Many species of *Tetramesa* (Eurytomidae) damage cereals and stems of crops by causing galls. As a result of gall formation, the conductive function of the stems is impaired and the stem becomes lignified and brittle, breaking easily by the action of wind. The development of chalcid galls on the leaves has been reported to cause a decrease in leaf size (Van Staden and Devy, 1978). A curious phenomenon was noted in North America during the first large outbreak of certain species of *Tetramesa* affecting wheat. Mattresses were fitted with hay containing larvae of this gall former. In the spring, as the adult insects began to emerge, the sleepers were stung (Westwood, 1882). An interesting report by Galil and Copland (1981) reveals that because of the heavy incidence of the pteromalid *Odontofroggatia galili* Wiebes, the syconia of *Ficus microcarpa* L. ceased dropping at an early stage and remained on the trees for several more weeks, producing abnormally larger, softer, juicy and almost black fruit which accumulated in large numbers underneath the trees or became crushed on cars parked along side. Additionally, frugivorous bats feeding on these fruits splattered their dark droppings on to the walls of the nearby buildings causing shabby appearance to the buildings. This happened because the sycophilic wasps prevented seed formation in the fruits as a result of their development and gall formation and so the fruits became seedless parthenocarpic. These fruits ceased dropping at an early stage and remained attached in the tree. The unusual developmental changes brought about by these wasps have been attributed to the contents of the acid glands of the reproductive system of female wasps (Grandi, 1961). The gall chalcids are also found to be very useful since many are pollinators of various plants. For instance, the agaonids or the fig-wasps are important pollinators of fig trees. According to Wiebes (1981b) the study of fig-wasps and the fig symbiosis may enable one to integrate flower ecological data into a testable hypothesis of reproductive strategy. No fig-wasp can propagate outside figs and no fig can thrive without its specific fig-wasps (Wiebes, 1979, 1982b). Thus the fig and the fig-wasp is believed to have evolved together. Related figs have related pollinators and both groups probably descended from a common ancestor fig and its pollinator wasp. The existence of two species of agaonids (Joseph, 1953b, 1954; Wiebes, 1964) in the receptacles of the same species of *Ficus*, however, makes the acceptance of the theory that figs and fig-wasps evolved together, with some reservation. Another interesting observation of ecological importance concerning fig-wasps has been reported by Galil and Eisikowitch (1968b) and Wiebes (1982b). According to these reports the exit hole made by the males in the walls of a fig receptacle, cause a replenishment of internal atmosphere of syconium, lowering the carbon dioxide content to less than 3–4 per cent enabling the females to actively leave the galls.

The host finding behaviour of gall chalcids is varied and complex. However, a general pattern of finding the hosts by chance encounter as described by Narendran (1975), Narendran and Joseph (1976, 1977) can be seen in many of the gall chalcids too. This chance encounter is often followed or combined with antennal chemoreception. However, exception to this type of behaviour in the gall chalcids has also been reported. The olfactory sense plays little or no part in attracting the pteromalids viz., *Mesopolobus fasciventris* Westwood and *M. jucundus* (Walker) to their host-galls even though the galls are probably distinguished from other objects by antennal chemorecption (Askew, 1961c). These two species of pteromalids are attracted to their hosts by visual stimuli in the first instance, although some inconspicuous galls are discovered apparently by chance encounter. Galls of one year or more old are not usually accepted by these pteromalids, probably because of the absence of their natural odour. If the galls are extremely hard, the parasite *Mesopolobus* usually fails to oviposit even after its repeated attempts to pierce the wall of the gall.

Ecological factors like climate have an important influence on the morphology of some gall chalcids. Differences in colour and ovipositor length in the two generations of *Torymus auratus* and in some other species of *Torymus* are common (Askew, 1965).

The existence of both phytophagy and entomophagy in chalcids associated with plant galls is another interesting phenomenon. A change from parasitism to phytophagy and *vice versa* can be observed in many species of chalcids associated with plant galls. Consumption of host plant tissue after feeding on the host larva is an important factor which is mainly due to the behavioural and ovipositional preference in the case of the genus *Torymus* (Grissel, 1976); and many species of *Torymus* are specific to niche, rather than specific to hosts (Askew, 1965; Townes, 1962 and Miller, 1970) in some parasitic wasps as well. The species *Syntomaspis eurytomae* Puz-Mal (Torymidae), oviposits within the black thorn fruits containing larvae of young almond seed-eater *Eurytoma amygdali* End. After killing the larvae of *E. amygdali*, the parasite larva *S. eurytomae* feeds on the seed, terminating its own development as a true seed-feeder. The species *Liodontomerus perplexus* Gah. usually parasitises the larvae of clover and sainfoin seed-eaters, but often changes its feed under certain conditions, to plant parts after destroying the young seed-feeding larvae. *Eurytoma parva* Gir. is another example of a chalcid developing on a mixed diet. This American species lays its eggs into the galls of *Tetramesa tritici* (Fitch) and when the parasite (*E. parva*) hatches from the egg, it first eats the young host larva and then begins to feed on the gall. In experimental conditions, *E. parva* developed quite normally on a vegetable diet (Philips, 1927). According to Philips (1927) "The larva of *Eurytoma parva* is at present in a highly plastic or adaptive condition, a fact which indicates to the writer that it is in the process of changing over from parasitism upon animals to phytophagy, while as yet entirely dependent upon *Harmolita tritici* for the stimulus leading to oviposition." Similar instances (change from parasitism to phytophagy) have also been reported in a few other species of chalcids, viz., *Eurytoma brevitergis* Bugbee *Torymus auratus* (Fourcroy) etc., (Wangberg, 1976; Askew, 1961b). In contrast to the above behavioural phenomenon of changing from the entomophagous diet to a phytophagous one, in several other chalcids like cleotoparasitic torymids, a change from phytophagy to entomophagy is doubted. According to Joseph (1958) the larva of *Philotrypesis caricae* (L.) develops along with the agaonid host in the gall-ovary of the fig sharing the same food till the available food is exhausted. Then the torymid kills the agaonid host. In the case of the torymid *Syntomaspis cyanea* (Boh.) a similar phenomenon is reported (Askew, 1961b). It may therefore be concluded that phytophagy in Chalcidoidea probably originated independently in a number of families associated with parasitism of various species on gall-inducers or seed-eaters (Niko'skaya, 1952; Gordh, 1979).

The habits and ecology of sawflies are highly varied and diverse. Unlike the chalcids, very few genera of sawflies are associated with plant galls. The adults of most species of sawflies usually emerge during spring and early summer. An adult sawfly has only a life span of a few days. The female oviposits within the plant tissue. The duration of larval stages vary, but generally last for about two weeks. At the end of the final larval stage, the mature larva moults and becomes a prepupa. If there is only a single generation, the prepupa will remain as a 'resting stage' till the next spring. Depending on the species and diapause requirement or latitude, there may be several generations in a year or it may take several years to complete the cycle (Smith, 1970). The emergence of adults of nematine sawflies inducing galls, is sychronized with daylight. Even after emergence the adults appear to rest in the immediate vicinity until the temperature rises later in the day.

Though the galls produced by sawflies rarely have an adverse effect on the host plants, heavy infestation by the stem-galling *Euura* causes marked branching around cecidia and this pruning effect provides additional sites for oviposition in the following year. The branches beyond the gall may wither. According to Smith (1970) "there are optimum ecologic sites for the gall makers as well as hosts. Sheltered areas overhanging water or humid swales or thickets are the most favoured, with galls often confined to the lee side of the plants. Sucker growth is preferred, and populations can be increased substantially by harvesting all the galls on the lower branches, leaving the hard-to-get galls in the overstory, and then carefully pruning the tree. While the majority of sawflies is thus removed, the following year's population will be larger, suggesting that a chief limiting factor is availability of suitable oviposition sites in meristematic tissue."

Acknowledgements

The author expresses his gratitude to Dr. G.L. Prinsloo (Pretoria), Dr. R.R. Askew (Manchester), Dr. D.R. Smith (Washington, D.C.), Dr. E.L. Smith (California), Dr. K.V. Krombein (Washington, D.C.), Dr. E.E. Grissel (Washington, D.C.), Dr. G. Gordh (California), Dr. J.T. Wiebes (Leiden), Dr. J. Galil (Tel Aviv), Dr. J.S. Noyes (London), Dr. B.R. Subba Rao (London), Dr. Hedqvist (Stockholm), Dr. C.M. Yoshimoto (Ottawa), Dr. Kamijo (Hokkaido), Dr. L. Mihatlovic (Belgrade), Dr. S. Nagarkatti (Bangalore), Prof. M.S. Mani (Madras), Prof. T.N. Ananthakrishnan (Madras) and Prof. K.J. Joseph (Calicut) for their immense help and assistance in the preparation of this chapter; they cooperated by supplying copies of their research papers, or by permitting the use of the information present in their publications and to quote various passages from their paper, or by supplying various relevant information concerning the gall insects, or by sending several specimens for the study.

The author is especially grateful to Dr. Z. Boucek (London) for his substantial help throughout the preparation of this chapter. The assistance of the authorities of the British Museum (Natural History), London is acknowledged for providing all facilities for the studies on chalcids during 1979–1980. The author has pleasure in thanking his research students Mr. V.V. Sudheendrakumar, Miss. P.A. Rosy and Miss. Mariamma Daniel for various kinds of assistance in the preparation of this chapter.

REFERENCES

1. Agarwal, B.D., Jain, M.K. 1981. Studies on the leaf and shoot galls of *Emblica officinalis* Gaertn. formed by Chalcidoidea (Hymenoptera). *Cecid. Internat.* 2 : 39–47.

2. Ananthakrishnan, T.N., Swaminathan, S. 1977. Host-parasite and host-predator interactions in the gall thrips *Schedothrips orientalis* Anan. (Insecta : Thysanoptera). *Entomon* 2 : 247–251.

3. Ashmead, W.H. 1887. Studies on the North American Chalcididae, with descriptions of new species, chiefly from Florida. *Trans. Am. Entomol. Soc.* 14 : 183–203.

4. Ashmead, W.H. 1894. Descriptions of thirteen new parasitic Hymenoptera bred by Prof. F.M. Webster. *J. Cincinnati Soc. nat. Hist.* 17 : 45–55.

5. Askew, R.R. 1961a. On the palaearctic species of *Syntomaspis* Forster (Hym. Chalcidoidea, Torymidae). *Ent. mon. Mag.* XC VI : 184–191.

6. Askew, R.R. 1961b. On the biology of the inhabitants of oak galls of Cynipidae (Hymenoptera) in Britain. *Trans. Soc. Brit. Ent.* 14 : 237–268.

7. Askew, R.R. 1961c. A study of the biology of species of the genus *Mesopolobus* Westwood (Hymenoptera : Pteromalidae) associated with cynipid galls on oak. *Trans. R. ent. Soc. Lond.* 113 : 155–173.

8. Askew, R.R. 1961d. *Eupelmus urozonus* Dalman (Hym., Chalcidoidea) as a parasite in Cynipid Oak Galls. The *Entomologist* (August) : 196–201.

9. Askew, R.R. 1961e. *Ormocerus latus* Walker and *O. vernalis* Walker (Hym. Pteromalidae), parasites in Cynipid Oak Galls. The *Entomologist* (August) : 193–195.

10. Askew, R.R. 1965. The biology of the British species of the genus *Torymus* Dalman (Hymenoptera : Torymidae) associated with galls of Cynipidae (Hymenoptera) on oak, with special reference to alternation of forms. *Trans. Soc. Brit. Ent.* 16 : 217–232.

11. Askew, R.R. 1966. Observations on the British species of *Megastigmus* Dalman (Hym., Torymidae) which inhabit Cynipid Oak Galls. The *Entomologist* (May) : 124–128.

12. Askew, R.R. 1980. The diversity of insect communities in leaf mines and plant galls. *J. Animal Ecology* 49 : 817–829.

13. Askew, R.R., Ruse, J.M. 1974. The biology of some Cecidomyiidae (Dpitera) galling the leaves of birth (Betula) with special reference to their chalcidoid (Hymenoptera) parasites. *Trans. R. ent. Soc. Lond.* 126 : 129–167.

14. Benson, R.B. 1950. An introduction to the natural history of British sawflies (Hymenoptera : Symphyta). *Trans. Soc. British Ent.* 10 : 45–142.

15. Benson, R.B. 1954. British sawfly galls of the genus *Nematus (Pontania)* on *Salix*. *J. Soc. British Entomology* 4 : 206–211.

16. Bhagat, R.C. 1982. Aphid galls and their parasitoids from Kashmir, India. *Entomon* 7 : 103–105.

17. Boucek, Z. 1964. In Peck, O., Boucek, Z. and Hoffer, A. Keys to the Chalcidoidea of Czechoslovakia (Insecta : Hymenoptera). *Mem. ent. Soc. Canada* 34 : 1–120.

18. Boucek, Z. 1970. Contributions to the knowledge of Italian Chalcidoidea, based mainly on a study at the Institute of Entomology in Turin, with descriptions of some new European species (Hymenoptera). *Mem. Soc. ent. ital.* 49 : 35–102.

19. Boucek, Z. 1974. On the Chalcidoidea (Hymenoptera described by C. Rondani). *Redia* 55 : 241–285.

20. Boucek, Z. 1977. A faunistic review of the Yugoslavian Chalcidoidea (Parasitic Hymenoptera). *Acta Entomol. Jugoslav.* 13 : 1–145.

21. Boucek, Z. 1978. A study of non-podagrionine Torymidae with enlarged hind femora, with key to the African genera (Hymenoptera). *J. ent. Soc. sth. Afr.* 41 : 91–134.

22. Boucek, Z. 1982a. Oriental chalcid wasps of the genus *Epitranus*. *Jour. Nat. Hist.* 16 : 577–622.

23. Boucek, Z. 1982b. Descriptions of a new *Hockeria* (Hymenoptera : Chalcididae) a parasite of lepidopterous gall-causer on *Tamarix*. *Israel J. Ent.* 16 : 49–51.

24. Boucek, Z., Narendran, T.C. 1981. Indian chalcid wasps (Hymenoptera) of the genus *Dirhinus* parasitic on synanthropic and other Diptera. *System. Ent.* 6 : 229–251.

25. Boucek, Z., Watsham, A., Wiebes, J.T. 1981. The fig wasp fauna of the receptacles of *Ficus thonningii* (Hymenoptera : Chalcidoidea). *Tijds. Entomol.* 124 : 149–235.

26. Brewer, J.W., Johnson, P.R. 1977. Biology and Parasitoids of *Contarinia coloradensis* Felt. a gall midge on ponderosa pine. *Marcellia* 39 : 391–398.

27. Bronner, R. 1981. Observations on cynipid galls modified by inquiline larvae. *Cecid. Internat.* 11 : 53–61,

28. Burks, B.D. 1958. In *Hymenoptera of America North of Mexico, Synoptic Catalog Agriculture Monograph.* 2, (Suppl.), ed. Krombein, K.V. 2 : 62–84. United States Government Printing Office, Washington, D.C. 305 pp.

29. Burks, B.D. 1971. A synopsis of the genera of the family Eurytomidae (Hymenoptera : Chalcidoidea). *Trans. Amer. ent. Soc.* 97 : 1–89.

30. Burks, B.D. 1979. In *Catalog of Hymenoptera in America North of Mexico, Symphyta and Apocrita* (Parasitica), ed. K.V. Krombein. *et al.,* 1 : 768–889, 961–1042. Smitsonian Institution Press. Washington, D.C. 1198 pp.

31. Caltagirone, L.E. 1964. Notes on the biology, parasites and inquilines of *Pontania pacifica*, a leaf gall incitant on *Salix* lasiolepis. *Ann. Entomol. Soc. Amer.* 57 : 279 : 291.

32. Carleton, M. 1938. The biology of *Pontania proxima*, the bean-gall of willows. *Jour. Linn. Soc, Lond.* 40 : 575–624.

33. Claridge, M.F. 1961. An advance towards a natural classification of Eurytomid genera (Hym., Chalcidoidea) with reference to British forms. *Trans. Soc. Brit. Ent.* 14 : 165–185.

34. Clausen, C.P. 1940. *Entomophagous Insects*. New York : Hafner Publishing Co., nc. 661 pp.

35. Crosby, C. 1909. Chalcis-flies reared from galls from Zumbo, East Africa. *Broteria* (Serie Zoologica) 8 : 77–90.

36. Domenichini, F. 1966. Palearctic Tetrastichinae. *Index entomoph. Ins.* 1: 1–101.

37. Doutt, R.L., Viggiani, G. 1968. The classification of the Trichogrammatidae (Hymenoptera: Chalcididae). *Proc. Calif. Acad. Sci.* (Ser. 4) 35: 377–586.

38. Forester, A. 1856. *Hymenopterogischen studien.* 2: 1–90.

39. Free, J.B. 1970. *Insect Pollination of Crops. Moraceae.* New York/London: Academic Press. Chapter 33: 374–379. 544 pp.

40. Fullaway, D.T. 1912. Gall-fly parasities from California. *J.N.Y. Entomol. Soc.* 20: 274-282.

41. Gahan, A.B., Ferriere, Ch. 1947. Notes on some gall inhabiting chalcidoidea (Hymenoptera). *Ann. ent. Soc. America* 11: 271–302.

42. Galil, J., Dulberger, R., Rosen, D. 1970. The effects of *Sycophaga sycomori* L. on

the structure and development of syconia in *Ficus sycomorus* L. *New Phytol.* 69: 103-111.

43. Galil, J., Eisikowitch, D. 1968a. On the pollination ecology of *Ficus sycomorus* in East Africa. *Ecol.* 49: 259-269.

44. Galil, J., Eisikowitch, D. 1968b. On the pollination ecology of *Ficus religiosa* in Israel. *Phytomorph.* 18: 356-363.

45. Galil, J., Copland, J. W. 1981. *Odontofroggatia galili* Wiebes in Israel, a primary fig wasp of *Ficus microcarpa* L. with a unique ovipositor mechanism (Epichryso-malinae Chalcidoidea). *Proc. Kon. Ned. Akad. Wetensch.* (C) 84: 183-195.

46. Galil. J., Meiri, L. 1981a. Number and structure of anthers in fig syconia in relation to behaviour of the pollen vectors. *New Phytol.* 88: 83-87.

47. Galil, J., Meiri, L. 1981b. Druplet germination in *Ficus religiosa* L. *Israel Jour. Bot.* 30: 41-47.

48. Girault, A.A. 1916. A new genus of Trichogrammatidae from Australia characterised by bearing a postmarginal vein. *Entomol.* 49: 102-103.

49. Girault, A.A. 1920a. Some insects never before seen by mankind. Privately published. 3 pp.

50. Girault, A.A. 1920b. New genera and species of Australian Trichogrammatidae (Hymenoptera). *Insecutor Inscit. menstr.* 8: 199-203.

51. Girault, A.A. 1923. Loves wooed and won in Australia. Privately published. 3 pp.

52. Gordh, G. 1979. In *Catalog of Hymenoptera in America North of Mexico. Symphyta and Apocrita (Parasitica)*, eds. K.V. Krombein *et al.*, 1: 743-748, 890-967. Washington, D.C. Smithsonian Institution Press. 1198 pp.

53. Gordh, G., Hawkins, B.A. 1982. *Tetrastichus cecidobroter* (Hymenoptera: Eulophidae), a new phytophagous species developing within the galls of *Aspondylia* (Diptera: Cecidomyiidae) on *Atriplex* (Chenopodiacea) in southern California. *Proc. Entomol. Soc. Wash.* 84: 426-429.

54. Graham, M.W.R. de V. 1969. The Pteromalidae of north-western Europe (Hymenoptera: Chalcidoidea). *Bull. Bt. Mus. nat. Hist. Ent. suppl.* 16:1- 908.

55. Graham, M.W.R. de V. 1974. A new species of *Tetrastichus* Walker (Hymenoptera: Eulophidae) from Britain. *Entomologist's Gaz.* 25: 365-367.

56. Grandi, G. 1929. Plasticita somatica, morphologia ed etologia in insetti viventi in particolari condizioni di segregazione. *Mem. Accad. Sc. Inst. Bologna* 6: 89-105.

57. Grandi, G. 1961. The hymenopterous insects of the superfamily Chalcidoidea developing within receptacles of figs. *Boll Ist. Ent. Univ. Bologna* 26: 1-33.

58. Grissel, E.E. 1973a. New species of North American Torymidae (Hymenoptera) *Pan.-Pac. Entomol.* 49: 232-239.

59. Grissel, E.E. 1973b. New species of *Eurytoma* associated with Cynipidae. *Pan-Pac. Entomol.* 49: 354-362.

60. Grissel, E.E. 1976. A revision of the Western nearctic species of *Torymus* Dalman (Hymenoptera: Torymidae). *Univ. Calif. Pub. Entomol.* 79: 1-120.

61. Grissel, E.E. 1979. In *Catalog of Hymenoptera in America North of Mexico. Symphyta and Apocrita (Parasitica)*. 1: 748-768, Washington D.C.: Smithsonian Institution Press. 1198 pp.

62. Hayat, M. 1983. The genera of Aphelinidae (Hymenoptera) of the World. *Systematic Entomology* 8: 63-102.

63. Hedqvist, K.J. 1968. Some Chalcidoidea (HYm) collected from the Philippine, Bismark and Soloman islands. 1. Leucospididae, Perilampidae and Ormyridae. *Entomologiske Meddelese* 36: 162-165.

64. Hedqvist, K.J. 1970. Result of the Lund University expedition in 1950-1951: Hymenoptera (Chalcidoidea): Eupelmidae. *South African Animal life* 14: 402-444.

65. Hoffer, A. 1964. In Peck, O., Boucek, Z., Hoffer, A. Keys to Chalcidoidea of

301

Czechoslovakia (Insecta: Hymenoptera). *Mem. ent. Soc. Canada* 34: 63-89.
66. Hovanitz, W. 1959. Insects and plant galls. *Sci. Amer.* 201: 151-162.
67. Ishii, T. 1931. Notes on phytophagous habits of some chalcidoids with descriptions of two new species. *Kontyu* 5: 132-138.
68. Joseph, K.J. 1953a. Contributions to our knowledge of fig-insects (Chalcidoidea: Parasitic Hymenoptera) from India 111. Descriptions of three new genera and five new species of Sycophagini, with notes on biology, distribution and evolution. *Agra Univ. J. Res.* 2: 53-82.
69. Joseph, K.J. 1953b. Contribution to our knowledge of fig insects (Chalcidoidea: Parasitic Hymenoptera) from India IV. Descriptions of three new and records of four known species of Agaonini. *Agra Univ. J. Res.* 2: 267-284.
70. Joseph, K.J. 1954. Contributions to our knowledge of fig insects (Chalcidoidea: Parastic Hymenoptera) from India VI. On six species of Agaoninae. *Agra Univ. J. Res.* 2: 401-416.
71. Joseph, K.J. 1958. Rescherches sur les Chalcidiens *Blastophaga psenes* (L) et *Philotrypesis caricae* (L.) du figuier *Ficus* carica (L.) Ann. Des. Sc. nat. Zool. 20:197-260.
72. Kamijo, K. 1963. A revision of the species of the Monodontomerinae occurring in Japan (Hymenoptera: Chalcidoidea). *Insecta Matsumurana* 26: 89–98.
73. Kamijo, K. 1976. Notes on Ashmead's and Crawford's types of Eulophidae (Hymenoptera: Chalcidoidea) from Japan. *Kontyu* 44: 482-495.
74. Kamijo, K. 1977. A new genus and three new species of Ormocerini (Hymenoptera: Pteromalidae) from Japan. *Kontyu* 45: 531-537.
75. Kamijo, K. 1979. Four new species of Torymidae from Japan with notes on two known species (Hymenoptera: Chalcidoidea). *Akitu* N.S. 24: 1-11.
76. Kamijo, K. 1981a. Pteromalid wasps reared from Cynipid galls on oak and Chestnut in Japan, with descriptions of 4 new species. *Kontyu* 49: 272-282.
77. Kamijo, K. 1981b. Three new species of *Spaniopus* (Hymenoptera: Pteromalidae) from Japan. *Akitu* n.s. 36: 1–9.
78. Kamijo, K. 1982. Two new species of *Torymus* (Hymenoptera: Torymidae) reared *Dryocosmus kuriphilus* (Hymenoptera: Cynipidae) in China and Korea. *Kontyu* 50: 505–510.
79. Lichtenstein, J.L. 1919. Note preliminaire an subject de *Philotrypesis* caricae Hass. *Bull. Soc. ent. Fr.* 17: 313–316.
80. Longo, B. 1909. Osservazioni e ricerche sul *Ficus carica* L. *Ann. di Botanica* 7: 234–256.
81. Mani, M.S. 1938. Catalogue of Indian Insects, Chalcidoidea. 23: 1–174. Delhi: Government of India Publication. 174 pp.
82. Mani, M.S. 1964. *Ecology of Plant Galls.* The Hague: W. Junk Publishers. 434 pp.
83. Mayer, P. 1982. Zur naturgeschichte der Feigninsekten. *Mitt. Zool. Stn.* Neapel 3: 551–590.
84. McCalla, D.R., Genthe, M.K., Hovanitz, W. 1962. Chemical nature of an insect gall growth factor. *Plant Physiol.* 37: 98–103.
85. Miller, C.D. 1970. The Nearctic species of *Pnigalio* and *Sympiesis. Mem. Entomol Soc. Can.* 68: 3-121.
86. Narendran, T.C. 1973. In Joseph, K.J., Narendran, T.C., Joy, P.J. *Oriental Brachymeria* (Hymenoptera: Chalcididae), Zoological Monograph, Calicut: Calicut University publication. 215 pp.
87. Narendran, T.C. 1975. Studies on Biology, Morphology and Host-parasite relationships of *Brachymeria lasus* (Walker) (Hymenoptera: Chalcididae). India: Calicut, Thesis submitted to Calicut University for doctoral degree in 1975.
88. Narendran, T.C. 1980. A new species and a new record of the interesting genus *Smicromorpha* Girault (Hymenoptera: Chalcididae) from Oriental region. *Bombay nat. Hist. Soc.* 73: 908-911.

89. Narendran, T.C., Joseph, K.J. 1976. Biological studies of *Brachymeria lasus* (Walker) (Hymenoptera: Chalcididae). *Entomon* 1: 31–38.

90. Narendran, T.C., Joseph, K.J. 1977. Studies on some aspects of host-specificity with references to *Brachymeria lasus* (Walker), a polyphagous chalcid parasite of Lepidopterous insects. In *Insects and Host-specificity*, 85–89. ed. T.N. Ananthakrishnan. India: The Macmillan Company of India Ltd. 127 pp.

91. Nikol'skaya, M.N. 1952. *The Chal. cid. fauna of USSR*. (Chalcidoidea): 1–593. Translated from Russian by A. Birron and Cole, Z.S. in 1963. Published for the National Science Foundation, Washington, D.C.: and the Smithsonian Institution by the Israel Program for scientific translations. 593 pp.

92. Noble, N.S. 1938. *Epimegastigmus* (*Megastigmus*) *brevivalus* Girault: A parasite of the citrus gall wasp (*Eurytoma fellis* Girault): with notes on several other species of Hymenopterous gall inhabitants *N.S. Wales Dept. Agr. Sci. Bul*. 65: 1-46.

93. Noyes, J.S. 1978. On the numbers of genera and species of Chalcidoidea (Hymenoptera) in the world. *Entomologist's Gazette* 29: 163–164.

94. Noyes, J.S. 1980. A review of the genera of Neotropical Encyrtidae (Hymenoptera: Chalcidoidae). *Bull. Br. Mus. nat. Hist. (Ent.)* 41: 107–253.

95. Peck, O. 1951. In *Hymenoptera of America north of Mexico.*, Synoptic Catalog. *Agricultural Monograph*. eds. C.W.F. Musebeck. Washington D.C.: U.S. Dept. of Agri. 410-594. 14420 pp.

96. Peck, O. 1963. *A catalogue of the Nearctic Chalcidoidea* (Insecta: Hymenoptera). *Can. Ent. Suppl. Ottawa*. 1092 pp.

97. Philips, W.J. 1927. *Eurytoma parva* (Girault) Philips and its biology as a parasite of Wheat jointworm *Harmolita tritici* (Fitch). *Journ. agric. Res, Wash*. 34: 743–758.

98. Prinsloo, G.L. 1980. An illustrated guide to the families of African Chalcidoidea (Insecta : Hymenoptera). *Sci. Bull. Dep. Agric. Fish. Repub. S. Afr*. 1–48.

99. Riek, E.F. 1970. Hymenoptera (Wasps, bees, ants): In *The Insects of Australia*. CSIRO, Melbourne University Press. 1029 pp.

100. Risbeck, J. 1952. Contribution a l'etude des Chalcidoides de Madagascar. *Mem. Inst. Scient. Madg*. (S.E.) 2: 1–449.

101. Roskam, J.C. 1982. Larval characters of some eurytomid species (Hymenoptera: Chalcidoidea). *Proc. K. Ned. Akad. Wet*. 85: 293–305.

102. Ross, H.H. 1932. The hymenopterous family Xyelidae in North America. *Ann. Entomol. Soc. Am*. 25: 153–169.

103. Schmitz. L.G. 1946. Exploration due Parc National Albert Park. *Inst. Parcs, Nat. Congo Belge*, Bruxelles 48: 1–192.

104. Smith, D.R. 1969. Nearctic Sawflies I. Blennocampinae: Adults and Larvae (Hymenoptera: Tenthredinidae). *Technical Bulletin* No. 1397: 1–179. Agricultural Research Services; USDA.

105. Smith, E.L. 1968. Biosystematics and morphology of Symphyta 1.: stem galling Euura of California Region, and a few female genitalic nomenclature. *Ann. Entomol. Soc. Amer*. 61: 1389–1407.

106. Smith, E.K. 1970. Biosystematics and Morphology of Symphyta. II. Biology of Gall-making Nematine Sawflies in the California Region. *Ann. Entomol. Soc. Wash*. 63: 36–51.

107. Smith, D.R. 1979. In Krombein *et al*. (Eds.), Catalog of Hymenoptera in America North Mexico. Smithsonian Institution Press, Washington, D.C. 2188 pp.

108. Subba Rao, B.R. 1976. *Narayana*, gen. nov. from Burma and some synonyms (Hym.: Mymaridae). *Oriental Ins*. 10: 87–91.

109. Subba Rao, B.R. 1978. New genera and species of Eurytomidae (Hymenoptera: Chalcidoidea). *Proc. Indian Acad. Sci*. 87: 293–319.

110. Tachikawa, T. 1981. Hosts of Encyrtid genera in the world (Hymenoptera: Chalcidoidea). *Mem. Ehime Univ*. 25: 85–110.

111. Townes, H. 1962. Host selection patterns in some Nearctic ichneumonids. *Proc. xi. Int. Congr. Entomol,* 1960: 2:738–741.

112. Triapitzin, V.A. 1973a. The classification of parasitic Hymenoptera (Chalcidoidea): Part I. Survey of the systems of Classification., The subfamily Tetracneminae Howard, 1892 (In Russian) Ent. Obozr. 52: 163–175 (English translation: *Ent. Rev. Wash.* 52: 118–125).

113. Triapitzin, V.A. 1973b. Classification of parasitic Hymenoptera of the family Encyrtidae (Chalcidoidea). Part. II. Subfamily Encyrtinae Walker, 1837 (In Russian) *Ent. Obozr.* 52: 416–429 (English translation: *Ent. Rev. Wash.* 52; 287–295).

114. Van Staden, J., Davey, J.E., Noel, A.R. 1977. Gall formation in *Erythrina lattissima. Z. Planzenphysiol. Bd.* 84: 283–294.

115. Van Staden, J., Davey, J.E. 1978. Endogenous cytokinins in the laminae of and galls of *Erythrina latissima* leaves. Bot. Gaz. 139: 36–41.

116. Varley, G.C. 1937. Descriptions of the eggs and larvae of four species of Chalcidoid Hymenoptera parasitic on the knapweed gall fly. *Proc. R. ent. Soc. Lond.* 6: 122–130.

117. Walker, F. 1871. *Notes on Chalcididae.* Part IV. Chalcididae, Leucospidae, Euchari-dae, Perilampidae, Oryridae, Eurytomidae., London. 70 pp.

118. Wangberg, J.K. 1975. Biology of the thimble berry gall maker *Diastrophes kincaidii.* The *Pan-Pacific Entomologist* 51: 39–48.

119. Wangberg, J.K. 1976. The insect community in galls of *Diatrophus kincaidii* Gillete (Hymenoptera: Cynipidae) on Thimble berry. *Dept., Entomol. Anniv. Pub.* 7: 45–50.

120. Wangberg, J.K. 1977. A new Tetrastichus parasitising Tephritid gall-formers on *Chrysothamnus* in Idaho. The *Pan-Pacific Entomologist* 53: 237–240.

121. Washburn, J.O., Cornell, H.V. 1981. Parasitoids, patches and phenology: Their possible role in the local extinction of cynipid gall wasp population. *Ecology* 62: 1597–1607.

122. Webb, F.E., Forbes, R.S. 1951. Notes on the biology of *Pleuroneura boreaiis* Felt (Xyelidae: Hymenoptera). *Can. Entomol.* 83: 181–183.

123. Westwood, J.O. 1882. Description of the insects infesting the seeds of Ficus *Sycomorus* and *Carica. Trans. ent. Soc. Lond.* 1–27.

124. Wiebes, J.T. 1964. Host specificity of fig wasps (Hymenoptera: Chalcidoidea, Agaonidae). *Proc. XII Int. Congr. Ent. Lond.* 1964–1965, Systematic Section 1: 95–96.

125. Wiebes, J.T. 1979. Co-evolution of figs and their insect pollinators. *Ann. Rev. Ecol. Syst.* 10: 1–12.

126. Wiebes, J.T. 1981a. The fig insects of La Reunion (Hymenoptera: Chalcidoidea). *Annls. Soc. ent. Fr.* (N.S.) 17: 543–570.

127. Wiebes, J.T. 1981b. Towards strategy concepts in flower ecology as exemplified by the fig wasp symbiosis. *Acta Bot. Neerl.* 30: 493–495.

128. Wiebes, J.T. 1982a. The phylogeny of the Agaonidae (Hymenoptera: Chalcidoidea). *Netherlands Journal of Zoology.* 32: 395–411.

129. Wiebes, J.T. 1982b. Fig wasps (Hymenoptera), 735–755. In *Monographiae Biologicae* Vol. 42. ed. Gressitt. Dr. W. Junk Publishers, The Hague.

130. Yoshimoto, C.M. 1965. Synopsis of Hawaiian Eulophidae including Aphilinidae (Hym.: Chalcidoidea). *Pacific Ins.* 7: 665–699.

131. Yoshimoto, C.M. 1971. Revision of the genera *Euderus* of America (Hym.: Eulophidae). *Can. Entomol.* 103: 541–578.

132. Yukawa, J., Ohsako, S., Ikenaga, H. 1981. Parasite complex of the Japanese species of the genus *Aspondylia* (Diptera: Cecidomyidae) including the soybean pod gall midge. *Proc. Assoc. Pl. Prot. Kyushu* 27: 113–115.

10. The Geography of Gall Insects

Raymond J. Gagné

Introduction

The gall-forming habit has arisen in insects many times and in many different ways. As has been outlined in previous chapters, differences abound among the various groups of gall-makers. These differences include the mechanisms for inducing galls, effects on the hosts, methods of feeding, and different life histories. Despite their peculiarities, the separate groups of gall-makers have been affected in similar fashion by geological and climatic processes. Since the Cretaceous period when flowering plants arose and began to flourish, whole continents have broken away from larger land masses, drifted apart as separate pieces and, in some cases, have collided together again. Parts of continents have been alternately dry and submerged. Glaciers have covered large areas, obliterating all life before them, and then retreated. These changes must be reflected in the present distribution of plants and animals and be major factors in understanding the evolution of gall insects. If unrelated groups have a congruent pattern of distribution, they were probably affected by the same geological and climatic processes.

As has often been pointed out, the various major groups of gall insects have generally different distributions. Gall-making thrips are restricted mostly to the warmer parts of the Eastern Hemisphere; cynipine gall wasps are mainly holarctic; scale insects are chiefly tropical or Australian; and gall midges are well distributed on all continents. But the situation becomes more interesting when one looks beyond the entire distribution of the various gall insects. One then finds that each group is made up of smaller, geographically restricted subdivisions that lend themselves better to direct comparisons. As an example, the oak-infesting Cynipidae are a holarctic group. They include not only the gall-forming *Neuroterus* (s. 1.) and *Cynips* (s. 1.) but the inquilinous *Synergus* and *Synophrus*. Each of these groups is

305

holarctic. The oak-infesting gall midge genera, *Polystepha* and *Macro-diplosis*, are also holarctic, as are oaks themselves. Instead of one general pattern of distribution, seven examples of congruent distribution have been listed here, each of which came about independently. The grass-infesting gall midges of the genus *Orseolia*, although not yet known from Australia, have otherwise the same general distribution in the warmer parts of the Eastern Hemisphere as the gall-forming thrips. And gall-forming thrips themselves have evidently developed the gall-making habit separately in the suborders Tubulifera and Terebrantia, and perhaps also in families within each suborder, resulting in multiple superimposed layers of distribution. The fact that very different gall-making insects and many subdivisions of the same group have the same range indicates that distribution is not random or entirely dependent upon the various strategies that the insects use to disperse, but must depend on extraneous physical factors that favour or hinder spread.

Understanding why insect gall-formers are distributed as they are is aided greatly if one considers the distribution of the plant groups upon which they live. Unlike their free-living insect relatives, gall-forming insects must be as restricted in their ultimate dispersal as are the host plants with which they are closely associated. Plant distribution thus adds yet another level of evidence to understanding the distribution of gall-formers. Fossil gall insects also provide corroborative evidence for age and geographical history of the various groups.

Correlation between geography and distribution is dependent upon correct, natural classifications of the biota. Before considering what is known about the distribution of gall-forming insects, the author first discusses how species that are recent immigrants distributed by man, and groups that are not really monophyletic but show superficial resemblances (homoplasy), could serve to cloud the analysis of distributional patterns.

Recent Immigration

Immigration by the agency of man is artificial dispersal that has to be recognized prior to determining the natural patterns of distribution. Many gallmakers are clearly recent immigrants. Two notable examples are the Hessian fly, *Mayetiola destructor* (Say) (Cecidomyiidae), which was brought from Europe to North America, probably on wheat straw sometime around 1776 and which subsequently spread throughout the continent wherever wheat was grown, and the grape phylloxeran, *Phylloxera vitifoliae* (Fitch) (Phylloxeridae), which was accidently tranported to Europe and Australia on potted American grape plants. Besides pest gall-makers, beneficial species have intentionally been introduced, e.g., *Zeuxidiplosis giardi* Kieffer (Cecidomyiidae) brought from Europe to California, Australia, and New Zealand to help suppress *Hypericum perforatum* L. (Hypericaceae).

Without direct evidence of immigration and subsequent spread, gall-makers can be assumed to be introduced species if they affect only immigrant plants in the adopted area. Some of these are *Contarinia acetosellae* (Rübsaamen) and *C. rumicis* (Loew) (Cecidomyiidae), whose sole hosts in North America are immigrant species of *Rumex* (Polygonaceae); neither is known from native North American *Rumex* spp. It is reasonable to assume that both species stowed away from Europe, possibly during colonial times. Another example of this situation is *Erosomyia mangiferae* Felt (Cecidomyiidae), a pest of mangoes that was described first from the West Indies and again from Brazil in the early part of this century. The species was discovered much later in India, where the mango is native, and given yet another name. The fact that an insect is an immigrant can be masked by its carrying a different name in the place of origin from that in the new home. *Janetiella siskiyou* Felt (Cecidomyiidae), known since 1917 as a pest of seeds of the Port Orford cedar, a tree native of western Oregon and California, was described as *Craneiobia lawsonianae* De Meijere in 1935 when the pest appeared on trees that had been introduced into the Netherlands. The synonymy was noticed only in 1973.

A gall-making species might incorrectly be considered a native or an immigrant until close attention is paid to it, as was done in a recent, comprehensive study of a genus of cecidomyiids across its whole range. From 1886, when first reported, to 1977, a species of *Semudobia* (Cecidomyiidae) found on native birches in the eastern United States was considered to be the European *S. betulae* (Winnertz). Roskam (1977) showed that *S. betulae* does occur in North America but only on the ornamental European weeping birch, *Betula pendula* Roth. *S. betulae* of American authors was actually one of two native species on American birches in eastern North America. The species differences were noticed only after thorough comparative study.

The above examples are fairly clear cases of immigration because the species live on immigrant hosts and have remained on those original hosts. But cases exist that are less definite, as when a species is a general feeder and already has a wide distribution. One of the five *Semudobia* species, *S. skuhravae* Roskam, is truly holarctic (Roskam, 1977). Unlike the other four more restricted species in the genus, it accepts a large number of birch species as hosts and forms galls on bracts as well as on seeds. *Contarinia baeri* (Prell) (Cecidomyiidae) is a well-known European pest of *Pinus sylvestris* L. It has been found in recent years on the European *P. sylvestris* in Canada and the United States as well as on the native *Pinus resinosa* Ait. The earliest report of damage by this species in North America was in Kearby and Benjamin (1963), but it has been known in Europe on *P. sylvestris* since Prell (1931). *Contarinia baeri* could well be a native pest of two-needle pines across the whole holarctic region.

A puzzling situation exists in the case of a predator of a gall-maker. *Adelges piceae* (Ratzeburg) (Adelgidae) is a pest of firs and is a European native that was carried to North America with nursery stock. During 1957 and 1959, *Aphidoletes thompsoni* Möhn (Cecidomyiidae), a known predator of this and other adelgids in Europe, was released in the Maritime Provinces of Canada, and in the northeastern United States, North Carolina, British Columbia, and the U.S. Pacific Northwest to suppress the adelgid. The predator is reportedly established now in Washington and specimens have recently been caught in British Columbia. Specimens have also been taken in 1971 in Ontario and Alberta, two places very distant from the introductions. *A. thompsoni* may therefore have been in North America before 1957. Perhaps a specimen taken prior to that date will be found in a collection some day. Further evidence that *A. thompsoni* is a native North American species is that the other two species of *Aphidoletes*, both of which feed on various aphids, are also holarctic.

In general, it appears that a gall-maker is not an immigrant unless its host plant is also an immigrant. When a gall-making species occurs on both an immigrant host plant and a native one, the question of whether it is an immigrant or an oligophagous native will require study and may be difficult to resolve.

Problems Due to Homoplasy

As unrecognized immigrants can distort the view of natural distribution, so too can artificial groups classified on the basis of shared similarity and not on common ancestry. Mistakes can involve various levels of classification. A case in point involves subtribes of gall midges about whose classification no general agreement now exists. The problem concerns the heavy weight assigned to superficially similar character states and the general unawareness of differences in a few very important and basic anatomical structures.

Until recently, the gall midge subtribes Lasiopterina and Alycaulina (without the inclusion of *Camptoneuromyia* and *Trotteria*) were considered to constitute a monophyletic tribe, Lasiopterini. They are no longer considered as so closely related (Gagné and Hawkins, in press). The Lasiopterina occur on all continents except South America and presently comprise 259 species. The Alycaulina are restricted to North and South America and comprise 188 species. Many more species of both subtribes remain to be described, but it is clear that both are large, diverse groups. Characters common to both subtribes include heavily scaled bodies, foreshortened antennae, a very short R5 wing vein, and a particular pattern of larval setation. Almost all species of Alycaulina and Lasiopterina have an associated fungus in their galls, and most are stem gall-makers. The two subtribes have been considered monophyletic (Gagné, 1969, 1976) even if they were not considered separate subtribes (Möhn, 1975; Wünsch, 1979). Both subtribes have developed along parallel lines. In Europe, grass culms are inhabited by a genus of Lasiopterina, while in North America the same niche is filled by a genus of Alycaulina. The European genus *Ozirhincus* of the subtribe Lasiopterina has an extreme development of the head and thorax that allows it to take sustenance from flowers of Asteraceae; in North America an undescribed species of *Alycaulina* has identical adaptations for life on the same plant family. Important differences between the two subtribes lie in the postabdomens of both sexes. Those of the females show two different modifications that presumably help to abrade plant tissue (Gagné and Hawkins, in press), so it is quite possible that the two subtribes are separately derived.

Still formally aligned to the Lasiopterini, according to Möhn (1975) and Wünsch (1979), are two further probable cases of homoplasy, the genera *Camptoneuromyia* and *Trotteria*. *Camptoneuromyia* is a genus of 16 species limited in distribution to the Western Hemisphere. *Trotteria* has a world-wide distribution, and 22 species are known. Yet the female postabdomens of both groups are unlike either of the types found in the subtribes. Based on those differences, it is reasonable to consider the four groups, the Lasiopterina, the Alycaulina (including *Meunieriella* and most inquilinous genera of Wünsch (1979)), *Camptoneuromyia*, and *Trotteria*, as each separately derived from the large paraphyletic remainder of the Lasiopteridi. Although the groups are superfically similar, certain important and telling characters weigh heavily against one of the groups being ancestral to another (Gagné and Hawkins, in press). With the additional evidence of different distributions to consider, it is easier to believe that the four groups are separately derived.

Continental Connections and Discontinuities of Gall-formers

1) FAUNAL CONNECTIONS BETWEEN THE NORTHERN CONTINENTS

The greatest number of examples of intercontinental biotic connections among continents are those between North America and Eurasia. This is probably because of the close proximity and recent connections between these continents and because their flora and fauna have been studied in greatest depth. Examples of similarities are especially numerous for those gall-makers on circumboreal genera of plants. To list a few: the sawfly genera *Euura* and *Pontania* and the gall midge genus *Rhabdophaga* on willows; the moth genus *Mompha* on *Epilobium* (Onagraceae); the adelgid genera *Adelges* and *Pineus* on spruces and other conifers; and the gall midge genera *Thecodiplosis* and *Cecidomyia* on pines. Continental faunal connections can be on the species or higher taxonomic level. The gall wasp genera *Neuroterus* (s. 1.) and *Cynips* (s. 1.) are found on both continents but their strict generic subdivisions are not. Pemphigine aphids use poplars as their primary host in both Eurasia and North America, but their secondary hosts are different in many cases from one continent to the other.

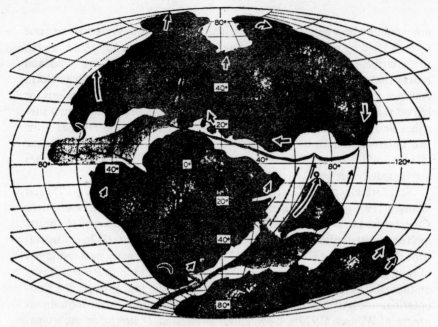

Figure 1. The continents ca. 135 million years ago (p. 36 of "The Breakup of Pangaea" by Robert S. Dietz and John C. Holden, October 1970, *Scientific American*).

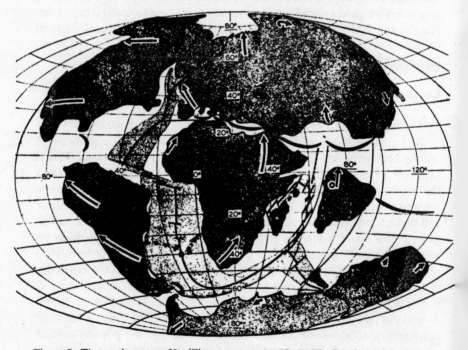

Figure 2. The continents ca. 65 million years ago (p. 37 of "The Breakup of Pangaea" by Robert S. Dietz and John C. Holden, October 1970, *Scientific American*).

Some of these biotic connections may be due to dispersal across Beringia, the intermittent Ice Age land connection between Asia and western North America. But explaining the great biotic congruence between the continents from this connection alone requires that all these fauna and flora dispersed separately in one direction or another over what was probably a very cold land bridge. An alternative explanation for their distribution on both continents is that they or their ancestors already existed on both Eurasia and North America when those two land masses made up the northern supercontinent Laurasia and were fragmented in the same way the land masses were. The great general biotic similarity that presently exists between the two continents is more easily explained by fragmentation than dispersal. Figures 1 and 2 illustrate successive stages of the breakup of Pangaea. Details of the presentations in these maps may change, but they generally show how broadly connected present day Europe and eastern North America once were.

In many groups of animals, greater affinities appear between eastern North America and Europe than between eastern and western North America, thus reflecting this ancient connection. Examples can be shown in fossil mammals of pre-midEocene age (McKenna, 1975), spiders (Platnick, 1976), carabid beetles (Noonan, 1979) and fungus gnats (Gagné, 1981). If these groups depended upon dispersal across Beringia, one would expect the European fauna to resemble more closely the western than eastern North American fauna. The connection between Europe and North America separated as late as the mid-Eocene, more or less 50 million years ago (McKenna, 1975), and might have been more southerly than at present, as indicated in figure 2. Ellesmere Island, the northernmost point of land in North America, presently within the Arctic Circle, at that time was covered with a temperate flora. During much of the Tertiary, the individual continents were also divided longitudinally by epicontinental seas that evidently served as barriers to faunal exchange. Since that time the midcontinental areas have been mostly dry, perhaps accounting for why certain groups, e.g., some species and genera of fungus gnats, do not seem to have been able to disperse latitudinally across the continents.

Although no examples of apparent close connections between gall insects of eastern North America in particular and Europe exist, excellent studies by Roskam (1977, 1979) on the gall midges associated with female birch catkins do illustrate an American-Eurasion dichotomy on the species level. Figure 3 shows the distribution of the five known species of *Semudobia*, a genus of gall midges that form galls in the bracts and seeds of birches. Figure 4 shows a hypothetical cladogram that reflects the cladistic argumentation scheme as outlined in detail in Roskam (1979). *Semudobia skuhravae* is the only naturally holarctic species and has the most generalized habits. It has the widest species range and can live in either seeds or bracts. The remaining four species live only in seeds and are restricted to

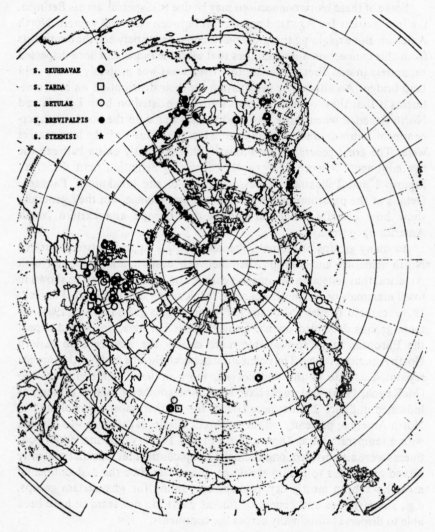

Figure 3. Distribution of *Semudobia* species.

either Eurasia or North America. The two American species are allopatric, one in the West, the other in the East. Discontinuity between Eurasia and North America may have resulted in the differentiation between *steenisi-brevipalpis* and *betulae* as the cooling of North America and subsequent isolation on eastern and western portions of the continent could have resulted in the division of *steenisi* and *brevipalpis* as well as their host birches. If evolution of the group did occur this way, it means that *skuhravae* is very old and has remained unchanged for at least 50 million years even though isolated on separate continents.

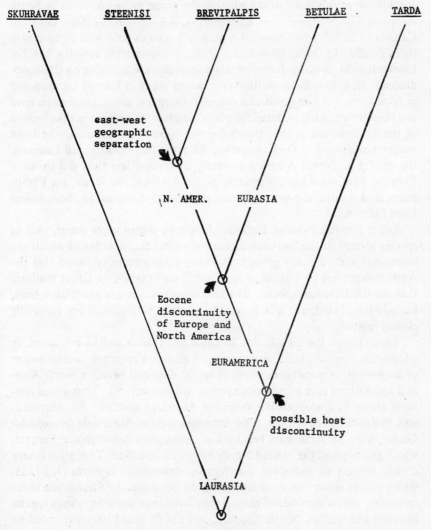

Figure 4. Cladogram of *Semudobia*.

Where *Semudobia* illustrates two speciation events following the break-up of Eurasia and North America, the gall midge subtribes Lasiopterina and Alycaulina involve hundreds of such events during the same period of time. The subtribe Alycaulina, touched upon earlier in the discussion of homoplasy, comprises 188 described species restricted to the Western Hemisphere. Since the subtribe does not occur elsewhere, one supposes that it appeared in North America after the mid-Eocene Euramerican separation. The group is very diverse in North as well as in South America and has radiated to many host plant families on both continents. It would be wild speculation at this point to say whether the group arose on North or South America. The other subtribe, Lasiopterina, is known in the Americas from Canada to El Salvador, where it is relatively species-poor with 39 species in three genera. In the Eastern Hemisphere, however, that subtribe has 220 known species in many distinct generic segregates, e.g., those on Chenopodiaceae. Most species in North America and many in Europe are classified in *Lasioptera*, but the genus is a catch-all category, and relationships have not been investigated in detail. A plausible explanation for the distribution of the Lasiopterina is that they were widespread on Laurasia, the large northern continent of the Cretaceous. After the fragmentation of Laurasia, the species on North America evidently diversified less than did those in Eurasia. That some Lasiopterina do occur in Africa, Indonesia, the Philippines, and Australia must mean that the groups dispersed to those places from Eurasia.

Other possible cases of Eurasian-American dichotomies occur, such as species groups of the gall midge genus *Asphondylia*, distributed on all the continents and on many groups of plants. I have recently found that the American species on Fabaceae have a different pattern of larval vestiture than do the Eurasian species. The two separated groups are thus distinct, but additional evidence is lacking as to whether both groups are especially closely related.

Even though the distribution of some gall insects can be explained by geographic fragmentation, dispersal must also be a common natural means of movement. A possible example of recent dispersal between North America and Europe is in a gall midge genus, *Mayetiola* (s.s.). That genus contains about 29 European but only one American species. All *Mayetiola* spp. live in culms of grasses. The American species, *Mayetiola ammophilae* Gagné, lives on American beachgrass, *Ammophila breviligulata* Fernald, which grows along the eastern North American coastline. That grass is very closely related to European beachgrass, *Ammophila arenaria* (L.) Link, which occurs along the northern European coastline. In Pleistocene times or earlier, the two species of grasses may have been one continuous species between Europe and North America, and the fly could have expanded its range from Europe. No *Mayetiola* species has yet been reported from European beachgrass, but one probably exists. It will be interesting to see how closely that species resembles *M. ammophilae*.

On the Asian and western North American extremities of the two continents, some similarities are apparent between the biota. The gall wasp genus *Cynips* (s.l.) of Europe appears to be more closely related to California *Cynips* (s.l.) than to those east of the Sierra Nevadas (Kinsey, 1930). *Rhopalomyia* (*Diarthronomyia*) gall midges associated with a tribe of asteraceous plants are well distributed in the drier parts of Asia, southern Europe, and North Africa, as well as in western North America.

2) FAUNAL CONNECTIONS AMONG THE SOUTHERN CONTINENTS

It appears that South America and Australia were joined via Antarctica during the Tertiary at a time when the area was more temperate than now. Biotic affinities between South America and Australia to support this are numerous. A good example of a three-layered system was outlined by Schlinger (1974). On Australia and South America, the hosts, southern beeches (Fagaceae), support aphids of the genus *Sensoriaphis*, which in turn support primary hymenopterous parasitoids. This distribution is plausibly explained as having followed directly from the fragmentation of the southern continents. Some eriococcid scale insects are gall-formers in South America and others, specifically those in the subfamily Apiomorphinae, are an endemic gall-forming group in Australia. I am advised that the gall-forming habit arose separately within the Coccoidea, but the distribution of the family suggests ties via the southern land connection.

Because of its relatively long period of isolation, Australia has many endemic groups of gall-makers, particularly in the Hymenoptera (Riek, 1970), but it also has groups with affinities to Asia. For example, a family of acalyptrate flies, the Fergusoninidae, thought until recently to be endemic to Australia, has now been reported in India (Harris and Joshi, 1980). Gall-making thrips are species-rich in both Australia and southeast Asia, but they exhibit endemism in both places. It needs to be pointed out here that gall midges have generally been considered to be poorly represented in Australia, but little work has been done on the cecidomyiid fauna of that continent, where they have perhaps been masked by the more conspicuous gall-making scale insects. A recent study of the grass and sedge gall midges by Harris (1979) and the fact that such wide-ranging genera as *Asphondylia* and *Lasioptera* are known from Australia indicate that the gall midge fauna there is diverse and numerous.

Pantropical connections are also evident. Gall-making Tanaostigmatine wasps are now known from Australia, Asia, Africa, and South America. The fig wasps and their hosts are also pantropical. Cecidomyiid predators of coccoids are another example of a pantropical group (Harris, 1968). It is possible that coccoids and their cecidomyiid predators co-evolved.

3) CONTINENTAL AND LOCAL PECULIARITIES

It is sometimes difficult to tell whether a local group has evolved since the breakup of the continents or is a relict of a once wide-ranging fauna. The gall midge genus *Macrolabis* contains 30 species, all restricted to Europe. Many are inquilines in the galls of other cecidomyiids, but some are gall-formers. The fact that they are not restricted to a particular plant group indicates that they would be more widespread had they had the chance. They may have arrived or evolved in Europe after the separation of Euramerica. An example of a possibly relict group is *Caryomyia*, a genus that contains many nearctic species, all forming galls on leaflets of hickories

(*Carya* spp.: Juglandaceae). Twelve species are described, but many more are known only from the galls that are specific to a species of *Caryomyia*. Figures 5 and 6 show 34 kinds of *Caryomyia* galls reported by Wells (1915). I have found 14 others for a total of 48 kinds of galls and probable species. The various types can be seen on many kinds of hickories, so no barrier

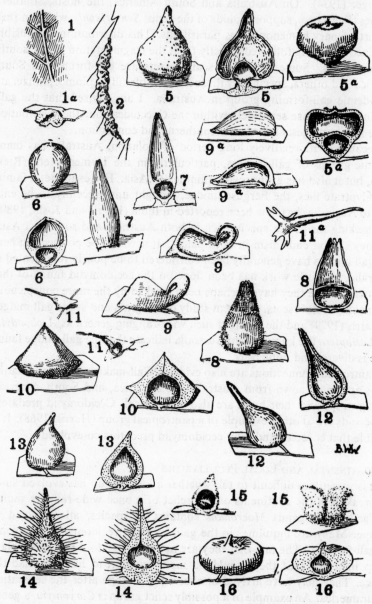

Figure 5. Hickory leaf galls; 5–16 caused by *Caryomyia* spp. (Wells, 1915).

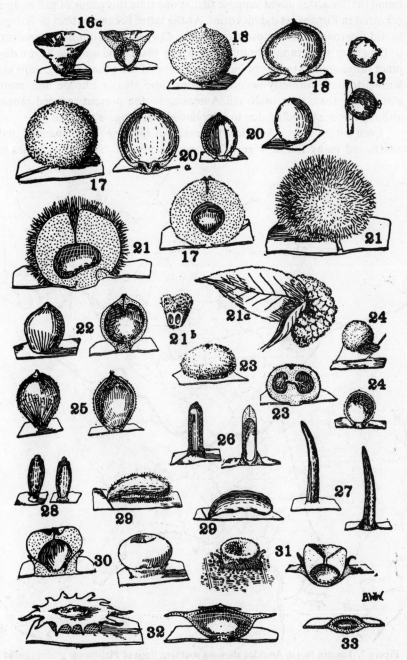

Figure 6. Hickory leaf galls; 16a–32 caused by *Caryomyia* spp. (Wells, 1915).

between host species is evident. Hickories now occur only in southeastern North America (24 species) and China (one species). If *Caryomyia* is ever found in China, one might suppose that at one time this genus of gall midges occurred in Europe, as did hickories. As the latter became extinct in Europe so did *Caryomyia*. The closest relative to *Caryomyia* is not now apparent, so the group's age cannot be inferred. Many plants and animals have a disjunct range similar to that of hickories. Some of the other plant groups are known to have formerly occurred at one time also in Europe and more extensively than now in North America, so the present reduced range, although now expanding due to interglacial warming, is relictual.

Even within North America, the range of hickories has expanded and contracted many times. Figure 7 shows the present range of hickories in

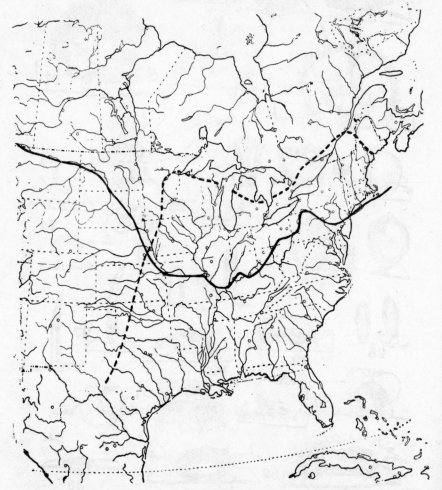

Figure 7. Eastern North America showing southern limit of Pleistocene glaciers (solid line) and present northern limit of hickory trees (broken line).

North America superimposed over the maximum extent of the glaciers during the Pleistocene. When the glaciers were at their southernmost extent, hickories must have been restricted to the far southern United States, perhaps even in isolated pockets. Their ranges then expanded during each interglacial period. This may be one reason for the diversity of *Caryomyia* species. During each glacial period, many species could also have become extinct, leaving now a small percentage of those species that ever lived and thereby adding greater difficulty to elucidation of the phylogeny of the genus.

Interpretation of other groups of gall insects may be less difficult. The gall wasp genus *Atrusca*, possibly because it is relatively young, shows, according to Kinsey (1936), a directional change in character states from southern Mexico northward (Fig. 8). Kinsey connected 43 living species of *Atrusca* in one phylogenetic train. A modern analysis of the available taxonomic characters of those gall wasps would be instructive in assessing the hypothesis that the taxon with the most primitive character states is closest to the place of origin. It is possible the species group in the eastern United States evolved during the present interglacial period. In North American oak gall wasps in general, the fact that the fauna west of the Sierra Nevada is almost entirely distinct from that in the rest of the U.S. and Mexico indicates that the group is fairly young. Even the subgenera of *Cynips* (ş.l.) are distinct from one side of that area to the other. Had oaks ever been continuous across what is now the Sierra Nevada and not been obliterated on either side, and if gall wasps once inhabited the entire area, one would expect to find some groups of these wasps in common. Kinsey (1936) did not believe that the fauna was ever shared between the two areas.

Another geographical aspect to explore is whether the number of gall-formers in a given taxon is correlated with the number of host species. This may be the reason that so many species of gall-formers have evolved on the very numerous *Eucalyptus* species in Australia and on the numerous *Acacia* species there and elsewhere. More oaks and more hackberries exist in south-central United States and Mexico than in northeastern United States, and more gall wasps and gall midges are known from those plants, respectively, from the former area than from the latter. It would be interesting to see whether more *Caryomyia* species are found where hickories are most diverse.

Fossil Gall Insects

Fossils are important for at least two major reasons: one, they are indicative of the general climate of when and where they lived; and two, they give us an estimate of the minimum time that a group has existed. Baltic amber was formed at different times between the early Eocene and Oligocene, as outlined by Larsson (1978). Northern Europe had a much warmer

Figure 8. Evolutionary tree and distribution map of genus *Atrusca* (Cynipidae)
(Kinsey, 1936).

climate than it does now over much of that time. It is easy to understand how plants such as the ancestors of present-day oaks and their associated gall insects that are now mainly temperate in distribution could have had a continuous distribution across Euramerica.

A few fossil galls have been found. Möhn (1960) reviews a cross-section of gall-forming groups: the oldest, on willow leaves (Upper Cretaceous), resembling those made by tenthredinid sawflies today; a poplar petiole gall (Miocene), presumably made by an aphid; an oak leaf gall (Pliocene) of probable gall wasp origin, and galls on leaves of poplar (Miocene), walnut (Miocene) and beech (Pliocene), presumbly made by gall midges. Möhn (1960) extracted from cone galls of a sequoia (?Miocene) gall midge larvae and pupae, which he described as *Sequoiomyia kraeuseli* Möhn. Later, Gagné (1968) assigned to that genus a living cone gall-maker of bald cypress in eastern North America.

Gall insects are not common as fossils. Although many cecidomyiids have been described from various amber deposits (Larsson, 1978; Gagné, 1973, 1977) most are not gall-makers. In a passive trap set out today, one will find that most cecidomyiids trapped are also not gall-makers but saprophages and predators, so fossils may reflect the same proportion. Gagné (1973) reported a *Contarinia* sp., probably a gall-maker, from Mexican amber (Oligocene-Miocene). That genus and some of its species groups are now well represented all over the world. Kinsey (1919) described three species (Miocene and ?Oligocene) of *Aulacidea*, a genus of gall wasps, two species from Colorado, one from Baltic amber. The age of these galls and gall-formers indicates that certain groups were well established long ago at a time when the face of the world appeared somewhat different from now.

REFERENCES

1. Gagné, R.J. 1968. Revision of the gall midges of Bald Cypress (Diptera: Cecidomyiidae). *Entomol. News* 79: 269–274.
2. Gagné, R.J. 1969. A tribal and generic revision of the Nearctic Lasiopteridi (Diptera: Cecidomyiidae). *Ann. Entomol. Soc. Amer.* 62: 1348–1364.
3. Gagné, R.J. 1973. Cecidomyiidae from Mexican Tertiary amber (Diptera). *Proc. Entomol. Soc. Wash.* 75: 169–171.
4. Gagné, R.J. 1976. New Nearctic records and taxonomic changes in the Cecidomyiidae (Diptera). *Ann. Entomol. Soc. Amer.* 69: 26–28.
5. Gagné, R.J. 1977. Cecidomyiidae (Diptera) from Canadian amber. *Proc. Entomol. Soc. Wash.* 79: 57–62.
6. Gagné, R.J., Hawkins, B.A. Biosystematics of the Lasiopterini (Diptera: Cecidomyiidae) associated with *Atriplex* spp. (Chenopodiaceae) in southern California. *Ann. Entomol. Soc. Amer.* In Press.
7. Harris, K.M. 1968. A systematic revision and biological review of the cecidomyiid predators (Diptera: Cecidomyiidae) on world Coccoidea (Hemptera: Homoptera). *Roy. Entomol. Soc. Lond., Transactions* 119: 401–494.
8. Harris, K.M. 1979. Descriptions and host ranges of the sorghum midge, *Contarinia sorghicola* (Coquillett) (Diptera: Cecidomyiidae), and of eleven new species of *Contarinia* reared from Gramineae and Cyperaceae in Australia. *Bull. Entomol. Res.* 69: 161–182.
9. Harris, K.M., Joshi, R.C. 1980. First record of *Fergusonina* galls in India. *Cecidol. Int.* 1: 123.
10. Kearby, W.H., Benjamin, D.M. 1963. A new species of *Thecodiplosis* (Diptera: Cecidomyiidae) on red pine in Wisconsin. *Can. Entomol.* 95: 414–417.
11. Kinsey, A.C. 1919. Fossil Cynipidae. *Psyche* 26: 44–49.

322

12. Kinsey, A.C. 1930. The gall wasp genus *Cynips*. *Indiana Univ. Stud.* 16: 1–577.
13. Kinsey, A.C. 1936. The origin of Higher Categories in *Cynips*. Indiana: Bloomington, Indiana University. 334 pp.
14. Larsson, S.G. 1978. Baltic Amber—A Palaeobiological Study. *Entomonograph* 1: 1–192.
15. McKenna, M.C. 1975. Fossil mammals and early Eocene North Atlantic land continuity. *Ann. Missouri Bot. Gardn.* 63: 335–353.
16. Möhn, E. 1960. Eine neue Gallmucke aus der niederrheinischen Braunkohle, *Sequoiomyia kraeuseli* n.g., n. sp. (Diptera, Itonididae). *Senckenbergiana lethaea* 41: 513–522.
17. Möhn, E. 1975. Gallmücken (Diptera, Itonididae) aus El Salvador. 8 Teil: Lasiopteridi. *Stuttgart. Beitr. Naturk.* (A) 276: 1–101.
18. Noonan, G.R. 1979. The science of biogeography with relation to carabids. In *Carabid Beetles: Their Evolution, Natural History, and Classification*, Ed. T.L. Erwin *et al.*, pp. 295–317. The Hague-Boston-London: W. Junk.
19. Prell, H. 1931. Die nadelknickende Kiefergallenmucke (Cecidomyia baeri n. sp.), ein verbreiteter neuer Kiefernschadling. *Tharandt. Forstl. Jahrb.* 82: 36–52.
20. Platnick, N.I. 1976. Drifting spiders or continents?: vicariance biogeography of the spider subfamily Laroniinae (Araneae: Gnaphosidae). *Syst. Zool.* 25: 101–109.
21. Riek, E.F. 1970. Hymenoptera. In *Insects of Australia*, pp. 867–957. Melbourne University Press, CSIRO.
22. Roskam, J.C. 1977. Biosystematics of insects living in female birch catkins. I. Gall midges of the genus *Semudobia* Kieffer (Diptera: Cecidomyiidae). *Tijdschkr. Entomol.* 120: 153–197.
23. Roskam, J.C. 1979. Biosystematics of insects living in female birch catkins. II. Inquiline and predaceous gall midges belonging to various genera. *Neth. J. Zool.* 29: 283–351.
24. Schlinger, E.I. 1974. Continental drift, *Nothofagus*, and some ecologically associated insects. *Ann. Rev. Entomol.* 19: 323–343.
25. Wells, B.W. 1915. A survey of the zoocecidia on species of *Hicoria* caused by parasites belonging to the Eriophyidae and the Itonididae (Cecidomyiidae). *Ohio J. Sci.* 16: 37–57, 2 pls.
26. Wünsch, A. 1979. Gallenerzeugende Insekten nordkolumbiens, speziell Asphondyliidi und Lasiopteridi (Diptera, Cecidomyiidae) aus dem Küsterbereich um Santa Marta. Waiblingen: Offsettdruck Kappenhöfer. 238 pp.

11. The Biology of Gall Mites

G.P. Channabasavanna and Neelu Nangia

Introduction

Mites have for long escaped detection because of their minute size, despite their occurrence for well over 50 million years. Phytophagous mites damage food crops and fruits, affect ornamental plants and, some are known to transmit viral diseases and also to induce galls. The nature of symptoms that mites induce on the host plants is so diverse and varied that no other group of parasitic organisms can compete with them in this regard. Gall formation by mites was first noticed by Reamur in 1737; he emphasised that 'worm-like' organisms were responsible for galling, although he believed them to be the immature stages of an insect. Nearly a hundred years later, Turpin (1833) reexamined these linden galls (Reamur 1737) and reported *Sarcoptes gallarum tiliae* Tur., as the gall maker. However, Duges (1834) believed that the adult mites in the galls to be 'larvae', and assumed that the adult insects laid their eggs within the galls and then escaped through the apertures. These larvae were described as organisms with two pairs of seven-segmented legs; reconsidering their 'larval' status, Siebold (1850) named them as *Eriophyes* and suggested that they propagated asexually, and the adult form was yet to be known. A further examination of the linden galls by Dujardin (1851) unveiled that these organisms were in fact adults, and he named them as *Phytoptus*. Later observations of Landois (1864) and Scheuten (1857) supported the belief these worm-like organisms were mites, maintaining that each mite possessed a pair of aborted limbs in addition to two pairs of normal legs; however, the presence of a third pair of rudimentary legs was negatived later (Low 1874).

The cecidogenous acarines mostly belong to the family Eriophyidae (recently raised to the level of a superfamily Eriophyoidea) and to a lesser extent to the Tenuipalpidae. The acarine-induced deformities including

323

galls that appear very specific and distinct to each species of mites. This led to the conclusion that growth regulators, specific to each causal organism, are involved in eliciting a characteristic growth response on plant parts that are modified into deformities from simple erinea to complex pouches, bladders, fingers, as well as bead-like galls on buds and leaves, providing food and shelter to the cecidogenous mites. In some of the host plants the normal elongation of floral axes or lateral or terminal shoots tends to become inhibited leading to contorted foliage, deformed flowers, and telescoped shoot axes. Another interesting example of mite action is the organisation of Witches' Brooms, which are a cluster of stunted branches of arborescent plants. Limited information on the diagnosis and bioecology of gall-inducing mites is available through the works of Felt (1940), Mani (1964) and Westphal (1977), and the purpose of this chapter is to highlight the more-important aspects of the biology of the gall mites examined in the perspectives of taxonomy, bioecology, and host relationships.

Taxonomy

1) ERIOPHYIDS

Modern eriophyid taxonomy originates from the pioneering work of Nalepa (1892, 1893, 1893a, 1893b, 1893c, 1894 and 1894a). His contributions to this field are monumental, many species described by him were determined from the dorsal, ventral and lateral views of mites, and on the host relations as well. Though such data are generally adequate to speculate the identity of eriophyids with some accuracy, additional considerations of the morphological details relating to legs, feather claw, skin and genitalia, are becoming important. The host relationships of eriophyids are intimate, and the species involved nearly show a high degree of host specificity. The host plant species, the plant organ affected, and the characteristics of the resultant injury are useful in determination; however the discovery of more species on a single host (Keifer 1952) points out the limitations of this. Yet, the symptoms of mite feeding and the host response continue to enjoy some purpose and merit for the identification of cecidogenous mites.

Earlier to the work of Nalepa, names were given to eriophyids based on the nature of host symptoms, shape and size of galls produced. Ever since the period of Nalepa, concerted efforts of Liro (1940, 1941, 1942, 1943), Liro and Roivainen (1951), Roivanen (1949, 1950, 1951, 1951a, 1953, 1953a) from northern Europe, Massee (1927, 1927a, 1937) from England, and Lamb (1952, 1952a, 1953, 1953a, 1953b, 1960) from New Zealand, and of a few others (Nemec 1924; Roivainen 1953a) have added information to Cecidology. Keifer (1942, 1944, 1959) and Keifer and Knorr (1978) and Keifer *et al.* (1982) have made substantial contributions to the knowledge of eriophyids of North America. Besides the major contributions of these, Farkas (1960, 1960a, 1961, 1961a, 1962), Boczek (1960, 1961, 1961a, 1962, 1968,

1969) on the mite fauna of Hungary and Poland, Hodgkiss (1930) on the apple mites of England, Massee (1927, 1927a, 1937) on the eriophyids of England, Niblett (1959) on the gall mites of London, Meyer (1968) on the gall mites of South Africa, Batchelor (1952) on the eriophyids of Washington, Natcheff (1966, 1966a, 1967, 1967a 1979, 1981) from Bulgaria, Hall (1967a) on Eriophyoidea of Kansas, Keifer and Knorr (1978) on eriophyids of Thailand, Channabasavanna (1966), Mohanasundaram (1982), Chakrabarti and Mondal (1979), Mondal and Chakrabarti (1980, 1982) on the eriphyid mites of India, Keifer *et al.* (1982) on the gall mites of the USA, and Briones and McDaniel (1976) on eriophyids of South Dakota, have added considerable information to gall-mite literature.

The eriophyids constitute the most preponderant group inducing different types of deformities on their host plants. Till 1964, all these mites were grouped under one family, *viz.*, Eriophyidae. Keifer (1964) recognized three distinct groups which could be considered as distinct families, which in turn could be grouped under the superfamily, Eriophyoidea. The families he recognized under the superfamily were: (1) Rhyncaphytoptidae, (2) Nalepellidae, and (3) Eriophyidae, which are recognizable by the following key:

KEY TO FAMILIES OF ERIOPHYOIDEA

1. Rostrum large and abruptly bent down near the base and tapering; oral stylet long. Dorsal setae present or absent; when present always pointing forward to some degree. Habits : rust mites or leaf vagrants......RHYNCAPHYTOPTIDAE Keifer, 1961
— Rostrum evenly bent down, oral stylet always short. Dorsal setae present or absent; if present, pointing forward or backwards....................................2
2. One or two anterior shield setae; anterior thanosomal (subdorsal abdominal) pair of setae often present. Internal female spermathecal tubes long or short; when short extending anteriorly first from centre near genital opening. Habits : bud mites, gall mites, rust mites and grass mites....NALEPELLIDAE Newkirk and Keifer, 1971
— Two or no shield setae, never with anterior shield seta. Internal spermathecal tubes always short and extending laterally or diagonally to rear from central opening. Never with lateral tibial spur or subdorsal abdominal setae. Habits; bud mites, erineum makers, gall mites, rust mites and leaf and green stem vagrants..........
....ERIOPHYIDAE Nalepa, 1898

The suprageneric taxa that are concerned in forming plant galls are distinguishable by the following key.

FAMILY NALEPELLIDAE

1. Genitalia relatively close to coxae. One central anterior shield seta, dorsal pair present or absent; internal spermathecal tubes 3 to 5 times longer than spermathecae. Body either wormlike and with subdorsal thanosomal seta pair present, or lacking subdorsal setae and usually more robust.............Nalepellinae Roivainen, 1953
— Four shield setae, rear pair rarely minute; spermathecal tubes short.............2
2. Body wormlike, with abdominal rings subequal dorsoventrally; subdorsal abdominal setae present.......................Phytocoptellinae Newkirk and Keifer, 1971
— Body more fusiform and often flattened; abdominal rings with lateral tergal—sternal

differentiation; subdorsal thanosomal setae present or absent.....................
....Sierraphytoptinae Keifer, 1944

FAMILY RHYNCAPHYTOPTIDAE

The two subfamilies of this family are distinguishable by the following key:

1. Featherclaw simple, undivided...............Rhyncaphytoptinae Roivainen, 1953
— Featherclaw divided, usually deeply so..Diptilomiopinae Newkirk and Keifer, 1971

FAMILY ERIOPHYIDAE

Of the five subfamilies four of them have one or more representatives which cause galls on their host plants. In all, 39 genera are concerned with galls in this family. The four subfamilies are distinguishable as follows:

1. Tibiae reduced or completely fused with tarsi; foretibia without seta. No spatulate projections for burrowing either on rostrum or legs. First setifarous coxal tubercles absent...Nothopodinae Keifer, 1956
— Tibia and tarsi distinct; foretibial seta nearly always present....................2
2. Female genitalia appressed to coxae and projecting from venter; anterior female internal apodeme appearing shortened; ribs on female coverflap typically in two uneven ranks. Most genera lacking dorsal tubercles and setae....................
....Cecidophyinae Keifer, 1966
— Genitalia not appressed to coxae and more or less on level with thanosome venter; internal anterior apodeme extending moderate distance forward. Coverflap variably embellished; dorsal setae rarely absent from shield............................3
3. Body wormlike, thanosomal rings subequal dorsoventrally, at least on anterior half or two-third shield typically lacking anterior lobe, or with slight projection over rostrum base..Eriophyinae Nalepa. 1898
— Body more fusiform and fitted for exterior living; shield usually with broad based and rigid anterior lobe over rostrum; thanosome typically with broad tergites and narrow soft sternites; if no dorso-ventral contrast present, then broad shield lobe present...Phyllocoptinae Nalepa, 1898

2) TENUIPALPIDS

The mites of the family Tenuipalpidae are commonly known as false spider mites. They are generally small, flat and reddish in colour. The most specialized members of this family, which are very few, form galls within which they feed, and among the two forms *Obdulia* sp. forming twig galls on *Tamarix marismortuii* and *Larvacarus transitans* causing stem galls on ber, are important.

Bioecology

ERIOPHYIDS

Detailed information on eriophyid life cycles, with observations on the morphology of the eggs, larvae and adults, including different types of females have been provided by Putman (1939, 1940) and Keifer (1942, 1952). The life histories of most eriophyids are simple. The gravid female lays eggs in the place where it feeds. However, the existence of two types of life cycles has been reported in some species (Keifer, 1952). Before developing into a sexually mature adult, the eriophyids display two nymphal instar stages and the complex life cycle involves two types of females: (1) the primary type or protogyne resembling the males, and (2) the structurally different secondary type or deutogyne. A third type of female is also known recently, wherein the protogyne female is ovoviviparous. Live larvae were observed (Shevtshenko, 1961) inside the body of the female of *Eriophyes laevis* (Nalepa), a deuterogynous species. A similar phenomenon was reported (Hall, 1967) in *Vasates quadripedes* Shimer.

Incidence of spermatophores indicates that there is no direct transfer of sperms from the male to the female mites, and that the fertilization occurs by the contact of females with sperm sacs laid on the host by males. Egglaying rates vary among species, but generally it is three or four a day under optimum conditions. A female may lay up to 80 eggs during a 30-day ovipositional period. The eggs are very small, and oval to spherical. The egg hatches into a first stage nymph which resembles the adult in many respects, but is much smaller without external genitalia, and fewer annulations and variations in the microtuberculation. Genital setae are always on the venter of these nymphs, and occur on each side of the midventral line, a few rings away from the second coxa. Before developing into an adult, the first nymph moults into the second nymph resembling the first in many respects, and undergoes a brief rest or pseudo-pupal stage. These I and II nymphs are very similar to adults, if not identical with respect to setal direction and microtuberculation. All the I-stage nymphs so far studied in the family Eriophyidae and Rhyncaphytoptidae have a peculiar dorsal discontinuity in the first four to six abdominal rings just behind the cephalothoracic shield.

The eriophyids which inhabit deciduous hosts exhibit alternation of generations. This alternation enables the species to have a primary female, which is capable of reproducing during favourable environmental conditions, and changing into a secondary female (which does not have a male counterpart) during unfavourable conditions displaying deuterogyny. The deutogynes appear to be restricted to temperate and arctic regions, though a few exceptions are known to occur in Thailand, viz., *Cisaberoptus kenyae* Keifer on mango foliage, and in India, *Aculus indicus* ChannaBasavanna on *Celtis wightii* Planch. The protogyne is an active egg layer and builds up the female population of the colony so long as the environmental conditions are favourable. The life span of a protogyne is four to five weeks. Deuterogyny occurs in all the three families of the Eriophyoidea. The Eriophyidae, which is the largest of the three families includes a majority of the gall formers, has the maximum number of deuterogynous species. Jeppson (1975) has classified the deutogynes into four principal types. The first type refers to those which are strongly microtuberculate on all sides of the abdomen. The California black walnut leaf pouch gall mites, *Eriophyes brachytarsus* Keifer, *Phytoptus sorbi* Canestrini, and *Eriophyes caulis* Cook, the eastern North American black walnut petiole gall mite are some of the well-known examples of this group. The second group comprises deutogynes which have microtubercles restricted around the abdomen, but the granules are thinned out along the dorsal surface as in the sugar maple finger gall mite, *Vasates aceriscrumene* (Riley). The third group includes deutogynes with microtubercles restricted to the lower parts of the abdominal ring, as in the red erineum mountain maple mite, *Eriophyes calaceris* (K.), Persian walnut leaf gall mite, *E. trisetacus* (Nal.) and the alder leaf bead gall mite, *Phytoptus laevis* Nal. The fourth group comprises those deutogynes which are devoid of microtubercles on the abdominal rings. A unique species that makes finger galls on the leaves of *Prunus* spp. *Phytoptus emarginatae* (K.) and the bladder gall mite *Vasates quadripedes* Shimer are some of the examples.

The biology of deutogynes appears attuned to leaf maturation or to the onset of decline in temperature (Jeppson, 1975). They do not seem to reproduce in the year that they develop; they feed on the leaves, then withdraw into the bark crevices of lalternal buds where they overwinter; on coming out of hibernation in the spring the deutogynes lay eggs on the developing leaf primordia, which hatch into males and protogynes. The primary females (protogynes) then lay eggs which produce primary females or both primary and secondary females (deutogynes), as well as males.

Despite increased interest in eriophyids during the past two decades, there are still many important aspects concerning the biology of eriophyids that remain unknown. An outline of the biology of some of the gall formers is presented as follows:

Information on the biology of some of the gall formers, *Eriophyes sheldoni* Ewing causing "big bud" galls on *Citrus sinensis*, *E. pyri* Pagenstecher inducing galls on *Pyrus communis* (Minder, 1957), *E. vitis* Pagenstecher including galls on *Litchi chinensis* (Allen and Wadud, 1963), *E. calaceris* (Keifer) causing erineum or *Acer glabrum*, *Acalitus phloecoptes* (Nalepa) causing bud and shoot galls on *Prunus domestica*, *Aceria jasmini* Channabasavanna causing erineum on *Jasminum pubescens*, *Trisetacus quadrisetus* Thomas inducing galls on *Juniperus communis*, *Vasates quadripes*, causing galls on *Acer saccharinum* and *A. dasycarpum*, *Phytoptus laevis* Nalepa inducing galls on *Alnus* spp., *Phyllocoptella avellanae* (Nalepa) causing galls on *Tilia* sp. and *Cecidophyopsis ribis* (Westwood) causing bud galls on *Ribes* spp. is available. For the sake of clarity the biology of *E. sheldoni* is provided here as a model system. In *E. sheldoni* hatching of eggs is at 25 °C, 98 per cent R.H. after an incubation period of three to 14 days, and the life cycle is generally extended over a period of 12 to 33 days. The males are known to deposit 25 to 100 stalked spermatophores at the rate of two to 15 per day which are picked up by virgin females (Sternlicht and Goldenburg, 1971).

Mites develop into adults through the first and second nymphal stages. The second moult period, the longest resting period of the two moults, is responsible for the development of a sexually mature individual. Neither of these nymphs has external genitals, but as the adult develops the genitalia are extruded between the coxae and the genital setae on the anterior part of the thanosomal venter.

In general, eggs are laid singly on the host in the bracts, leaves, flowers and within buds, and a female lays six to 50 eggs. The incubation period is variable for different species ranging from two to 14 days. Embryonic development is fast during temperatures regimes of 18–24 °C, while slow at lower temperatures (10–17 °C). However, high temperature, high humidity and heavy rainfall are unfavourable for the development of *E. litchi*, although *C. ribis* has been reported to resist temperatures as low as −7 to −12 °C. There are two nymphal stages in most eriophyids but three nymphal stages are occasionally known in *A. jasmini*. The total life cycle in general, extends from 12–36 days. In most species two to three generations occur per year but *E. emerginatae* has a univoltine life cycle, the females overwintering in crevices of old buds near the base of branches move to newly developing buds in spring (Oldfield, 1969).

In rare instances such as *Acalitus phloecoptes* (Nalepa) causing bud and shoot galls on *Prunus domestica* only one female eriophyid occurs during August and the galls become packed with 4000–5000 mites. From the beginning of autumn until the period of leaf-fall, mites migrate to the leaf axils and crawl under the first layer of bracts of the buds. In *Trisetacus quadrisetus* Thomas inducing galls on *Juniperus communis*, the individuals are found only from early spring to late winter. *Vasates quadripes* Shimer and *Eriophyes calaceris* (Keifer) have a life cycle which show alternation of generation. The latter colonises the erineum on Rocky Mountain Maple, *Acer glabrum* until September and then migrate to the stems, twigs and bark crevices, and developing buds in spring. The eriophyid *Phytoptus lavevis* Nalepa exhibits deuterogyny. Generally after hibernation, the deutogynes induce formation of galls and deposit eggs that develop into protogynes and males of the first generation. Each gall contains over 400 mites and the emergence of mites from galls is facilitated by pronounced positive phototaxi of the deutogynes (Jeppson, 1975). In *C. ribis* which reproduces on black currants, the migration of individuals occurs in April/May and several generations of mites develop each year (Jeppson, 1975).

2) TENUIPALPIDS

The biology of tenuipalpid mites is simple. The egg after the incubation period hatches into a larva. After a short quiescent period, the larva hatches into a protonymph followed by another moult into deutonymph. The deutonymph finally moults into an adult. The total life cycle may take about 18–25 days depending on the environmental conditions. Galls are generally caused on the stem or leaves of the host.

Larvacarus transitans (Ewing) oviposits within the galls or the stem of *Zizyphus jujuba*. The rate of oviposition is correlated with the temperature. In the winter season (November to March), it is retarded due to low temperature and acclerated towards the end of March when the temperature rises. A minimum of two eggs per gall to a maximum of 397 has been observed. The mite passes through the usual number of instars before attaining sexual maturity. Emergence of adults takes place towards the onset of monsoon and continues during the rainy season. It is known that 2 inches rainfall which provides 95–98 per cent RH and temperature between 90.8 and 93.6 °F is suitable for emergence. The mites after emergence, settle singly on the tender shoots. After 10–15 days the mite is covered with a scale which keeps enlarging. In April a slit appears on one side of the scale, facilitating the mite emergence during monsoon (Latif and Wali, 1961).

Host Relationships

1) ERIOPHYIDS

Among the varied kinds of deformities induced by eriophyid mites on their host plants, the greatest diversity of galls occurs on the leaves. The galls may be elongate, hemispherical, rounded or bead like; they may be scattered irregularly over the lamina, or may be restricted to the venal angles between the veins. They may be solitary, coalesced or in clusters, and may be visible on the upper or lower surface of the leaf, or on both. The galls may be formed on other parts of the plant, viz., leaf petiole, catkin flowers, flower buds, leaf buds and leaf axils. The interior gall structures may provide important diagnostic characters. The other types of abnormal growth induced by these mites are erineum, big bud, witches' broom, etc. The erineum has several diagnostic aspects which include position on the leaves, density of hairs, colour, type of hairs which may be elongate, clavate, capitate, unicellular or multicellular.

A wide morphogenetic range exists in galls produced by acarines. Many of the plants damaged by eriophyid mites show structural and morphological modifications (Chandrasekaran and Balasubramanian, 1972). These result from variations caused in the parenchyma cells which get transformed into meristem as a result of the feeding by mites. The cells start dividing and in the long run form galls on the tissue. The damaged tissues contain a large quantity of free amino acids, indicating a disturbed protein synthesis. With the ageing of galls, red pigment material increases in certain cases and extends gradually into the other tissues. Each host plant reacts in a different manner to the feeding activity by mites. Auxin and phenolic contents are also generally higher (Balasubramanian and Purushothamanan, 1972; Purohit et al., 1979) in the galls than the normal leaf tissue.

Eriophyids neither suck out the cell sap entirely nor remove subcellular particles so as to kill it, but salivary injection by mites actually promotes tissue succulence. Feeding punctures caused by gall-forming eriophyids are partially callosified by the host, permitting the restoration of cell wall integrity following puncturing (Westphal, 1968, 1968a, 1977). The nutritive tissue is characterized by hypertrophy and lobing of the nucleus, hypertrophy of the nucleolus, fragmentation of vacuoles, and abundance of organelles in the cytoplasm, higher density of ribosomes and endoplasmic reticulum and hypertrophy of the plastids (Westphal, 1977). The differentiation of the nutritive tissue appears to show a gradual gradation from a mild effect to a dramatic one as in the simple erinea to complex galls (Westphal, 1977).

Further the feeding damage and gall initiation by eriophyid mites indicate a series of physiological changes involving a higher rate of nucleic acid metabolism and higher quantity of growth promoting substances (Tandon *et al.*, 1976).

Structural modifications and chemical changes associated with the formation of some of the common types of host tissue responses to mite feeding injury are discussed briefly.

Erineum: Unlike galls with escape holes, erinea are not closed growths, but enable mites to live within dense trichomatous tissue. Erinea vary from small localized patches to extensive areas covering much of the foliar and petiolar surfaces. Erineal patches may be seen on any tender parts and even flowers. This suggests that the growth regulators engendering erineal development differ from gall regulators by being able in most instances to be translocated laterally. Sometimes the dense growth of hairs on the undersurface of leaves forms hemispherical dome which projects out through the upper surface, causing leaf distortions.

The actively growing hairs of erineum of *Tilia intermedia* D.C. caused by *Phytoptis leisoma* Nal. show enlarged nucleus with dispersion of chromation indicating strong synthesis of RNA (Westphal, 1977). The cell beside the one pierced by mouth parts turns necrotic and grows into hairs. The hairs are more lignified than the normal cells. Further the intercellular spaces increase in size.

The eriophyid, *Eriophyes paderineus* Nalepa causes on leaves of *Prunus padus* L. felting of pedicellate globular hairs which are nutritive and unicellular, with one nucleus and a hypertrophied nucleolus. The histological development of erinea on grapevine (*Vitis* sp.) caused by *Eriophyes vitis* (Pagenstecher) (Slepyan *et al.*, 1969) results in the juvenile healthy grape leaves being covered with dense, long but narrow and bent, thin walled, unicellular trichomes plaited together. On mature leaves, the trichomes are two- to six-celled and hornlike. Erineous trichomes are commonly thin walled, elongate, unicellular and rounded at the apex. When leaves have ceased to unfold, fat granules are formed in the trichomes of the erinea, followed by deposition of tannic substances which imparts a rusty brown colour to it. The trichomes finally dry out and fall. The epidermal cells are also modified during erineum formation. The cuticle either does not develop fully or is wanting, while the walls are thickened. The epidermal stomata may either remain normal or be deformed. Significant changes occur in the mesophyll chlorenchyma in the erinea finally resulting in the local invagination of leaf blade towards the upper leaf surface.

Witches' broom: The basic organisation of this gall is that of modified shoot, somewhat branched and compacted with shortened internodes. The shortening of internodes produces a bud-like cluster of overlapping cottony leaves which are paler and thicker than normal branchlets. One feature which distinguishes gall leaves from the normal is the presence of swollen epidermal cells which stain more intensely than the normal cells. Swollen cells have dense cytoplasm with enlarged nuclei and nucleoli. The salivary factor which induces brooming travels in conductive tissue, being more of an inhibitor than one that diverts cell growth.

The Witches' brooms caused by *Aceria victoriae* Ramsay on *Haloragis erecta* (Banks ex. Murr.) vary considerably in appearance, being composed of bunches of discoloured and distorted leaves. The entire gall may be red or partly green and partly red. Portions of the galls may have a tumour-like fleshiness while other parts are leaf-like. Meristematic areas in these galls are generally pronounced but lack the symmetry of a normal shoot apex. Disorganisation of meristem in the gall is greatest where mite feeding has caused maximum cell destruction. While promoting meristematic activity, the gall mites impede differentiation so that air spaces do not develop as in uninfested shoots, and the gall leaves have a rudimentary appearance. In place of properly formed epidermis, the deformed leaf is bound by a surface of callus-like loosely adherent red cells. The apical dominance in gall is greatly reduced, thereby producing a branched effect on the gall leaves (Jeppson, 1975).

Pin or needle galls: *Eriophyes tilia* var. *rudis* injects a salivary secretion on the leaf of *Tilia cordata* Mill (Linden). This results in the formation of elongated rod-shaped galls on the upper surface of the leaf. Galls become reddish-brown in colour and 4–6 mm in length. Qualitative and quantitative differences in the carotenoid content of gall and lamina have been observed. The mite causes increased synthesis of some carotenoids and inhibits synthesis of others. This results in selective accumulation of certain carotenoids in the galls of linden. B-carotene comprises 27.3 per cent of all carotenoids in normal lamina whereas in galls it is only 1.7 per cent. Dihydroxy carotene comprised 19.2 per cent of all carotenoids in lamina, but this carotene was not found at all in the galls. The reverse holds as regards isocrypto-xanthia and 4-keto-carotene. In the galls these two carotenoids comprise 20 per cent whereas in the lamina, the former carotenoid is absent and the latter constitutes 0.6 per cent of all carotenoids (Czeczuga, 1975).

Pouch galls: A gall is basically a depressed area that further develops to acquire a definite shape that is characteristic of the host and the mite species. Generally an escape hole is present at the bottom with a protruded structure on the upper surface. This has been designed because mites are incapable of forcing their exit. The interior surface of the galls is lined by succulent or turgid host tissue which vary from hair-like papillae to hypertrophied cells.

Histologically the galls are not very different from normal leaves. Normal mature leaves of *Pavetta hispidula* possess one layer of palisade cells and six to eight layers of spongy cells with large air spaces; the lower epidermal layers are more elongated horizontally than the upper epidermal cells. In the gall lamina, palisade cells are recognizable and spongy mesophyll cells appear hypertrophied and devoid of air chambers. The galls have erineal hairs developed by modification of abaxial epidermal cells. Proliferation of spongy cells into callus-like growth is evident and very often the lower epidermis is ruptured. Due to persistent feeding by the mites, the peripheral collenchyma cells are hypertrophied and the epidermal cells develop into erineal hairs (Raman and Swaminathan, 1978).

The pouch galls caused by *Eriophyes tiliae* on *Tilia platylla* are generally formed on the abaxial side of undifferentiated leaf. The nutritive epidermis in gall has smaller cells, but larger nuclei, and in the gall these cells have dense cytoplasm with well developed endoplasmic reticulum and ribosomes. Starch which is present in the early phase of cecidogenesis disappears during later stages (Thompson, 1975). Elongate pouch galls on the dorsal surface of tender leaves of *Pongamia glabra* caused by *Aceria pongamiae* Channabasavanna have hairy outgrowths lining the inner surface of the galls.

Filz galls: Morphological and cytological observations show that filz gall hairs induced by *E. leiesoma* look like hairs which normally cover very young leaves of *Tilia intermedia* (Westphal, 1977). The pathological hairs show larger nuclei and hypertrophied nucleoli than normal hairs. Both kinds of hairs become progressively lignified. Observations on feeding punctures revealed that lignification of cell wall restricts nutrition of the mite.

2) TENUIPALPIDS

Only two tenuipalpids are important as gall formers: *Larvacarus transitans* causing stem galls *Zizyphus jujuba* Mill (Rhamnaceae) and *Obdulia* sp. causing stem galls on *Tamarix marismortuii*. Information on this group of mites is very meagre.

REFERENCES

1. Allen, M.Z., Wadud, M.A. 1963. On the biology of the litchi mite, *Aceria litchii* (Eriophyidae, Acarina) in East Pakistan. *Pakistan J. Sci.* 15 (5) : 231–240.

2. Balasubramanian, M., Purushothaman, D. 1972. Phenols in healthy and galled leaves of *Pongamia glabra* Vent caused by an eriophyid mite *Eriophyes cheriani* Masses (Eriophyidae : Acarina). *Indian J. exp. Biol.* 10 : 394–395.

3. Batchelor, G.S. 1952. The eriophyid mites of the state of Washington. *State College Wash. Tech. Bull.* No. 6, 32pp.

4. Boczek, J. 1960. A new genus and three new species of Eriophyid mites (Acarina). *Jour. Kansas ent. Soc.* 33(1) : 9–14.

5. Boczek, J. 1961. Badania nad ratoczamiz rodziny Eriophyidae (Sp Szpecia-lowata) W. Poloce 1. (Studies in eriophyid mites of Poland-1) (English Summary). *Prace 10R, Poznan,* 3(2) : 5–85.

6. Boczek, J. 1961a. Studies on eriophyid mites of Poland. II Acarologia t. III. fasc. 4 : 360–370.

7. Boczek, J. 1962. Nalepella haarlovi, N. sp. (Acarina : Eriophyidae) *Ent. Meddl.* 31 : 195–197.

8. Boczek, J. 1968. Studies on mites (Acarina) living on plants in Poland. VIII. *Bull. l'Acad. Polon. Sci.* Cl. 5, 16 : 631–636.

9. Boczek, J. 1969. Studies on mites (Acarina) living on plants in Poland. XI. *Bull. l'Acad. Polon. Sci.* Cl. 5, 17 : 393–398.

10. Briones, M.L., McDaniel, B. 1976. The eriophyid plant mites of South Dakota. *South Dakota State. Univ. Tech. Bull.* 43 : 123pp.

11. Chakrabarti, S., Mondal, S. 1979. Studies on the eriophyid mites (Acarina : Eriophyidae) of India III. Descriptions of five new species. *Acarologia* 21 : 396–407.

12. Chandrasekaran, V., Balasubramanian, M. 1972. Morphology, height diameter relationship and reduction in leaf area of leaf galls on *Eugenia jambolana* Lam. and on *Pongamia glabra* Vent. *Madras agric. J.* 59 : 411–414.

13. Channabasavanna, G.P. 1966. A contribution to the knowledge of Indian eriophyid mites (Eriophyoidea : Trombidiformes : Acarina). *Univ. Agri. Sci. Hebbal, Bangalore* 153pp.

14. Czeczuga, B. 1975. The carotenoid content of galls produeed by *Eriophyes tiliae* var. *rudis*. Nal. (Acarina) on *Tilia Cordata* Mill., leaves. *Marcellia* 38 : 223–225.

15. Duges, A.L. 1934. Nouvelles observations sur les Acariens. *Ann. Sci. Nat. Ser.* 2, 2 : 104–106.

16. Dujardin, F. 1851. Sur des acariens a Quarte Pied, Parasites des vegeteaux, et. Qui, Devient Former un genre particulier (Phytoptus). *Ibid. Ser.* 3, 15 : 166–169.

17. Farkas, H.K. 1960. Uber die Eriophyiden (Acarina) Ungarns 1. Beschreibung neuer und wenig bekannter Arten. *Acta Zool. Hung.* VI (3–4).

18. Farkas, H.K. 1960a. Afrikanische Gallamilben (Acarina : Eriophyidae) aus den Material des cecidologischen Herbarium des Ungarischen naturwissenschoftlichen Museum. *Ann. Hist. Nat. Mus. Nation Hung.* 52 : 429–435.

19. Farkas, H.K. 1961. Two new American gall mites (Acarina, Eriophyidae). *Ibid.* 53 : 507–509.

20. Farkas, H.K. 1961a. Uber die Eriophyiden (Acarina) Urgarns 11. Beschreibung einer neuer Gattung und zweinen Arten. *Acta. Zool. Hung.* 7(1–2) : 73–76.

21. Farkas, H.K. 1962. On the Eriophyids of Hungary III. The descriptions of two new species (Acarina : Eriophyidae). *Ann. Hist. Nat. Mus. Nation. Hung. Pars. Zool.* 52 : 429–431.

22. Felt, E.P. 1940. *Plant galls and gall makers.* New York : Comstock Pub. Co. 364 pp

23. Hall, C.C. Jr. 1967. A look at eriophyid life cycles. *Ann. Entomol. Soc. Am.* 60 : 91–94.

24. Hall, C.C. Jr. 1967a. The Eriophyoidea of Kansas. *Univ. Kansas Sci. Bull.* XL VII (9) : 601–675.

25. Hodgkiss, H.E. 1930. Mites on silver maple. *New York Agr. Exp. Sta. Tech. Bull.* 163 : 16–21.

26. Jeppson, L.R. 1975. Mites injurious to economic plants. In Jeppson, L.R., Keifer, H.H., Baker, E.W. Berkeley : Univ. Calif. Press. 614pp.

27. Keifer, H.H. 1942. Eriophyid studies XII. *Bull. California Dept. Agr.* 31 : 117–129.

28. Keifer, H.H. 1944. A review of North American economic eriophyid mites. *J. Econ. Entomol.* 39 : 563–570.

29. Keifer, H.H. 1952. The Eriophyid mites of California (Acarina : Eriophyidae). *Bull. Calif. Insect Surv.* 123pp.

30. Keifer, H.H. 1959. Eriophyid studies XXVI. *Bull. Calif. Dep. Agric.* 46 : 244pp.

31. Keifer, H.H. 1964. Eriophyid studies B-II. *Bureau Ent., Calif. Dept. Agr.* 20 pp.

32. Keifer, H.H., Baker, E.W., Kono, T., Delfinado, M., Styer, W.E. 1982. An illustrated guide to plant abnormalities caused by eriophyid mites in North America. *USDA, ARS, Agri. Hand Book* No. 573.

33. Keifer, H.H., Knorr, L.C. 1978. Eriophyid mites of Thailand. *Dept. Agric. Min. Agric. Co-op. Thailand. Plant Prot. Service Tech. Bull.* No. 38; 31pp.

34. Kendall, J. 1930. The structure and development of certain eriophyid galls. *Ztschr. f. Parasitenk.* 2 : 478–501.

35. Lamb, K.P. 1952. New plant galls. 1. Mite and Insect galls. *Trans. roy. Soc. N. Z.* 79 : 349–362.

36. Lamb, K.P. 1952a. A preliminary list of New Zealand Acarina. *Trans. roy. Soc. N. Z.* 79 : 370–375.

37. Lamb, K.P. 1953. Tomato galls from Morocco. *Bull. ent. Res.* 44(3) : 401–404.

38. Lamb, K.P. 1953a. A new species of *Diptilomiopus* Nalepa (Acarina : Eriophyidae), together with a key to the genus. *Trans. roy. Soc. N. Z.* 80 : 367–382.

39. Lamb, K.P. 1953b. New plant galls II. Descriptions of seven new species of gall mites and the galls which they cause. *Ibid.* 80 : 371–382.

40. Lamb, K.P. 1960. A check list of New Zealand plant galls (Zoocecidia). *Trans. Roy. Soc. New Zealand.* 88 : 121–139.

41. Landois, H. 1864. Eine Milbe (*Phytoptus vitis* milni Land.) also ursache des Trauben—Misswachses *Zeitschr. Wiss. Zool.* 14 : 351–363.

42. Latif, A., Wali, M. 1961. Distribution, Bionomics and description of *Larvacarus transitans* (Ewing). *Pak. J. Sci. Res.* 13(2) : 77–87.

43. Liro, J.I. 1940. Neue Eriophyiden aus Finland, *Ann. Zool. Soc. Zool. Bot. Fenn. Vanamo* 8 (1) : 1–67.

44. Liro, J.I. 1941. Neue und seltene Eriophyiden (Acarina), *Ibid.* 8 (7) : 1–54.

45. Liro, J.I. 1942. Neve finnische Eriophyiden (Acarina). *Ann. ent. Fenn.* 8 : 71–79.

46. Liro, J.I. 1943. Uber neve order sonst bemerkenswerte finnische Eriophyiden (Acarina). *Ann. Zool. Bot. Fenn. Vanamo.* 9 (3) : 1–50.

47. Liro, J.I., Roivainen, H. 1951. Eriophyidae of Finland. *Animalaia Fennica*, No. 6, Helsinki 281p.

48. Low, F. 1874. Beitrage zur Naturgcschichte der Gall-Milben. *Verhandl. Zool-Bot. Ges. Wien.* 24 : 3–14.

49. Low, F. 1878. Beitrage zur kenntniss der Milbengalles (Phytoptocecidien). *Wien. Verh. Zool. bot. Ges.* 28 : 127–150.

50. Mani, M.S. 1964. *Ecology of plant galls.* The Hague, W. Junk Publ. 400pp.

51. Massee, A.M. 1927. Descriptions of three new species of gall mites (Eriophyidae) from Sudan. *Ann. Mag. Nat. Hist.* 20 (118) : 372–375.

52. Massee, A.M. 1927a. A contribution to the knowledge of the species of gall mites (Eriophyidae) of Sussex. *Ibid.* 20 (119) : 375–379.

53. Massee, A.M. 1937. A species of gall mite (Eriophyidae) injurious to tomato. *Bull. Ent. Res.* 28 : 403.

54. Meyer, M.K.P. 1968. The grass stunt mite, *Aceria neocynodonis* Keifer in South Africa. *So. Afr. J. Agric. Sci.* 11 : 803–804.

55. Minder, I.F. 1957. Some information on biology of *Eriophyes pyri. Zoologitshesky zhurnal.* 36. 1007–1015 (in Russian).

56. Mondal, S., Chakrabarti, S. 1980. Studies on the eriophyid mites (Acarina : Eriophyoidea) of India. *VI. Orient. Insects* 14 : 453–459.

57. Mondal, S., Chakrabarti, S. 1982. Studies on the eriophyid mites (Acarina : Eriophyoidea) of India. *V. Acarologia* 23 (1).

58. Mohanasundaram, M. 1982. Four new species of phyllocoptine mites (Eriophyidae : Acarina) from Tamil Nadu, India. *Entomon* 7 : 23–30.

59. Nelepa, A. 1892. Neve Gallmilben. 5. *Fortsetzung. Anz. Acad. Wien.* 29 : 190–192.

60. Nalepa, A. 1893. Neve Gallmilben. 6. *Fortsetzung. Anz. Akad. Wien.* 30 : 31–32.

61. Nalepa, A, 1893a. Ueber neve Gallmilben. 6. *Fortsetzung. Botanisches centralblatt.* 53 : 342–343.

62. Nalepa, A. 1893b. Neve Gallmilben 7. *Fortsetzung. Anz. Akad. Wien.* 30 : 105.

63. Nalepa, A. 1893c. Neve Gallmilben. 8. *Fortsetzung. Anz. Akad. Wien.* 30 : 190–191.

64. Nalepa, A. 1894. Neve Gallmilben. 9. *Fortsetzung. Anz. Akad. Wien.* 31 : 38.

65. Nalepa, A. 1894a. Neve Gallmilben. 10. *Fortsetzung. Anz. Akad. Wien.* 31 : 179–180.

66. Natcheff, P.D. 1966. Studies on Eriophyid mites of Bulgaria. II. *Acarologia* 8 (3) : 415–420.

67. Natcheff, P.D. 1966a. Studies of Eriophyid mites of Bulgaria *V. C.R. Acad. Bulgar. Sci.* 19 : 1175–1178.

68. Natcheff, P.D. 1967. Studies on Eriophyid mites of Bulgaria. IV. *Naucn. Trud. Ser. Rast.* 18 : 385–386.

69. Natcheff, P.D. 1967a. Etude sur des acares Eriophydes en Bulgaria. Wissencheflliche Arbeiten 18 (46) *Ser. Pflanzeban* 315–335.

70. Natcheff, P.D. 1979. Eriophyid studies in Bulgaria. XIII. *Plant Science* 16 : 116–119.

71. Natcheff, P.D. 1981. *Etiophyid mites of Bulgaria.* Doctoral Thesis. 423pp.

72. Nemec, B. 1924. Unterschungen uber Eriophyiden gallen. *Studies from the Plant Physiological Laboratory of Charles University,* Prague Vol. 5 : 47–94.

73. Niblett, M. 1959. The gall mites (Eriophyidae) of London area. *London Nat.* No. 38 : 51–54.

74. Oldfield, G.N. 1969. The biology and morphology of *Eriophyes emarginatae,* a prunus finger gall mite, and notes on *E. prunidemissae. Ann. Entomol. Soc. Am.* 62 : 269–277.

75. Pagenstecher, J. 1857. Uber Honing producrirende *Ameisen. Verh. Natr. Med. Ver. Heiderlburg.* 1 : 46–53.

76. Purohit, S.D., Ramawat, K.G., Arya, H.C. 1979. Phenolics and phenolase as related to gall formation in some arid zone plants. *Curr. Sci.* 48 : 714–716.

77. Putman, W.L. 1939. The plum nursery mite. *Seventh Annual Rpt. Ent. Soc. Onterio* p. 33.

78. Putman, W.L. 1940. The plum nursery mite (*Phyllocoptes fockeni* Nal. and Trt.) *Rept. Entomol. Soc. Ont.* 70 : 33–40.

79. Raman, A., Swaminathan, S. 1978. On the histopathology of mite galls on leaves of *Pavetta hispidula* (Rubiaceae). *Ind. J. Acar.* 3 : 67–70.

80. Reamur, M.de. 1737. Memoires pour servir a l'historie des insects tome troisieme, de, l'imprimerie royal. *Academie Royale des Sciences, Paris.* 421–423; 511–515.

81. Riley, C.V. 1970. Mite gall on sugar maple. *Amer. Ent. and Bot.* 2 : 339.

82. Roivainen, H. 1949. Eriophyid news from Denmark. *Ann. Ent. Fenn.* 15 (1) : 22–32.

83. Roivainen, H. 1950. Eriophyid news from Sweden. *Acta. Ent. Feen.* 7 : 1–51.

84. Roivainen, H. 1951. Contributions to the knowledge of the eriophyids of Finland. *Acta. Ent. Fenn.* 8 : 72.

85. Roivainen, H. 1951a. Contributions to the knowledge of Eriophyidae of Finland. *Ibid.* 8 : 1–72.

86. Roivainen, H. 1953. Some gall mites (Eriophyidea) from Spain. *Arch. Inst. Acelon.* 1 : 9–43.

87. Roivainen, H. 1953a. Subfamilies of European Eriophyid mites. *Ann. Ent. Fenn.* 19 : 83–87.

88. Scheuten, A. 1857. Einiges uber Milben. *Arch. Naturg.* 23 (1) : 104–112.

89. Shevtshenko, V.G. 1961. Postembryonic development of the four legged gall mites (Acariformes, Eriophyidae) and observations on the classification of *Eriophyes laevis. Zool. Jour.* 40 : 1143–1158.

90. Siebold, G.R. von. 1850. Bemerkumgen uber Psychiden. *Fahresber. Schfer. Ges. Kult.* 28 : 84–89.

91. Slepyan, E.I., Landsberg, G.S., Malchenkova, N.I. 1969. The gall of the mite *Eriophyes vitis* Pgst. (Acarina : Eriophyidae) and its ecological niche. *Entomol. Rev.* 48 : 67–74.

92. Sternlicht, M., Goldenburg, S. 1971. Fertilization, Sex ratio, and post embryonic stages of the citrus bud mite, *Aceria sheldoni* (Ewing). *Bull. Entomol. Res.* 60 (3) : 391–397.

93. Tandon, P., Vyas, G.S., Kant, U., Arya, H.C. 1976. Nucleic acid metabolism in Eriophyes induced Zizyphus gall and normal stem calli in culture. *Indian J. Expt. Biol.* 14 : 211–213.

94. Thompson, J. 1975. Development and histology of galls on *Tilia platyphylla* caused by *Eriophyes tiliae tiliae. Bot. Tidsskrift.* 69 : 262–270.

95. Turpin, P.J. 1833. Sur le developpement des galles corniculées de Tilleul. *Nouveau Bull.* des se. par la Soc. Philomatique de Paris, 163–165, dans Nalepa 191.

96. Westphal, E. 1968. Observations sur la morphologis et l'histocytologie des virescences et aultres modifications organoidea produites par *Eriophyes cladophtirus* Nal. sue *Solanum dulcamara* L. *Marcellia* 35 : 83–103.

97. Westphal, E. 1968a. Observations morphologiques et histologiques concernant les galles d'*Eriophyes peucedoni* Can. sur les feuilles de *Pimpinella saxifraga* L. *Marcellia* 35 : 247–263.

98. Westphal, E. 1977. Morphogenese, ultrastructure et etiologie de quelques galles d'eriophyes (Acariens). *Marcellia* 39 : 193–375 .

Index

354